Signals and Systems

Signals and Systems

E. Sreenivasa Reddy

Principal
University College of Engineering
Acharya Nagarjuna University.

BS Publications

An imprint of **BSP Books Pvt. Ltd.**
4-4-309/316, Giriraj Lane, Sultan Bazar,
Hyderabad - 500 095.

Signals and Systems
by E. Sreenivasa Reddy

Published by:

 **BS Publications**

An imprint of **BSP Books Pvt., Ltd.**

4-4-309/316, Giriraj Lane, Sultan Bazar,
Hyderabad - 500 095
Phone : 040 - 23445605, 23445688
e-mail : info@bspbooks.net
www.bspbooks.net

ISBN: 978-93-89354-15-7 (Hardback)

PREFACE

This book is mainly written for undergraduate engineering courses. This course is now taken by CSE & IT people also as it is needed to understand basic concepts of deep learning. Also now a days this course has gained tremendous importance due to the advances in digital signal processing, where study of signals is a prerequisite. Even where analog methods are still used the knowledge of frequency domain description of analog signals is needed. The study of systems is required for computer simulation and implementation. It encompasses analysis, design and implementation of systems as well as problems involving signal system interaction.

This textbook was written with goal of providing not only knowledge of signals and systems but about to prepare for competitive exams like GATE, IES etc. It cover both continuous and discrete signals and systems. All previous GATE Paper Questions were solved and a comprehensive assignment questions with answers were provided at the end of each chapter.

-Author

Contents

CHAPTER 3: Fourier Series

CHAPTER 4: Fourier Transforms

CHAPTER 5: Linear Time Invariant Systems

CHAPTER 6: Convolution and Correlation

CHAPTER 7: Sampling

CHAPTER 8: Laplace Transform

CHAPTER 9: Z-Transform

CHAPTER 1

Introduction to Signals & Systems

In this chapter we discuss about basic concepts of signals and systems.

Objectives of the chapter

Definition of signals and systems, classification of signals, classification of systems, operations on signals, exponential, complex exponential and sinusoidal signals, exponential and sinusoidal signal concepts of impulse function, unit step function, signum function. etc.

1.1 SIGNAL

A signal is defined as any physical quantity that varies with time, space or any other independent variable. A signal may be also treated as set of data. A signal contains a pattern of variation of some form. The variation may be with respect to time or space.

A signal is a physical variable which is a function of time and associated with a system. Some signals like electric charge is function of space.

Signals are represented as mathematical functions of one or more independent variables. This representation is known as signal modeling.

Ex: when a torch light is kept on continuously there is no signal present in it. When it is switched on and off w.r.t time then one will feel there is some signal in it. Hence any variable element constitutes some information in it.

A signal is a variable that carries information. It is a physical quantity which conveys information

Examples:

   (a)   Electrical Signals – Voltages and currents
   (b)   Acoustic Signals – Acoustic pressure vs time
   (c)   Mechanical Signals – Velocity of car vs time
   (d)   Video Signals – Intensity levels vs time
   (e)   Biological Signals – Sequence of bases in a gene

1.1.1 REPRESENTATION OF SIGNAL

Signals are represented as a function that describes the evolution of one or more independent variables such as time, space etc.

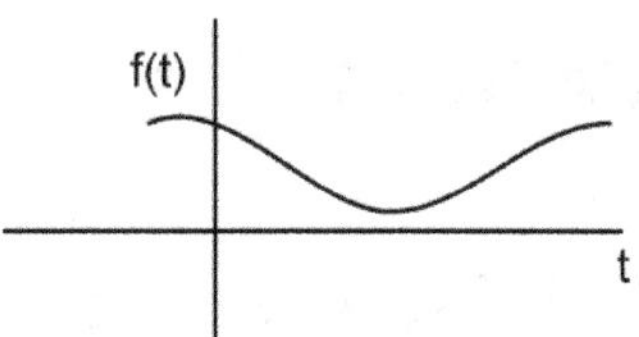

Ex: E C G signal, images and video signals

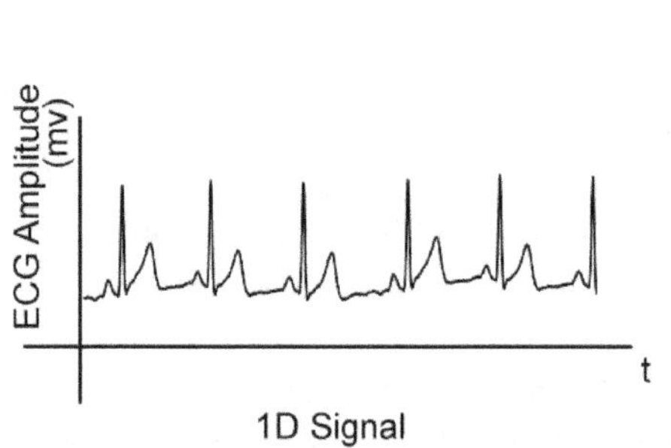

1D Signal
Independent variable : time

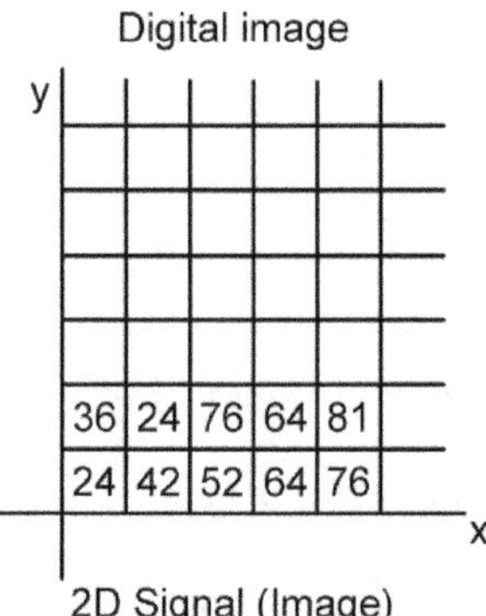

2D Signal (Image)
Independent variable : space

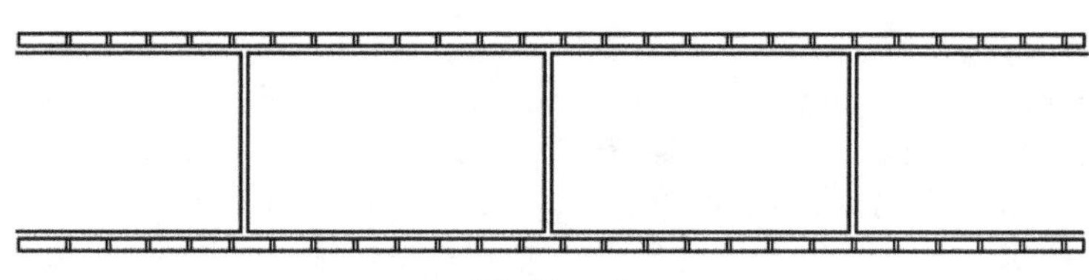

3D Signal
Independent variable : Time & Space

1.1.2 DIFFERENT TYPES OF SIGNALS

There are several classes of signals

1. **Deterministic Signal:** Signals that are modeled or defined completely by specified functions of time are known as deterministic signals.

   Ex: Unit pulse

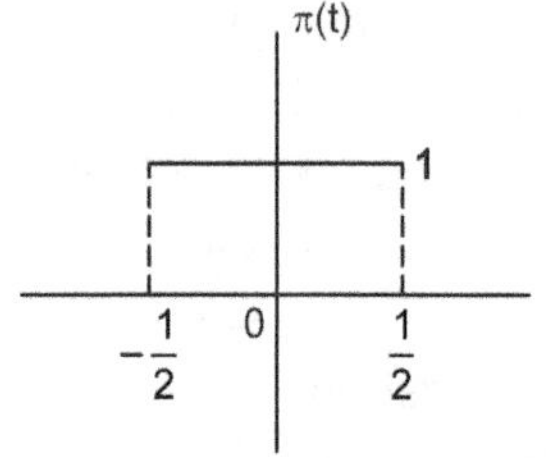

Fig. 1.1

$$\pi(t) = \begin{cases} 1 & |t| \leq \tfrac{1}{2} \\ 0 & \text{otherwise} \end{cases}$$

A deterministic signal physical description must be known completely in any form.

Even it is not continuous function of time it is deterministic.

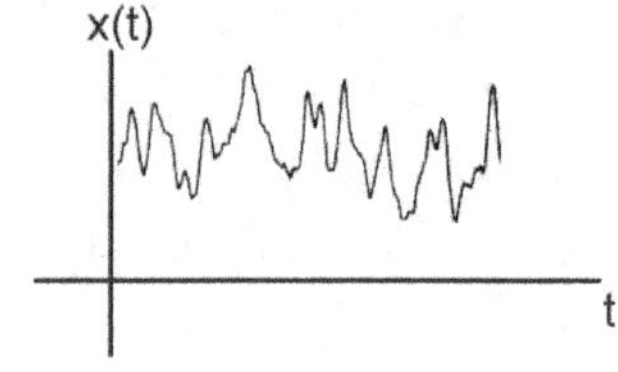

Fig. 1.2

2. **Random signal:** Signals that take random values at any given time and modeled probabilistically are known as random signals.

   Ex: Noise

   These signals are only known by mean, mean square values which are probabilistic.

3. **Continuous time Signal:** Signals that are defined for all values of time t are continuous time signals. It is a function of continuous time variable but may not be mathematically continuous function.

   Ex: E C G, AC wave forms

   $$x(t + T) = x(t)$$

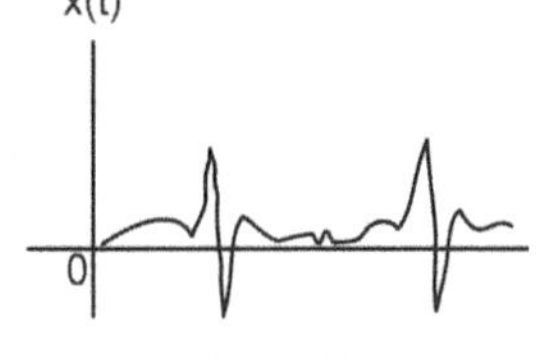

Fig. 1.3

4. **Discrete time signal:** Signals that are represented at discrete values of the independent variable such as time are known as discrete time signals.

   Ex: $x(nT) = x(t)\big|_{t=nT}$

   Unit pulse

   $$\delta(n) = \begin{cases} 1 & n = 0 \\ 0 & \text{otherwise} \end{cases}$$

Fig. 1.4

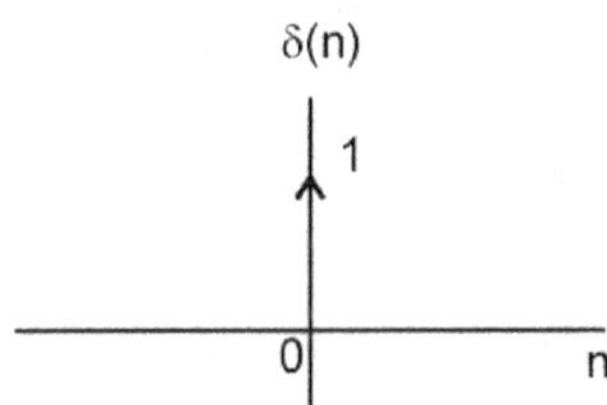

Fig. 1.5

Unit step

$$\mu(n) = \begin{cases} 1 & n \geq 0 \\ 0 & n < 0 \end{cases}$$

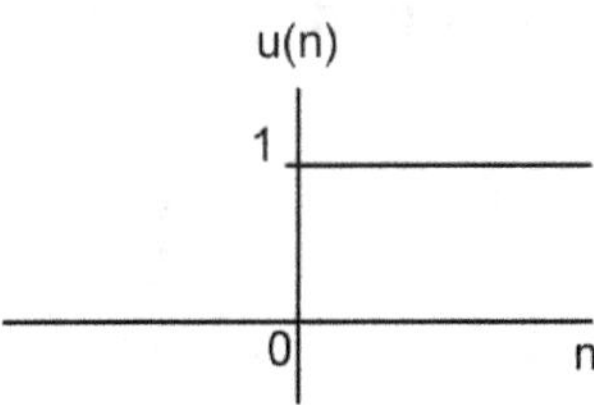

Fig. 1.6

5. **Periodic Signals:** A signal $x(t)$ is periodic if and only if $x(t + T_o) = x(t)$ for $-\alpha < t < \alpha$ where T_o is constant, and its smallest value is known as fundamental period.

   Ex: Sinusoidal

   $$x(t) = \Delta \sin (2\pi f_o t + \theta) \quad -\alpha < t < \alpha$$

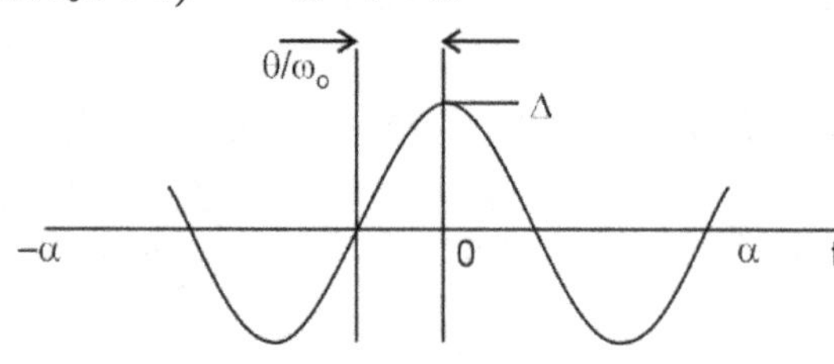

Fig. 1.7

   A periodic signal must start at $t = -\alpha$ because if it starts some finite instant, it may not be same as $x(t)$. A periodic signal can be generated by periodic extension of any segment of $f(t)$.

6. **Aperiodic Signals:** Any deterministic signal that does not satisfy the condition

   $x(t + T_o) = x(t)$ is known as aperiodic signal.

   The above condition is not satisfied for at least one value of t, then it is aperiodic

7. **Causal Signal and Non-Causal Signal:** A continuous time signal $x(t)$ is said to be causal if

   $$x(t) = 0 \quad \text{for} \quad t < 0, \text{ a signal that doesn't start before } t = 0.$$

   Other wise it is non-causal.

   Anti Causal Signal is defined as

   $$x(t) = 0 \quad \text{for} \quad t > 0$$

8. **Symmetric and anti symmetric Signals:** A signal $x(t)$ is referred as a symmetric (even) if it is identical to its time reversed counterpart.

   $$x(-t) = x(t) - \quad \text{continuous signal}$$

   $$x(-n) = x(n) - \quad \text{discrete signal}$$

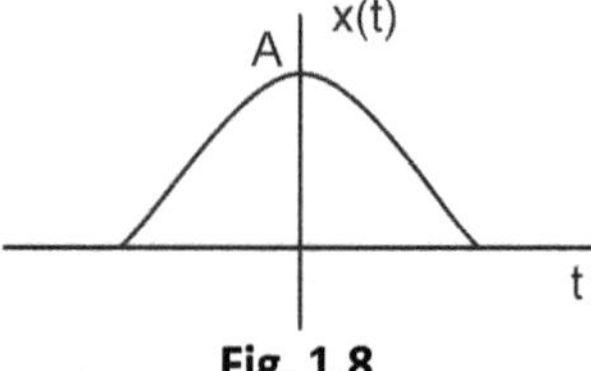

Fig. 1.8

   Ex: $x(t) = A \cos t$

   A signal is said to be anti symmetric (odd) if

   $$x(-t) = -x(t)$$

   Any signal $x(t)$ can be represented as sum of even and odd components.

   $$x(t) = x_e(t) + x_o(t)$$

   $$x(-t) = x_e(-t) + x_o(-t)$$

   $$= x_e(t) - x_o(t)$$

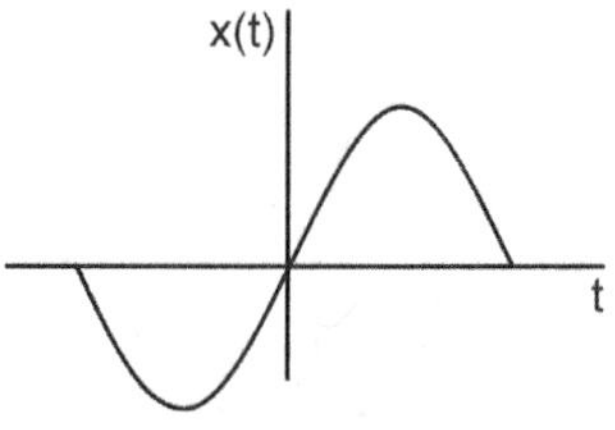

Fig. 1.9

$$2x_e(t) = x(t) + x(-t)$$

$$x_e(t) = \frac{1}{2}\{x(t) + x(-t)\}$$

$$x_o(t) = \frac{1}{2}\{x(t) - x(-t)\}$$

1.1.3 SIGNAL SIZE

The size of a signal depends on not only amplitude but also on its duration.

Hence the size of a signal is measured in terms of different parameters like energy, power etc. but not by its amplitude.

Signal Energy: Let us consider a signal x(t), its energy is given by

$$E_x = \int_{-\alpha}^{\alpha} \left|x^2(t)\right| dt$$

Signal size is a measure of area under x(t). But for large signals both positive and negative areas could cancel each other and looks like of small size. Hence always signal size is measured using $x^2(t)$.

Signal Power: If the energy is finite. Means signal, amplitude $\rightarrow 0$ as time $\rightarrow \alpha$

If signal energy is finite then we say measure of signal size is correct. If the amplitude doesn't become zero as time $\rightarrow \alpha$, we must measure power P_x, since energy becomes infinite. Power is the time average of energy.

$$P_x = \underset{T\rightarrow\alpha}{Lt} \frac{1}{T} \int_{-T/2}^{T/2} \left|x^2(t)\right| dt$$

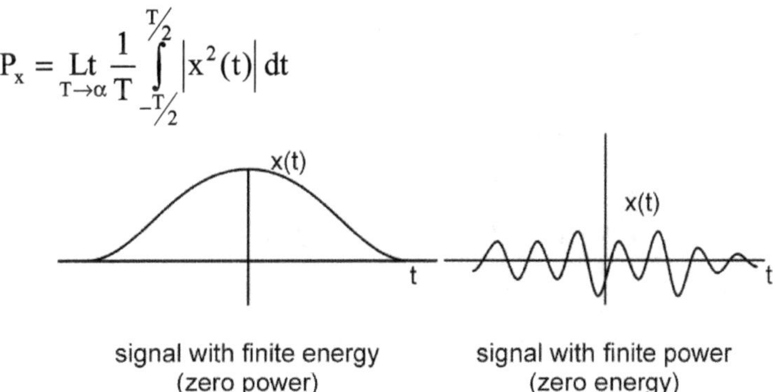

signal with finite energy   signal with finite power
(zero power)   (zero energy)

Fig. 1.10

- A signal is an energy signal if $E_\alpha < \alpha$
- A signal is a power signal if $0 < P_\alpha < \alpha$
- Most of the signals are both energy or power signals but not both

Note: Units of energy and power of a signal will not be in conventional manner like jouls and watts, as we are using them to represent size of signal.

1.1.4 OTHER TYPES OF SIGNALS

(a) **Classification of signals based on their energy and power:**

(i) *Energy signals*: Signals with finite total energy $E_\alpha < \alpha$ are known as energy signals.

These energy signals have zero average power $P_\alpha = 0$. Any signal with finite amplitude and finite duration are called as energy signals.

(ii) *Power signals*: Signal with finite average power $P_\alpha > 0$ are known as power signals

Power signals have finite total energy

$$E_\alpha = \alpha \quad \text{if } P_\alpha > 0$$

A signal with finite energy has zero power, but signal with finite power has infinite energy.

Ex: Periodic signals such as

$$x(t) = \cos(t) \text{ and } x[n] = \sin 5n$$

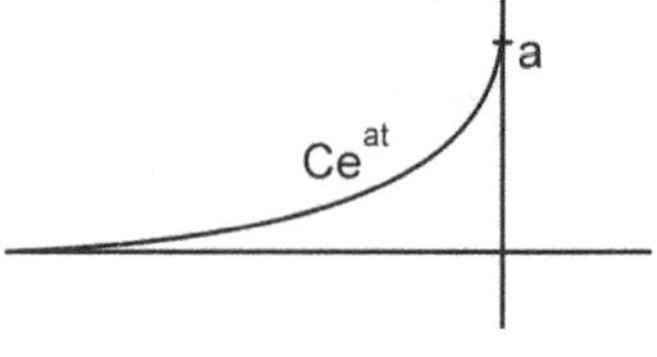

Fig. 1.11

(b) **Exponential and sinusoidal signal:**

A signal is real exponential if it can be represented by

$$x(t) = C\, e^{at}$$

where C and a are real

If a is imaginary, i.e., $a = j\omega_0$ it is called periodic complex exponential.

(c) **Bounded and Unbounded Signals:**

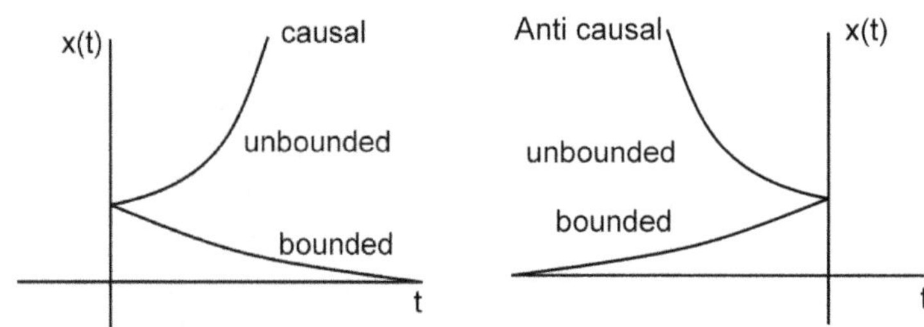

Fig. 1.12

(d) **Real and complex exponential signals:**

A real exponential signal is defined as

$$x(t) = A\, e^{at}$$

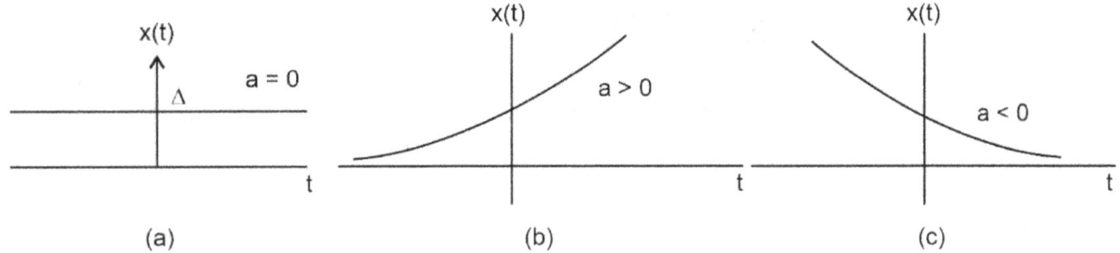

Fig. 1.13

Complex exponential signal is given by

$$x(t) = e^{st} \quad \text{where} \quad s = \sigma + j\omega$$

$$x(t) = e^{(\sigma + j\omega)t} = e^{\sigma t} \cdot e^{j\omega t}$$

$$= e^{\sigma t}\{\cos \omega t + j \sin \omega t\}$$

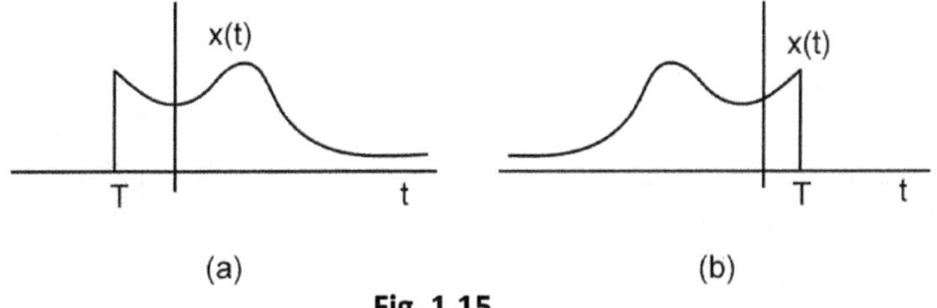

Fig. 1.14

(d) Right and left Sided Signals:

Right Sided Signal: If a signal is zero for $t > T$ then it is right sided signal.

Left Sided Signal: If a signal is zero for $t < T$ then it is left sided signal.

Fig. 1.15

1.2 ELEMENTARY SIGNAL MODELS

Some of the basic signals like unit step, impulse and exponential are needed to build other signals. Unit impulse is a rectangular pulse with a width that has become infinitesimally small, a height that has become infinitely large and overall area is unity.

1. **Unit Step function:** It begins at $t = 0$ (causal) and is constant up to ∞

$$u(t) = 1 \qquad \text{for} \quad t \geq 0$$
$$= 0 \qquad \text{for} \quad t < 0$$
$$u(t - a) = 1 \qquad \text{for} \quad t \geq a$$
$$= 0 \qquad \text{for} \quad t < a$$

If we need a signal to start at $t = 0$, we must multiply signal without x(t).

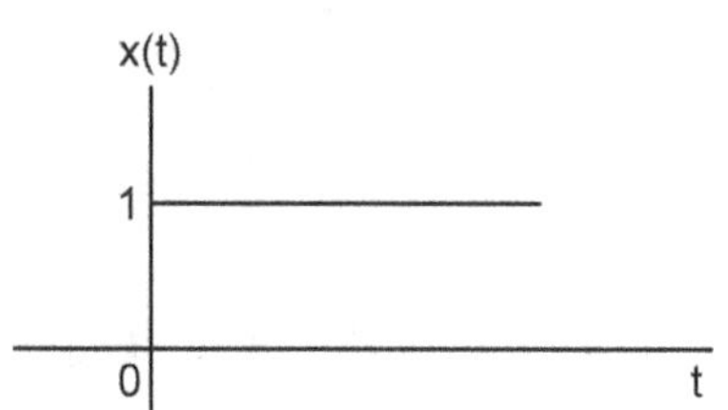

Fig. 1.16

2. **Impulse function:** It is first defined by P.A.M Dirac as

$$\int_{-\alpha}^{\alpha} \delta(t)\, dt = 1 \text{ for } \quad t = 0$$

and

$$\delta(t) = 0 \qquad \text{for} \quad t \neq 0$$

It has 0 amplitude everywhere except at $t = 0$. At $= 0$ its amplitude is infinite such that area under the curve is equal to 1

$$\delta(t) = \lim_{\Delta \to 0} x(t) \qquad \text{where } \Delta \text{ is the width of the}$$

signal and height is $1/\Delta$.

$$= \lim_{\Delta \to 0} \frac{1}{\Delta}\{u(t) - u(t - \Delta)\}$$

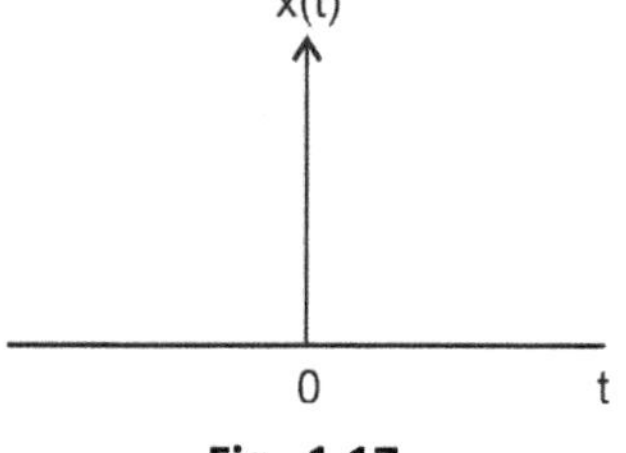

Fig. 1.17

delayed impulse

$$\int_{-\alpha}^{\alpha} \delta(t - a) = 1 \quad \text{for} \quad t = a$$

and

$$\delta(t - a) = 0 \quad \text{for} \quad t \neq a$$

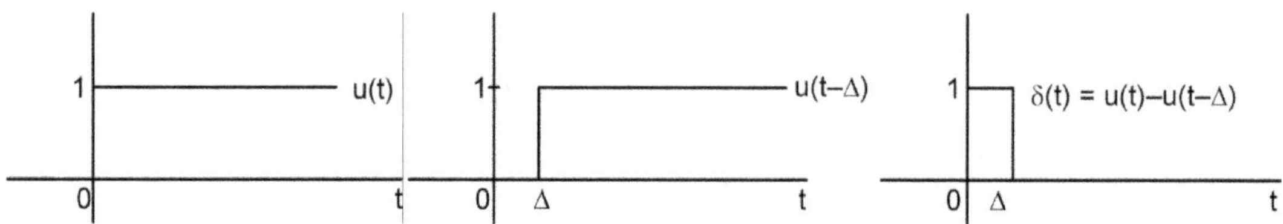

Fig. 1.18

3. **Rectangular pulse function:** A rectangular function can be described by unit step function as shown below.

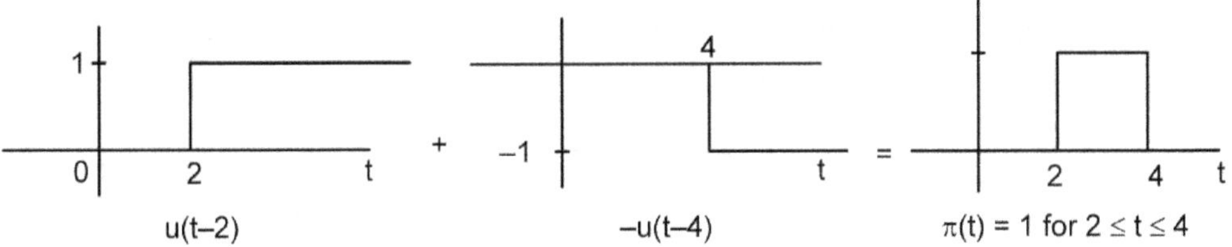

Fig. 1.19

In general a rectangular function is defined by

$$\pi(t) = 1 \quad \text{for} \quad |t| \le \frac{1}{2}$$

$$= 0 \quad \text{otherwise}$$

4. **Signum function**

Unit signum function is defined by

$$\text{sgn}(t) = \begin{cases} 1 & t > 0 \\ 0 & t = 0 \\ -1 & t < 0 \end{cases}$$

$$\text{sgn}(t) = -1 + 2u(t)$$

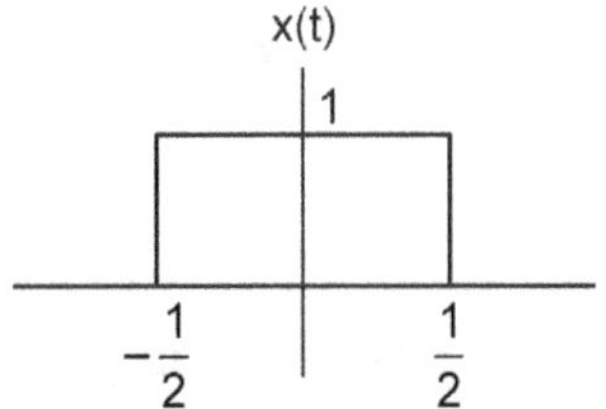

Fig. 1.20

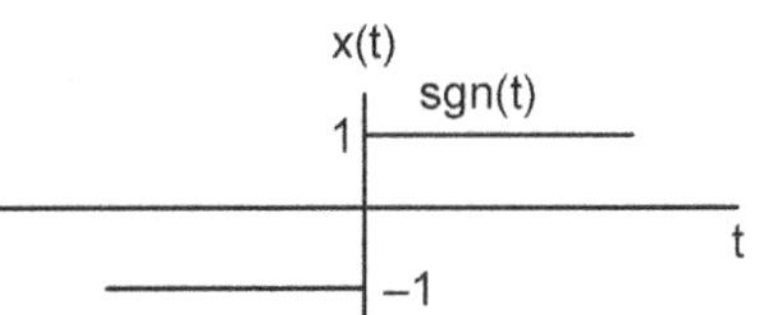

Fig. 1.21

5. **SinC function**

Unit sinC function is given by

$$\text{sinC}(t) = \frac{\sin \pi t}{\pi t} \quad -\alpha < t < \alpha$$

sinC function oscillates with period 2π and decays with increasing t.

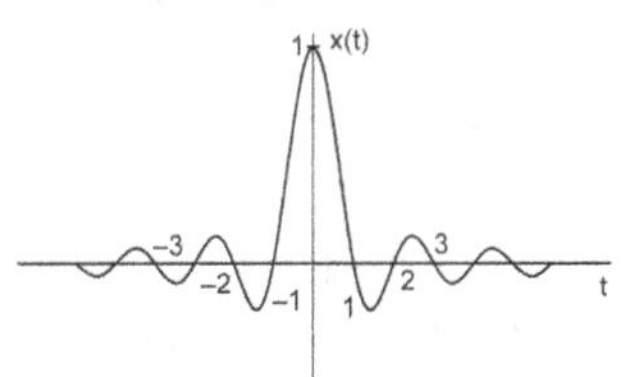

Fig. 1.22

6. **Unit ramp function**

Unit ramp function is defined by

$$r(t) = t \quad \text{for all } t$$

7. **Unit parabolic function**

Unit parabolic function is defined by

$$P(t) = \frac{t^2}{2} \quad \text{for } t \ge 0$$

$$= 0 \quad \text{for } t < 0$$

$$\text{or } P(t) = \frac{t^2}{2} u(t)$$

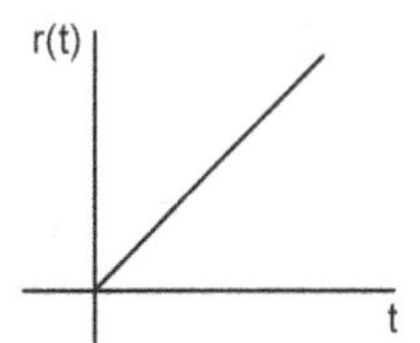

Fig. 1.23

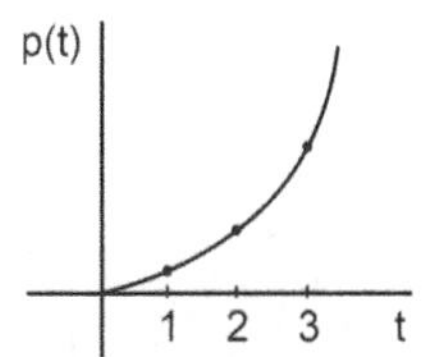

Fig. 1.24

1.3 BASIC OPERATIONS ON SIGNALS

The basic operations on signals are shifting, scaling and inversion.

1. **Time shifting:** Time shifting may delay or advance the signal in time domain.

$$y(t) = x(t - T) \quad \text{or} \quad y(n) = x(n - k)$$

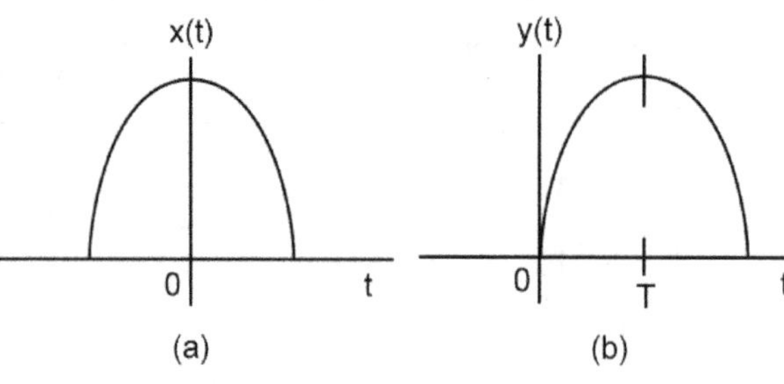
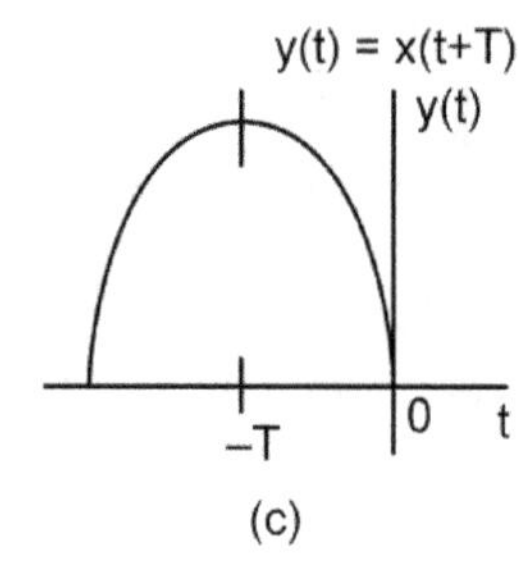

(a)　　　　　　　　(b)　　　　　　　　(c)

Fig. 1.25

2. **Time reversal:** Time reversal of a signal x(t) can be obtained by folding the signal w.r.t
   t = 0. Time reversal is obtained by rotating the frame by 180°.

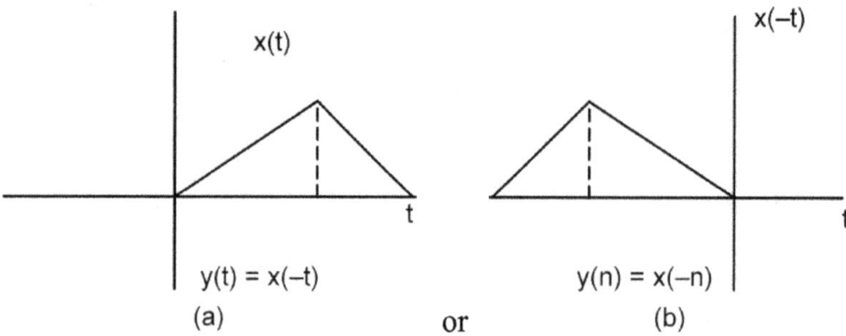

(a)　　　　　　　or　　　　　　　(b)

Fig. 1.26

3. **Time scaling:** Time scaling of a signal can be done by replacing t by at compression or
   expansion of a signal in time is known as time scaling.

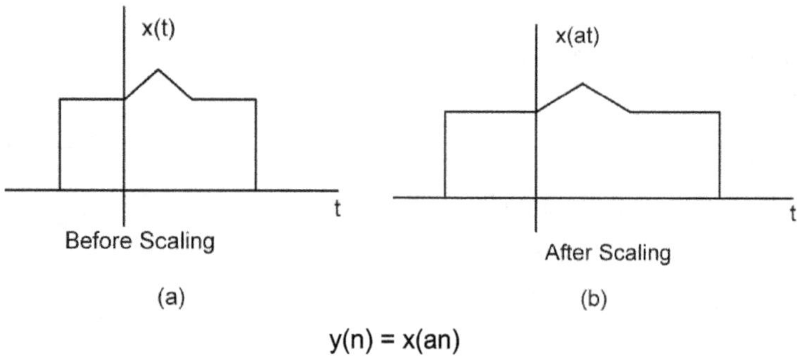

(a)　　　　　　　　　　　　　(b)

$$y(n) = x(an)$$

Fig. 1.27

4. **Amplitude scaling:** Amplitude scaling of a signal can be done by multiplying, amplitude by a constant

$$y(t) = ax(t)$$

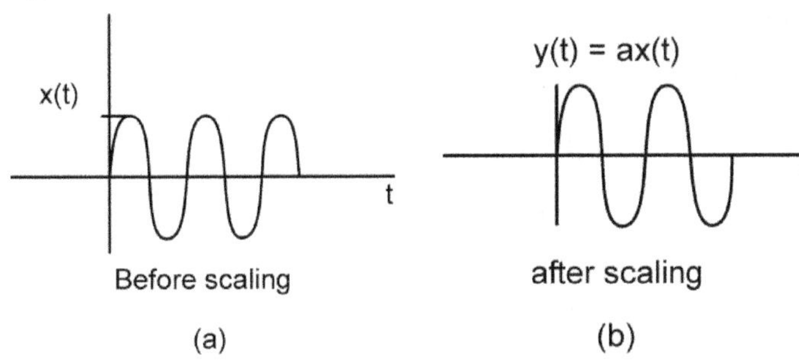

Fig. 1.28

1.4 PROPERTIES OF UNIT IMPULSE

We can assume impulse as a tall, narrow rectangular pulse of unit area. Exponential pulse, triangular pulse, gaussian pulse etc, can be used for impulse approximation.

Ex: exponential pulse ae^{-at} becomes narrow as a increases $\displaystyle\int_{-\alpha}^{\alpha} ae^{-at}dt = 1$

1. $\displaystyle\int_{-\alpha}^{\alpha} x(t)\delta(t)\,dt = x(0)$

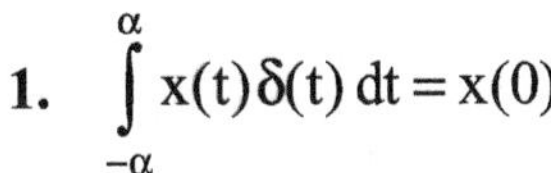

Let us consider the product of $x(t)$ and $\delta(t)$ which is $x(t)\,\delta(t)$ impulse exists at $t = 0$ only thus $x(t)$ exists only at $t = 0$.

$\therefore \quad x(t)\,\delta(t) = x(0)\,d(t)$

$\displaystyle\int_{-\alpha}^{\alpha} x(t)\,\delta(t)\,dt = x(0)\int \delta(t)\,dt = x(0)$

2. $\displaystyle\int_{-\alpha}^{\alpha} x(t)\,\delta(t-t_0)\,dt = x(t_0)$

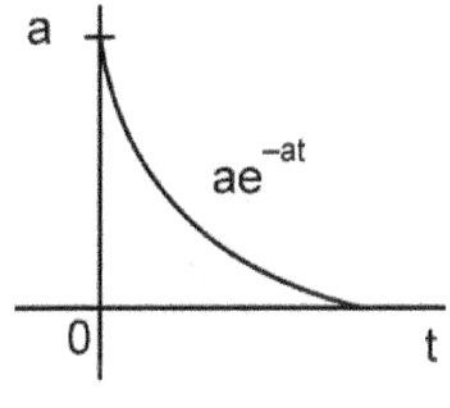

Since impulse is existing at t_0, $x(t)$ will exist at t_0 i.e., $x(t_0)$.

3. $\displaystyle\delta(at) = \frac{1}{|a|}\delta(t)$

$\displaystyle\int_{-\alpha}^{\alpha} x(t)\,\delta(at)\,dt$

Let at $= \lambda$

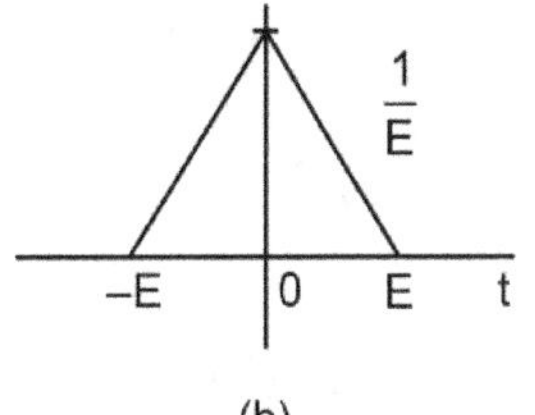

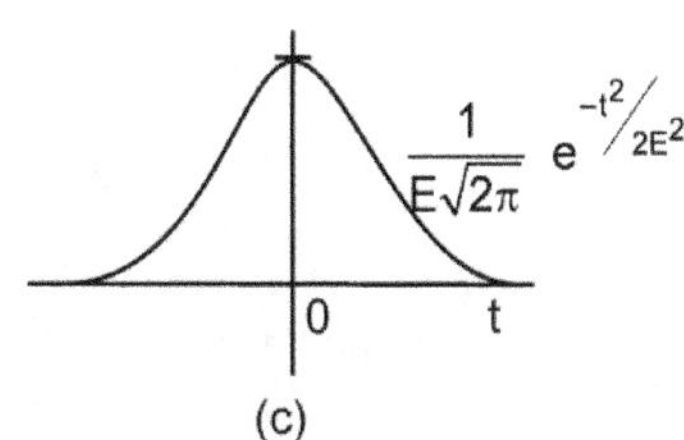

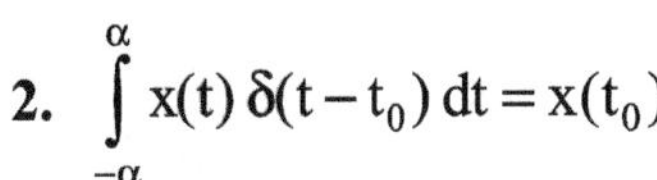

$$\text{adt} = d\lambda$$

$$dt = \frac{dT}{a}$$

$$= \frac{1}{a} \int_{-\alpha}^{\alpha} x\left(\frac{\lambda}{a}\right) \delta(\lambda)\, d\lambda$$

at $\lambda = 0$

$$= \frac{1}{|a|} x(0) = \frac{1}{a} \int_{-\alpha}^{\alpha} x(t)\, \delta(t)\, dt$$

$$= \frac{1}{a} \delta(t)$$

4. $\displaystyle \int_{-\alpha}^{\alpha} x(T)\, \delta(t-T)\, dt = x(T)$

replace t by T and t_0 by t in below equation

$$\int x(t)\, \delta(t-t_0)dt = \int x(T)\, \delta(T-t)dT$$

as impulse has even property

i.e., $\delta(t-T) = \delta(T-t)$, hence $\displaystyle \int x(t)\delta(t-t_0)dt = x(t_0)$

1.4.1 EXAMPLES USING IMPULSE FUNCTION

1. $\displaystyle \int_{-\alpha}^{\alpha} e^{-at^2} \delta(t-10)dt$

we have $\displaystyle \int x(t)\delta(t-t_0)dt = x(t_0)$

$\therefore$ $\displaystyle \int e^{-at^2} \delta(t-10)dt = e^{-a(t-10)^2}$

at $t = 0 = e^{-100a}$

2. $\displaystyle \int_{0}^{\alpha} t^2 \delta(t-3)\, dt = t^2 \big|_{t=3} = 9$

3. $\displaystyle \int_{-\alpha}^{\alpha} \delta(t)\cos t + \delta(t-1)\sin t\, dt$

$$= \int_{-\alpha}^{\alpha} \delta(t)\cos t\, dt + \int_{-\alpha}^{\alpha} \delta(t-1)\sin t\, dt$$

$$= 1 + \sin 1$$

4. $\displaystyle \int_{-\alpha}^{\alpha} \delta(t)e^{-j\omega t}\, dt = e^{-j\omega t}\big|_{\omega=0} = 1$

1.5   SOME OTHER PROPERTIES OF SIGNALS

(a)   Even and odd functions of signals: Every signal has even and odd components.

Even function × odd function = odd function

Odd function × odd function = even function

Even function × even function = even function

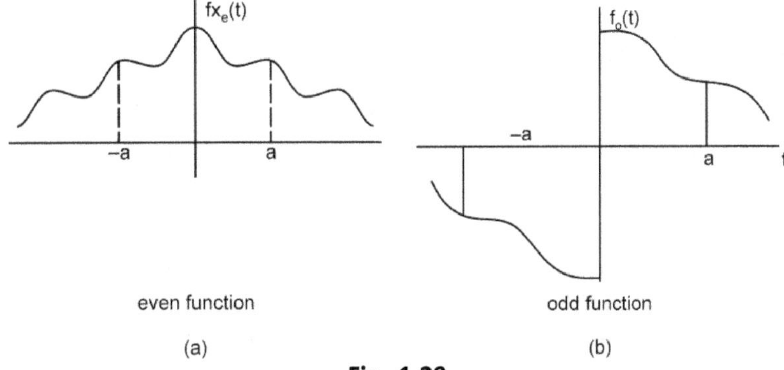

Fig. 1.29

Even signal is symmetrical about the vertical axis

$$\int_{-a}^{a} x_e(t)dt = 2\int_{0}^{a} x_e(t)dt$$

and for odd signal

$$\int_{-a}^{a} x_0(t)dt = 0$$

Any signal x(t) can be expressed as sum of even and odd components

$$x(t) = \frac{1}{2}\left|x(t) + x(t)\right| + \frac{1}{2}x(t) - x(-t)\left|\right.$$

$$= x_e(t) + x_0(t)$$

$$x_e(t) = \frac{1}{2}\left|x(t) + x(-t)\right|$$

$$x_0(t) = \frac{1}{2}\left|x(t) - x(-t)\right|$$

Examples:

1.   Find the even and odd components of e^{jt}

$$e^{jt} = x_e(t) + x_0(t)$$

$$x_e(t) = \frac{1}{2}\left\{e^{jt} + e^{-jt}\right\} = \cos t$$

$$x_0(t) = \frac{1}{2}\left\{e^{jt} - e^{-jt}\right\} = j\sin t$$

SOLVED PROBLEMS

EXERCISE 1

1. **Find the following signals are periodic or not.**

    (a) $2u(t) + 2\sin 2t$

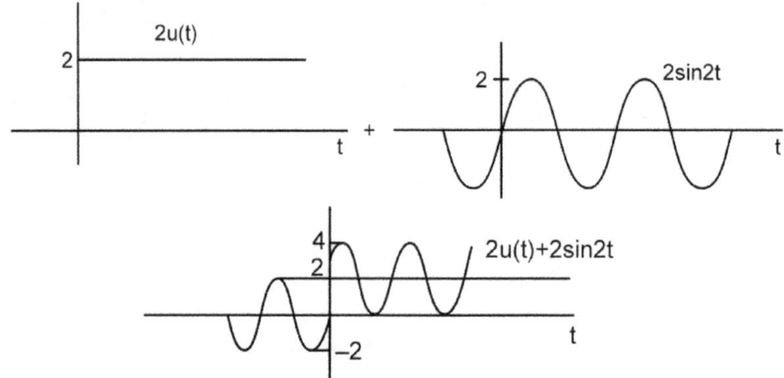

Fig. 1.30

∴ $2u(t) + 2\sin 2t$ is aperiodic

(b) $2\cos(10t + 1) - \sin(4t - 1)$

Time period of $2\cos(10t + 1)$ is $\dfrac{2\pi}{10} = \dfrac{\pi}{5}$

Time period of $2\sin(4t - 1)$ is $\dfrac{2\pi}{4} = \dfrac{\pi}{2}$

T_1/T_2, ratio of two periods $= \dfrac{2}{5}$ is a rational number, the sum of two signals is periodic

$$T = 2T_2 = 5T_1$$

(c) $x(t) = \cos 60\pi t + \sin 50\pi t$

$\cos 60\pi t$ time period $T_1 = \dfrac{2\pi}{60\pi} = \dfrac{1}{30}$

$\sin 50\pi t$ time period $T_2 = \dfrac{2\pi}{50\pi} = \dfrac{1}{25}$

∴ $\dfrac{T_1}{T_2} = \dfrac{5}{6}$ hence it is periodic

$T = 6T_1 = 5T_2$

2. Find the even and odd components of

(a) $x(t) = \cos t + \sin t + \cos t \sin t$

$x(-t) = \cos t - \sin t - \cos t \sin t$

$$x_e(t) = \frac{1}{2}\{x(t) + x(-t)\}$$

$$= \frac{1}{2}\{2\cos t\} = \cos t$$

$$x_o(t) = \frac{1}{2}\{x(t) - x(-t)\}$$

$$= \sin t + \cos t \sin t$$

(b) $x(t) = \cos t + 2 \sin t + 2 \sin^2 t \cos t = \cos t + 2 \sin t + (1 - \cos 2t) \cos t$

$x(-t) = \cos t - 2 \sin t + \cos t - \cos 2t \cos t$

$$x_e(t) = \frac{1}{2}\{2\cos t + 2\cos t - 2\cos 2t \cos t\}$$

$$= 2\cos t - \cos 2t \cos t$$

$$x_0(t) = \frac{1}{2}\{4\sin t\}$$

(c) $x(n) = \{-2, 1, 2, -1, 3\}$

$(n) = \{-2, -1, 0, -1, 2\}$

$$x_e(-n) = \frac{1}{2}\{x(n) + x(-n)\}$$

$$x_e(0) = \frac{1}{2}\{x(0) + x(-0)\} = \frac{1}{2} \times (2 + 2) = 2$$

$$x_e(1) = \frac{1}{2}\{x(1) + x(-1)\} = \frac{1}{2}\{-1 + 1\} = 0$$

$$x_e(2) = 0.5$$

$$x_e(n) = \{0.5, 0.2, 0, 0.5\}$$

EXERCISE 2

1. Determine the power for the signal

(a) $x(t) = A \cos(\omega_0 t + \theta)$

$$P = \underset{T \to \alpha}{Lt}\, \frac{1}{T} \int_{-T/2}^{T/2} A^2 \cos^2(\omega_0 t + \theta)\, dt$$

$$= \underset{T \to \alpha}{Lt}\, \frac{1}{T} \int_{-T/2}^{T/2} A^2 \left\{\frac{1 + \cos(2\omega_0 t + 2\theta)}{2}\right\} dt$$

$$= \underset{T\to\alpha}{Lt} \frac{A^2}{2T} \left\{ \int_{-\frac{T}{2}}^{\frac{T}{2}} 1\,dt + \int_{-\frac{T}{2}}^{\frac{T}{2}} \cos(2\omega_0 t + 2\theta)\,dt \right\}$$

$$= \underset{T\to\alpha}{Lt} \frac{A^2}{2T} \cdot T = \frac{A^2}{2}$$

(b) $x(t) = e^{j\omega t} \cos\omega_0 t$

$$P = \underset{T\to\alpha}{Lt} \frac{1}{T} \int_{-T/2}^{T/2} \left(e^{j\omega t} \cos\omega_0 t \right)^2 dt$$

$$= \underset{T\to\alpha}{Lt} \frac{1}{T} \int_{-T/2}^{T/2} \frac{1+\cos 2\omega_0 t}{2} dt$$

$$= \underset{T\to\alpha}{Lt} \frac{1}{2T} \left[\int_{-T/2}^{T/2} dt + \int_{-T/2}^{T/2} \cos 2\omega_0 t\, dt \right]$$

$$= \underset{T\to\alpha}{Lt} \frac{1}{2T} \cdot T$$

$$= \frac{1}{2}$$

(c) $x(t) = A\, e^{j\omega_0 t}$

$$p = \underset{T\to\alpha}{Lt} \frac{1}{T} \int_{-T/2}^{T/2} \left(A e^{j\omega_0 t} \right)^2 dt$$

$$= \underset{T\to\alpha}{Lt} \frac{A^2}{T} \int_{-T/2}^{T/2} \left| e^{j2\omega_0 t} \right| dt$$

$$= \frac{A^2}{T} \cdot T = A^2$$

2. Find which of the following are energy signals, power signals

(a) $x(t) = e^{-3t}\, u(t)$

$$\text{Energy of signal } E = \underset{T\to\alpha}{Lt} \int_{-T}^{T} \left(e^{-3t} \right)^2 u(t)\,dt$$

$$= \underset{T\to\alpha}{Lt} \int_{-T}^{T} e^{-6t} u(t)\,dt = \underset{T\to\alpha}{Lt} \int_{0}^{T} e^{-6t}\, dt = \frac{1}{6}$$

Power of the signal

$$P = \underset{T\to\alpha}{Lt}\ \frac{1}{T}\int_{-T/2}^{T/2} e^{-6t}\,dt$$

$$= \underset{T\to\alpha}{Lt}\ \frac{1}{T}\int_{0}^{T/2} e^{-6t}\,dt$$

$$= \underset{T\to\alpha}{Lt}\ \frac{1}{-6T}e^{-6t}\Big/_{0}^{T/2} = 0$$

As energy is finite and power is 0,

It is energy signal

(b) $x(t) = \cos(t)$

$$E = \underset{T\to\alpha}{Lt}\ \int_{-T}^{T} \cos^2 t\,dt$$

$$= \frac{1}{2}\underset{T\to\alpha}{Lt}\ \int_{-T}^{T} 1dt + \int_{-T}^{T}\cos 2t\,dt$$

$$= \alpha$$

$$P = \underset{T\to\alpha}{Lt}\ \frac{1}{T}\int_{-T/2}^{T/2} \cos^2 t\,dt$$

$$= \underset{T\to\alpha}{Lt}\ \frac{1}{2T}\int_{-T/2}^{T/2} 1 + \cos 2t\,dt$$

$$= \frac{1}{2T}\cdot T = \frac{1}{2}$$

As energy is infinite and power is finite

It is power signal

3. Find which of the signals are causal or non causal?

(a) $x_1(t) = e^{at}\,u(t)$

        for $t < 0$ $x_1(t) = 0$

        $\therefore$ it is causal

(b) $x_2(t) = e^{-2t}\,u(-t)$

        for $t < 0$ $x_2(t) \neq 0$

        $\therefore$ it is non causal

(c) $x_3(t) = \text{sinc } t$

        for $t < 0$, $x_3(t) \neq 0$

        $\therefore$ it is non causal

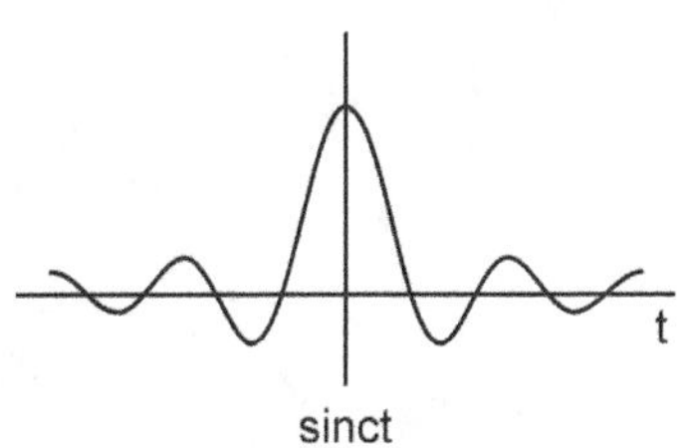

Fig. 1.31

4. Sketch the following signals

(a) $u(-t+1)$ (b) $-2u(t-1)$ (c) $3r(t-1)$ (d) $\pi(t+3)$

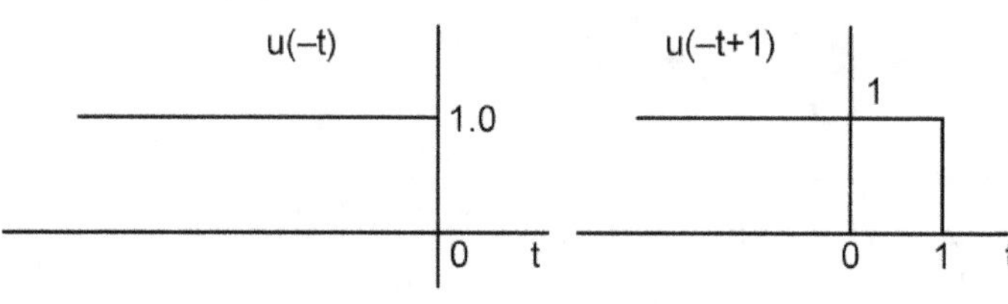

Fig. 1.32

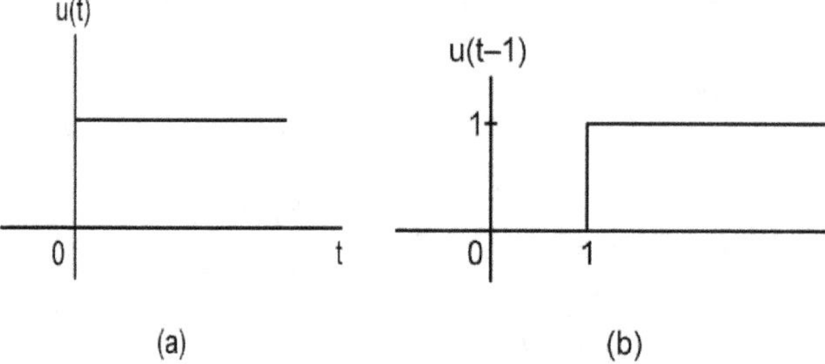

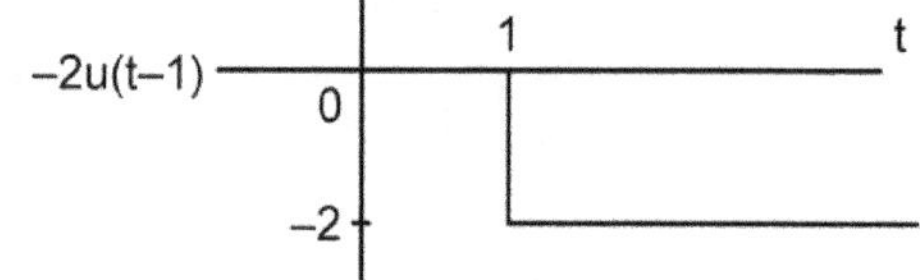

Fig. 1.33

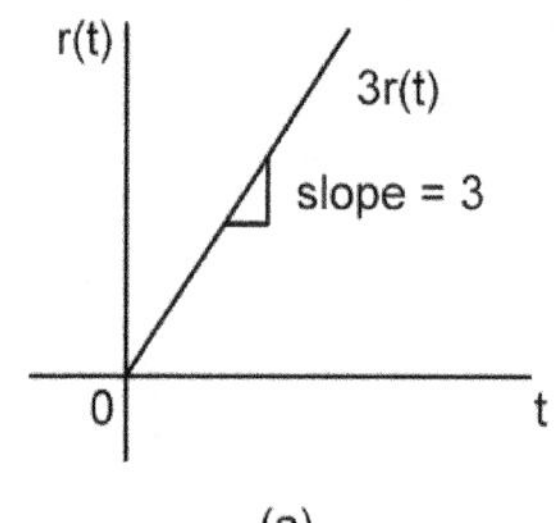

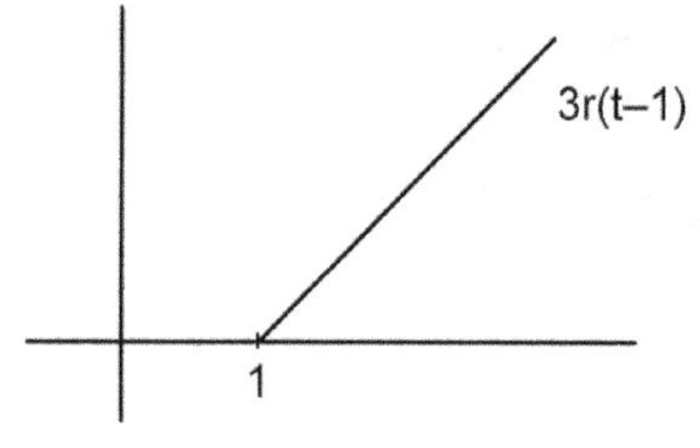

Fig. 1.34

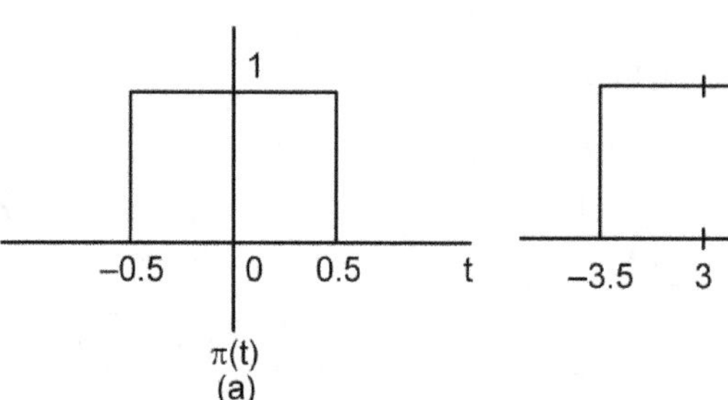

Fig. 1.35

EXERCISE 3

1. Find the even and odd components of below signals

Ans:

    (a) $x(t) = \cos t + \sin t + \sin t \cos t$

        $x(-t) = \cos t - \sin t - \sin t \cos t$

        $\therefore\ x_e(t) = \cos t$

        $x_o(t) = \sin t \cos t + \sin t$

    (b) $x(t) = 1 + t \cos t + t^2 \sin t + t^3 \sin t \cos t$

        $x_e(t) = 1 + t^3 \sin t \cos t$

        $x_o(t) = t \cos t + t^2 \sin t$

2. Consider $x(t) = A \cos(\omega t + \phi)$ find average power

Ans:

$$P = \frac{1}{T} \int_{-T/2}^{T/2} \left| x^2(t) \right| dt \qquad\qquad T = \frac{2\pi}{\omega}$$

$$P = \frac{\omega}{2\pi} \int_{-\pi/\omega}^{\pi/\omega} A^2 \cos^2(\omega t + \phi)\, dt = \frac{A^2}{2}$$

3. Find average power of a discrete signal $\quad x(n) = A \cos(\omega_0 n + \phi)$

$$P = \frac{1}{N} \sum_{n=0}^{N+1} x^2[n]$$

$$N = \frac{2\pi}{\omega_0}$$

$$P = \frac{\omega_0}{2\pi} \sum_{n=0}^{\frac{2\pi}{\omega_0}-1} A^2 \cos^2(\omega_0 n + \phi)$$

$$= \frac{A^2 \omega_0}{4\pi} \{1 + \cos 2\phi + 1 + \cos(2\phi - 2\omega_0)\}$$

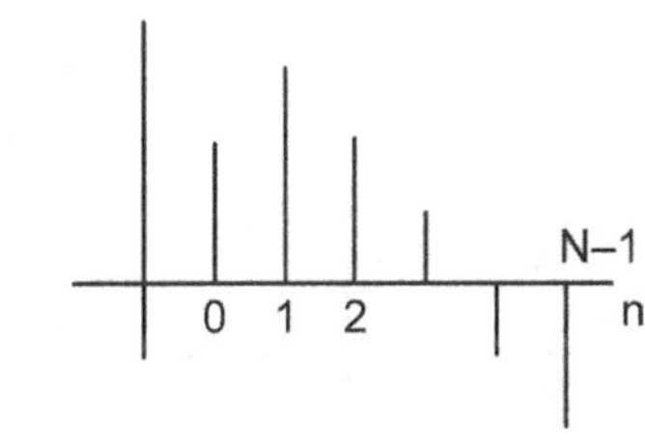

Fig. 1.36

4. Find the energy of signal

$$x(t) = \frac{1}{2}(\cos \omega t + 1) - \frac{\pi}{\omega} \le t \le \pi/\omega$$

$$E = \int_{-\alpha}^{\alpha} x^2(t)\, dt = \int_{-\pi/\omega}^{\pi/\omega} \frac{1}{2}(\cos \omega t + 1)^2\, dt$$

$$= \frac{3\pi}{2\omega}$$

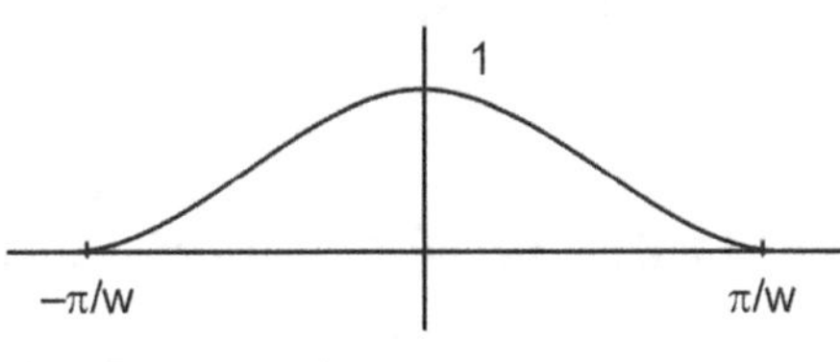

Fig. 1.37

5. Sketch the wave form

    (a) $x(t) = u(t + 1) - 2u(t) + u(t - 1)$

(b) $x(t) = r(t + 1) - r(t) + r(t - 2)$

(c) $x(t) = r(t + 2) - r(t + 1) - r(t - 1) + r(t - 2)$

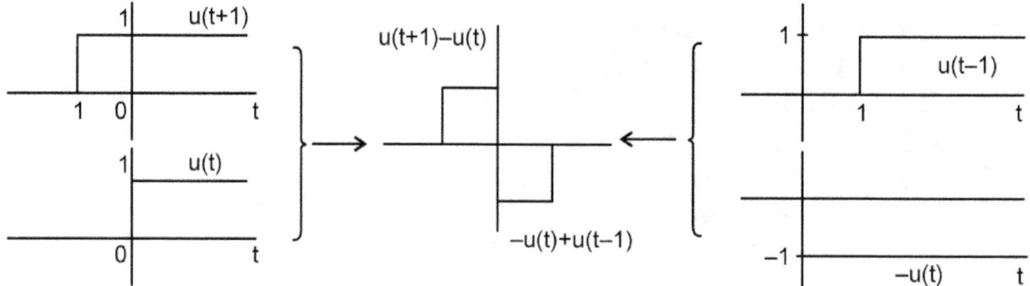

Fig. 1.38

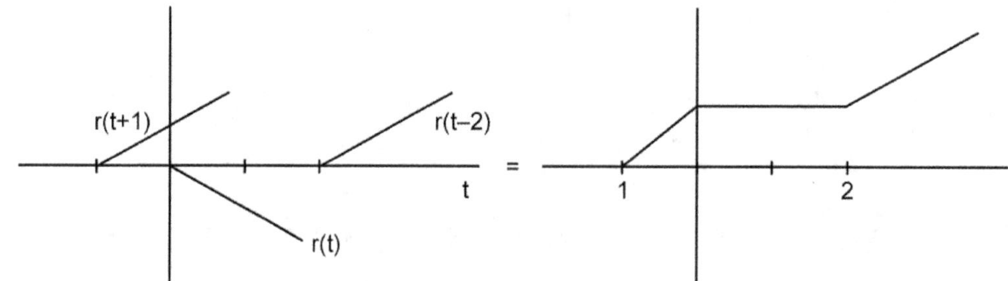

Fig. 1.39

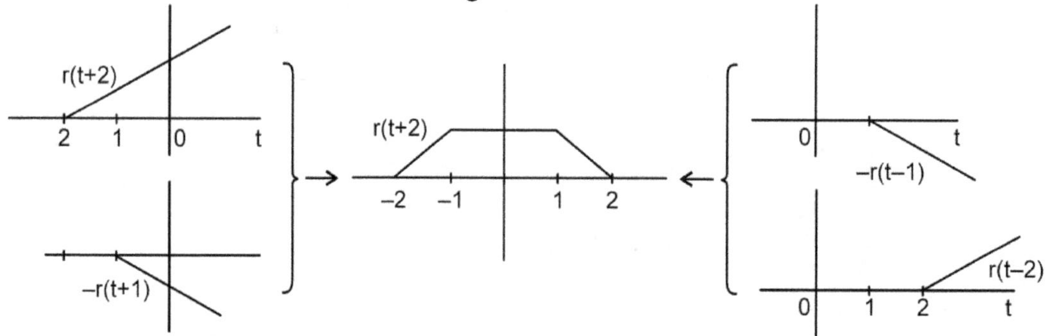

Fig. 1.40

6. Find the below signals are periodic or non periodic

(a) $x[n] = \displaystyle\sum_{k=-\alpha}^{\alpha} \left\{ \delta(n - 3k) + \delta(n - k^2) \right\}$

(b) $x(t) = \displaystyle\sum_{k=-\alpha}^{\alpha} (-1)^k \, \delta(t - 2k)$

(a) It is not periodic, since impulse is not repeating at the same interval

(b) It is periodic, its period is $(-1)^k$

SOLVED PROBLEMS FROM PREVIOUS QUESTION PAPERS

UNIT – 1   NOV - 2008

1. Sketch the following signal

(a) $\pi\left(\dfrac{t-1}{2}\right) + \pi(t-1)$

We have $\pi(t) = 1$ for $|t| \le \dfrac{1}{2}$

$\qquad\qquad = 0$ otherwise

(i) $\pi(t-1) = 1$

For $|t-1| \le \dfrac{1}{2}$

$\therefore\ t \le \dfrac{3}{2}$ and $t \ge 1/2$

(ii) $\pi\left(\dfrac{t}{2} - \dfrac{1}{2}\right) = \left|\dfrac{t-1}{2}\right| \le \dfrac{1}{2}$

$\therefore\ t \le 2$

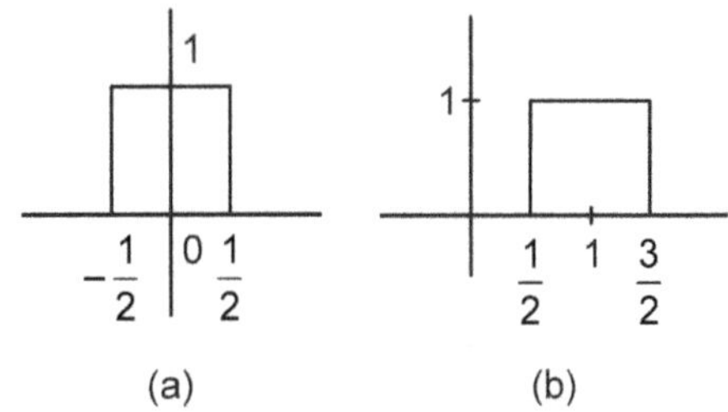

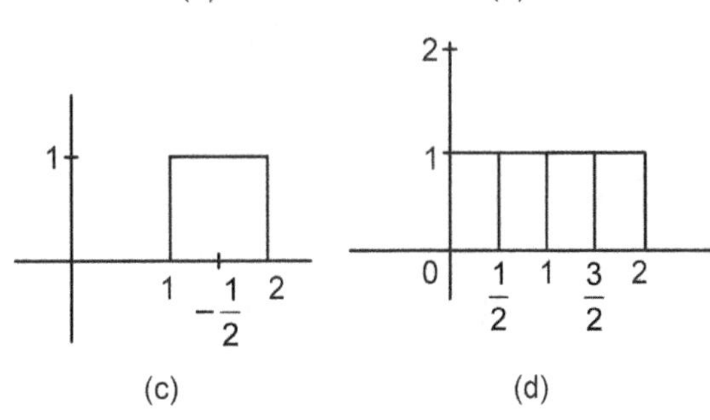

Fig. 1.41

2. $3u(t) - u - (t-1) - 5u(t-2)$

NOV – 2008

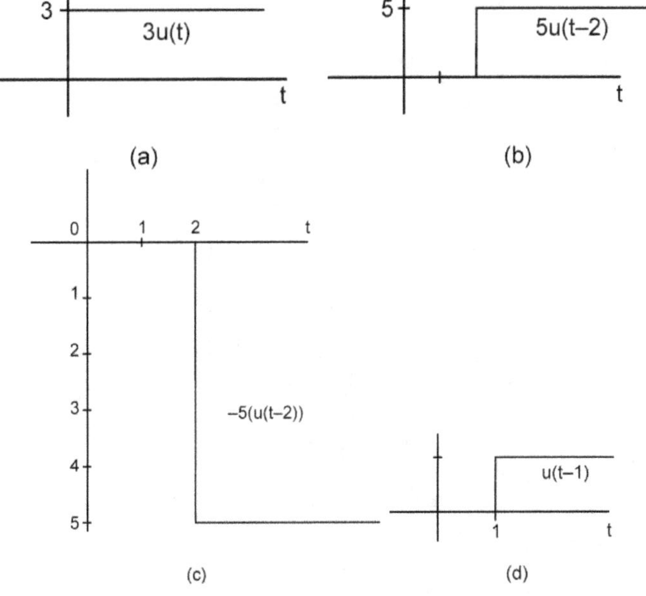

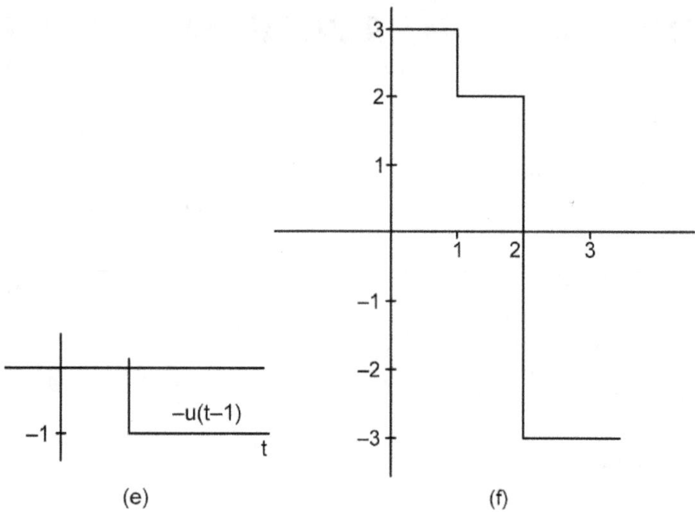

Fig. 1.42

3. Sketch (a) x(t − u) and (b) x(−t) for the signal shown below

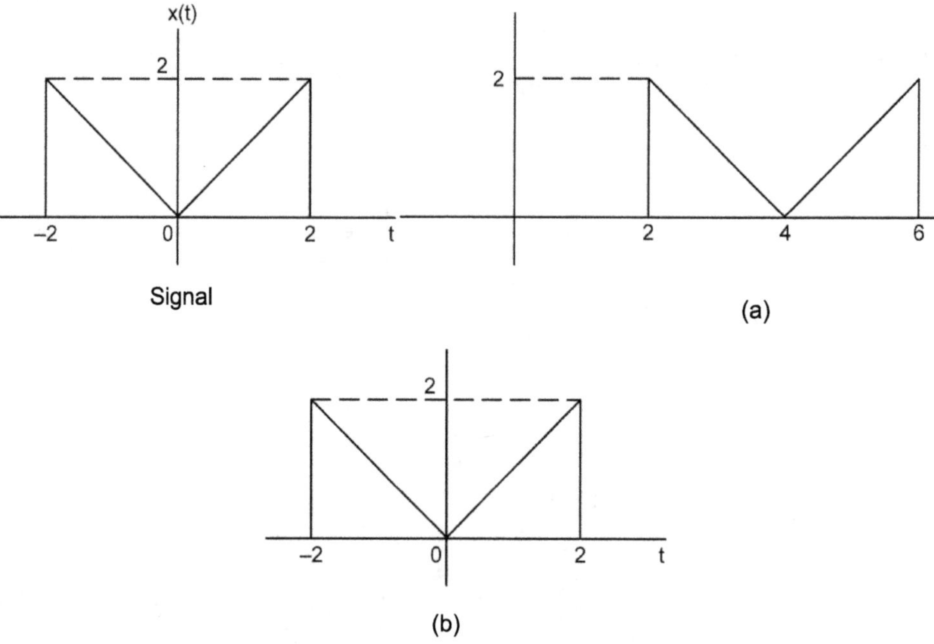

Fig. 1.43

(b) Unit parabolic function

$$P(t) = \frac{t^2}{2} u(t)$$

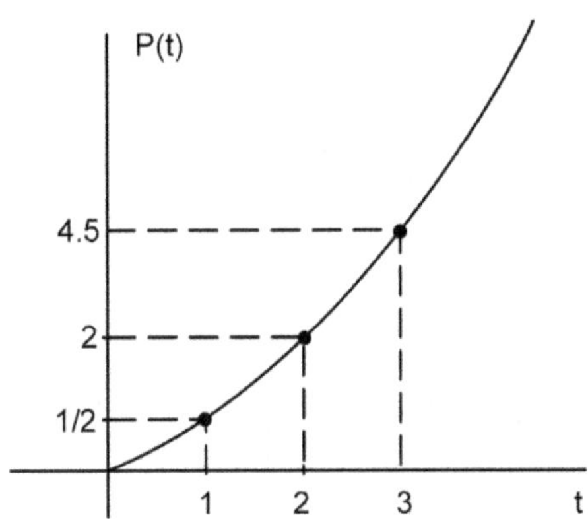

Fig. 1.44

(c) Delayed unit impulse $\delta(t - T) = 1$

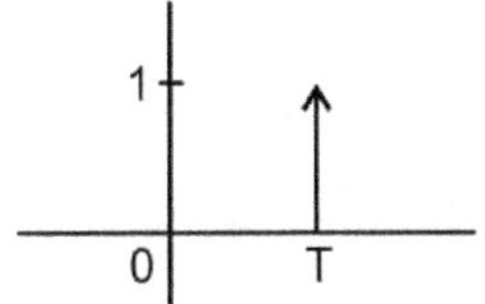

Fig. 1.45

4. Evaluate the integral

(a) $\displaystyle \int_{0}^{\alpha} \delta(t) \sin 2\pi t \, dt = \sin 2\pi t \int_{t=0} \delta(t) dt = 0$

(b) $\displaystyle \int_{-\alpha}^{\alpha} e^{-\alpha t^2} \delta(t - 10) dt = e^{-\alpha t^2} \int_{t=10} \delta(t) dt = e^{-100\alpha}$

5. Prove the following   Feb 2008

(a) $\delta(n) = u(n) - u(n - 1)$

$$u(n) = \sum_{n=0}^{\alpha} \delta(n)$$

$$u(n - 1) = \sum_{n=1}^{\alpha} \delta(n)$$

$\therefore \ u(n) - u(n - 1) = \delta(n) \quad$ at $n = 0$

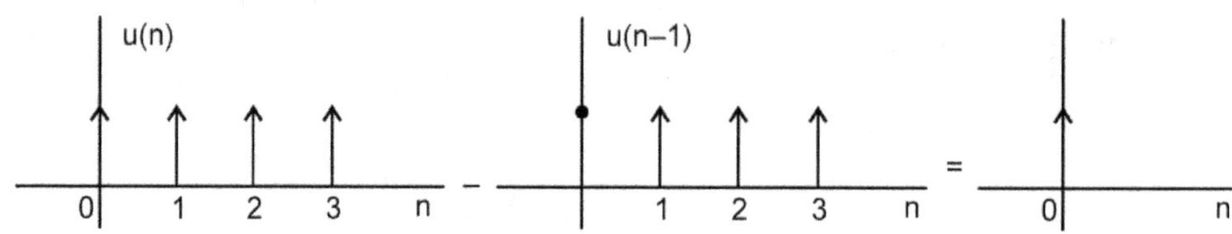

Fig. 1.46

(b) $u(n) = \sum\limits_{n=0}^{\alpha} \delta(k)$ Feb 2008

as $\delta(k)$ exists at $k = 0$

$\sum\limits_{k=0}^{n} \delta(k)$ means $\delta(k)$ exists at $K = 0, 1, 2, \ldots\ldots$

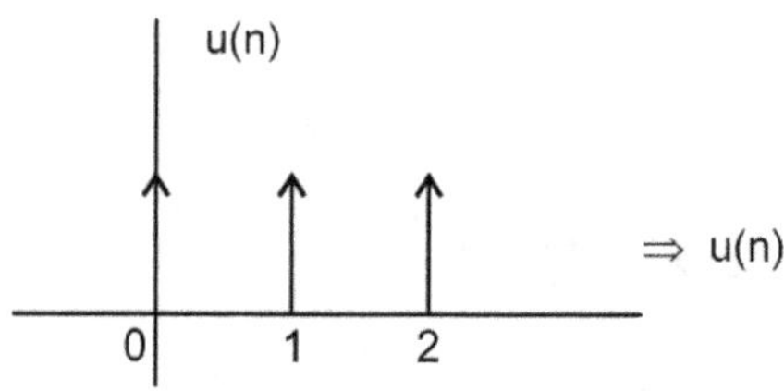

$\Rightarrow u(n)$

Fig. 1.47

6. Prove that $\text{sinc}(0) = 1$ and plot sin c function   Feb 2008

$$\text{sinc}(t) = \frac{\sin \pi t}{\pi t} \qquad -\alpha \le t < \alpha$$

$$\text{sinc}(0) = \frac{\sin 0}{0} = 1$$

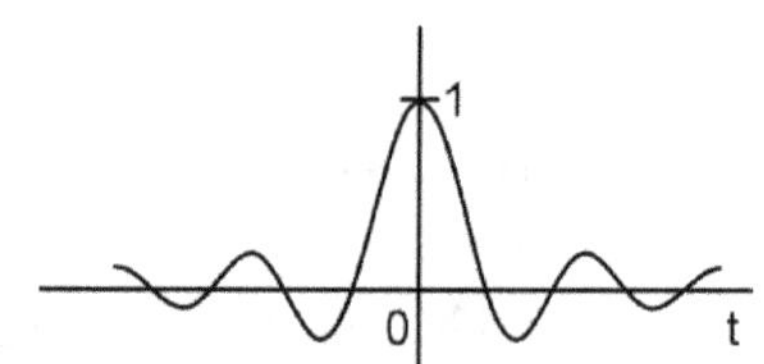

7. Prove that   Aug 2007

(a) $\int\limits_{-\alpha}^{\alpha} x(T)\delta(t-T)\,dt = x(t)$

impulse function has even property $x(t) = (-t)$

$\therefore \ \delta(t-T) = \delta(T-t)$

$\int x(T)\delta(t-T)\,dt = \int x(T)\delta(T-t)\,dt$

$\qquad\qquad = x(t)$

(b) $\int\limits_{-\alpha}^{\alpha} \delta(t)x(t)\,dt = -x(0)$ Aug 2007 where $\delta(t) = \dfrac{d}{dt}\delta(t)$

let $u = x(t) \quad dv = \delta(t)$

$\rightarrow v = \delta(t) \quad \int u\, dv =$

$du = x(t) = uv - \int v\, du$

$\int \delta(t)x(t)dt = \int u\, dv$

$\qquad = x(t)\,\delta(t) - \int x(t)\delta(t)dt$

at $\quad t = 0, \quad x(t) = 0$

$\qquad = -\int x(t)\,\delta(t)\, dt = -x(0)$

8. Show that whether $x(t) = Ae^{-at}$ is energy signal or not?   Nov 2008

Energy $\quad E = \displaystyle\int_{-\alpha}^{\alpha} \left|x^2(t)\right| dt$

$$E = \underset{T\to\alpha}{\text{Lt}} \int_{-T}^{T} (Ae^{-at})^2\, dt$$

$$= \underset{T\to\alpha}{\text{Lt}}\, A^2 \int_{-T}^{T} e^{-2at}\, dt = \underset{T\to\alpha}{\text{Lt}}\, \frac{A^2}{-2a} e^{-2at}\bigg|_{-T}^{T}$$

$$= \underset{T\to\alpha}{\text{Lt}}\, \frac{A^2}{-2a}\left\{e^{-2aT} - e^{+2aT}\right\} = \infty \quad \text{it is not energy signal}$$

9. Find whether the following signals are periodic or not?

(a) $\quad 3\cos\left(17\pi t + \dfrac{\pi}{3}\right) + 2\sin\left(19\pi t - \dfrac{\pi}{3}\right)$

Time period of $3\cos\left(17\pi t + \dfrac{\pi}{3}\right)$

$$T_1 = \frac{2\pi}{17\pi} = \frac{2}{17}$$

Time period of $2\sin\left(19\pi t - \pi/3\right)$

$$T_2 = \frac{2\pi}{19\pi} = \frac{2}{19}$$

$$\therefore \ \frac{T_1}{T_2} = \frac{2/17}{2/19} = \frac{19}{17}$$

$$\therefore \ T = 17\, T_1 = 19\, T_2$$

$\dfrac{19}{17}$ is a rational number. The sum of two signals is periodic

$$T = \cancel{17} \times \dfrac{2\cancel{\pi}}{\cancel{17}\,\cancel{\pi}} = 2\,\text{sec}$$

(b) $12e^{j(4t+\pi/2)} + 9e^{j(3\pi t+\pi/3)}$

$$T_1 = \dfrac{2\pi}{4}$$

$$T_2 = \dfrac{2\pi}{3\pi}$$

$$T_1 = \dfrac{2\pi}{\omega_1} = \dfrac{2\pi}{4} = \dfrac{\pi}{2}$$

$$T_2 = \dfrac{2\pi}{3\pi} = \dfrac{2}{3}$$

$$\dfrac{T_1}{T_2} = \dfrac{3\pi}{4}$$

$$4T_1 = 3\pi T_2$$

$\therefore$ it is not periodic

(c) $\cos(20n)$

$$\omega_0 = 20$$

$$N = \dfrac{2\pi}{20} = \dfrac{\pi}{10}$$

Since it is an integer, it is aperiodic

10. Which of the following signals are energy signals   or power signals?

(a) $r(t) - r(t-2)$

$$E = \lim_{T \to \alpha}\left[\int_{-T}^{2} |x(t)|^2\,dt + \int_{2}^{T} |x(t)|^2\,dt\right]$$

$$= \lim_{T \to \alpha}\left[\int_{0}^{2} t^2\,dt + \int_{2}^{T} 4\,dt\right]$$

$$= \lim_{T \to \alpha}\left[\dfrac{8}{0} + 4(t-2)\right] = \alpha$$

$$P = \lim_{T \to \alpha}\dfrac{1}{2T}\left[\int_{0}^{2} t^2\,dt + \int_{2}^{T} 4\,dt\right]$$

$$= \lim_{T \to \alpha}\dfrac{1}{2T}\left[\dfrac{8}{3} + 4(T-2)\right] = 2\omega$$

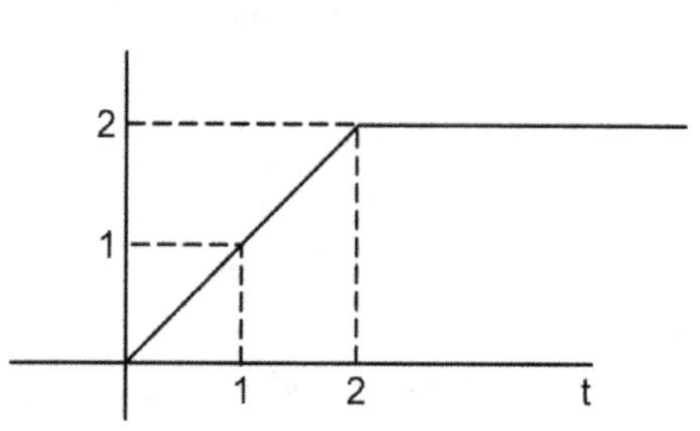

The energy of the signal is infinite and power is finite. Hence it is power signal.

(b) $u(t) - u(t-1)$

$$E = \lim_{T \to \alpha} \int_{-T}^{T} |x_1(t)|^2 \, dt$$

$$P = \lim_{T \to \alpha} \frac{1}{2T} \int_{0}^{1} 1^2 \, dt = 0$$

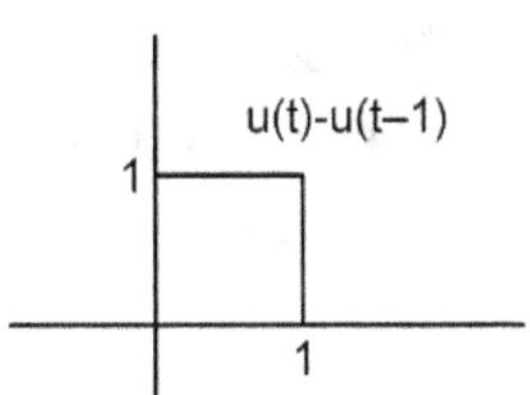

Hence it is energy signal

EXERCISE PROBLEMS (SHORT QUESTIONS)

1. Find the even and odd components of the below signals

(a) $x(t) = \cos t + \sin t + \sin t \cos t$

> **Ans:** $\cos t$ is even, $\sin t + \sin t \cos t$ is odd

(b) $x(t) = 1 + t + 3t^2 + 5t^3 + 2t^4$

> **Ans:** $1 + 3t^2 + 2t^4$ is even, $t + 5t^3$ is odd

(c) $x(t) = 1 + t \cos t + t^2 \sin t + t^3 \sin t \cos t$

> **Ans:** $1 + t^3 \sin t \cos t$ is even, $t \cos t + t^2 \sin t$ is odd

(d) $x(t) = (1 + t^3) \cos^3 10t$

> **Ans:** $\cos^3 10t$ is even, $t^3 \cos^3 10t$ is odd

(e) $x(t) = e^{-2t} \cos t = (\cosh 2t - \sinh 2t) \cos t$

> **Ans:** $\cosh 2t \cos t$ is even, $-\sinh 2t \cos t$ is odd

Hint for above problems:

$$x_e(t) = \frac{1}{2}\{x(t) + x(-t)\}$$

$$x_0(t) = \frac{1}{2}\{x(t) - x(-t)\}$$

2. For the below signals find fundamental period.

(a) $x(t) = \cos^2 2\pi t$ — **Ans:** Periodic, $T = 0.5s$

(b) $x(t) = \sin^3 2t$ — **Ans:** Periodic, $T = \dfrac{1}{\pi}$ s

(c) $x(t) = e^{-2t} \cos 2\pi t$ — **Ans:** Non periodic

(d) $x(n) = (-1)^n$ — **Ans:** Periodic, $N = 2$

(e) $x(n) = (-1)n^2$ — **Ans:** Periodic, $N = 2$

(f) $x(n) = \cos 2n$ — **Ans:** Non periodic

Hint for above Problems:

If the signal is periodic $x(t) = x(t + T)$

Fundamental period

$$T = \frac{2\pi}{\omega}$$

or $$N = \frac{2\pi}{\Omega}$$

3. Classify below signals as energy or power signals

(a) $x(t) = \begin{cases} t & 0 \le t \le 1 \\ 2-t & 1 \le t \le 2 \\ 0 & \text{otherwise} \end{cases}$

Ans: Energy signal $E = \dfrac{2}{3}$

(b) $x(n) = \begin{cases} n & 0 \le n \le 5 \\ 10-n & 5 \le t \le 10 \\ 0 & \text{otherwise} \end{cases}$

Ans: Energy signal $E = 85$

(c) $x(t) = 5\cos \pi t + \sin 5\pi t, -\alpha \le t \le \alpha$

Ans: Power signal $P = 13$

(d) $x(t) \begin{cases} 5\cos t\pi t & -1 \le t \le 1 \\ 0 & \text{otherwise} \end{cases}$

Ans: Energy signal $E = 25$

(e) $x(t) = \begin{cases} 5\cos t\pi t & -0.5 \le t \le 0.5 \\ 0 & \text{otherwise} \end{cases}$

Ans: Energy signal $E = 12.5$

(f) $x(n) = \begin{cases} \sin \pi n & -4 \le n \le 4 \\ 0 & \text{otherwise} \end{cases}$

Ans: Zero signal

(g) $x(t) = \begin{cases} \cos \pi n & n \ge 0 \\ 0 & \text{otherwise} \end{cases}$

Ans:

Hint for above problems:

$$\text{Energy} = \lim_{T \to \alpha} \int_{-T/2}^{T/2} x^2(t)\,dt = \int_{-\alpha}^{\alpha} x^2(t)\,dt$$

$$\text{Power} = \lim_{T \to \alpha} \frac{1}{T} \int_{-T/2}^{T/2} x^2(t)\,dt$$

for discrete signal

$$E = \sum_{n=-\alpha}^{\alpha} x^2(n)$$

$$P = \lim_{N \to \alpha} \frac{1}{2N} \sum_{n=-N}^{N} x^2(n)$$

4. Sketch the below signals [x (t) is triangular pulse]

(a) $x(3t)$

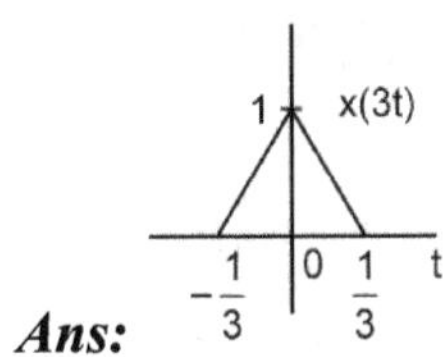

Ans:

(b) $x(3t + 2)$

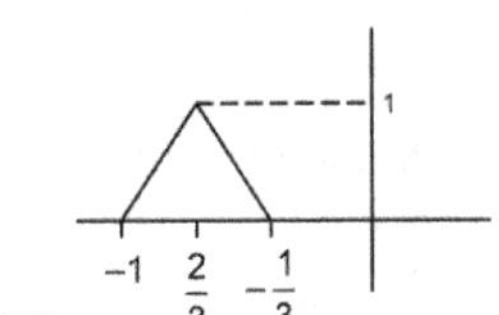

Ans:

(c) $x(-2t - 1)$

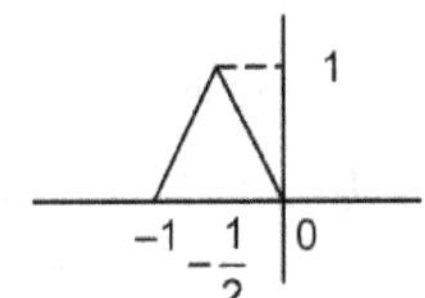

Ans:

(d) $x(2(t + 2))$

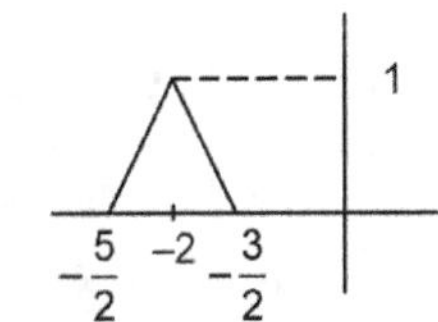

Ans:

5. Find the linear, time invariant signals among the below given signals.

(a) $y(t) = t^2 x(t - 1)$ *Ans:* Linear but not time invariant

(b) $y(n) = x(n + 1) - x(n - 1)$ *Ans:* Linear, time invariant

(c) $y(n) = x^2(n - 2)$ *Ans:* Linear, time invariant

6. Determine which of the below properties hold for each signals

(a) Memory less

(b) Time invariant

(c) Linear

(d) Causal

(i) $y(t) = x(t - 2) + x(2 - t)$ *Ans:* Non causal, time invariant

(ii) $y(t) = \dfrac{dx(t)}{dt} = \dfrac{x(t) - x(t-h)}{h}$ ***Ans:*** Non causal, linear

(iii) $y(t) = x(t) \cos 3t$ ***Ans:*** Memory less, non-causal, linear

(iv) $y(t) = \displaystyle\int_{-\alpha}^{+\alpha} x(t)\,dt$ ***Ans:*** Non causal, linear

EXERCISE PROBLEMS

1. Sketch the following signals

   (a) $\pi(t - \tfrac{1}{2})$

   (b) $\pi\big((t-1)/2\big) + \pi(t-1)$

   (c) $r(-2t+3)$

   (d) $r\big[(t+1)/3\big]$

   (e) $u\big[(t-2)/4\big]$

   (f) $y(t) = \displaystyle\sum_{n=0}^{\alpha} u(t-2n)u(1+2n-t)$

2. Sketch the following signals

   (a) $r(t+2) - 2r(t) + r(t-2)$

   (b) $u(t)\,u(10-t)$

   (c) $2u(t) + \delta(t-2)$

   (d) $2\,u(t)\,\delta(t-2)$

   (e) $e^{-10t}u(t)$

   (f) $\cos 10\pi t\, u(t)\, u(2-t)$

3. Express the following signals in terms of singularity functions.

   (a) $r(t)\,u(2-t)$

   (b) $r(t) - r(t-1) - r(t-2) + r(t-3)$

   (c) $\pi\big[(t-3)/6\big] + \pi\big[(t-4)/2\big]$

4. Find the fundamental periods of signals given below.

   (a) $\sin 50\pi t + \cos 60\pi t$

   (b) $2\cos\big(10\pi t + \tfrac{\pi}{6}\big)$

   (c) $\sin 2t + \cos\pi t$

   (d) $\cos 60\pi t$

5. Express the signals shown in terms of singularity functions.

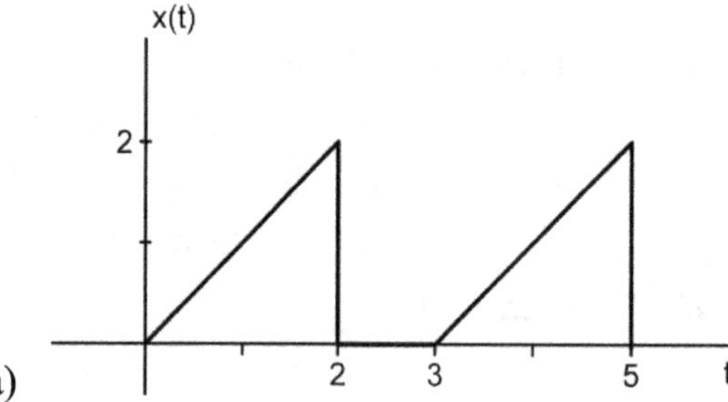

(a)

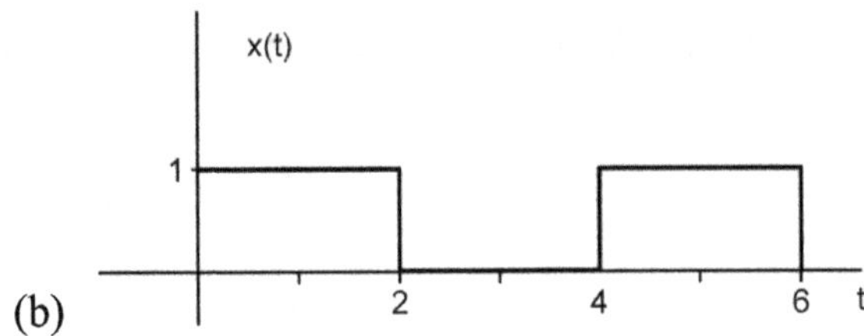

(b)

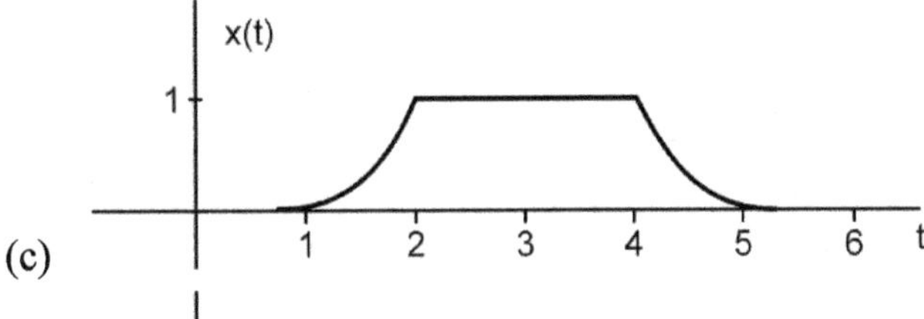

(c)

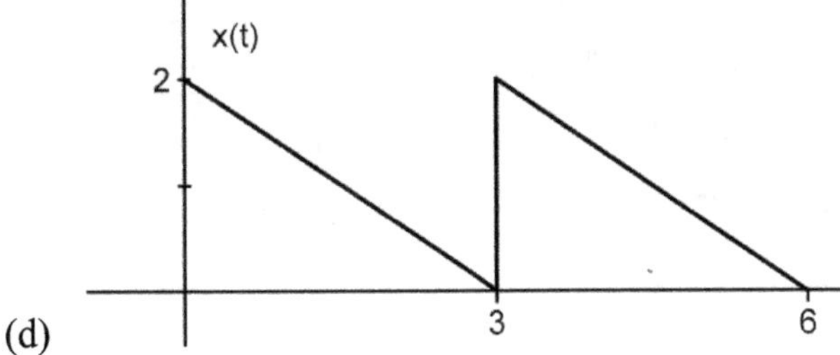

(d)

6. Which of the following are energy signals and which are power signals.

(a) $\left(1-e^{-5t}\right)u(t)$

(b) $t^{-1/4}u(t-3)$

(c) $t\,e^{-2t}\,u(t)$

(d) $u(t)-\dfrac{1}{3}u(t-10)$

(e) $\cos 10\pi t\; u(t)\, u(2-t)$

7. Evaluate the following integrals

(a) $\displaystyle\int_{0}^{\alpha}\cos 2\pi t\,\delta(t-2)\,dt$

(b) $\displaystyle\int_{-\alpha}^{\alpha}(t-2)^2\,\delta(t-2)\,dt$

(c) $\displaystyle\int_{-\alpha}^{\alpha}t^2\delta(t-2)\,dt$

(d) $\displaystyle\int_{-\alpha}^{\alpha}e^{3t}\delta(t-2)\,dt$

(e) $\displaystyle\int_{-\alpha}^{\alpha}\left[e^{-3t}+\cos(2\pi t)\right]\delta(t)\,dt$

8. Show that $\delta_\varepsilon(t)=\dfrac{e^{-t/\varepsilon}}{\varepsilon}u(t)$ in the limit $\varepsilon\to 0$

9. Given the signal $x(t) = \sin2\,(7\pi t - \pi/6) + \cos(3\pi t - \pi/3)$ sketch signal sided amplitude and double sided amplitude.

10. Given the signal $x(t) = 16\cos(20\pi t + \pi/4) + 6\cos(30\pi t + \pi/6) + 4\cos(40\pi t + \pi/3)$ find the power spectral density?

OBJECTIVE QUESTIONS

1. The energy E of a discrete signal is given by ___________________

2. If the energy E is finite, then $x(n)$ is called as ___________________

3. The average power of a discrete signal is ___________________

4. The signal $x(n)$ is periodic if and only if ___________________

5. The discrete time complex exponential signal is defined as ___________________

6. The fundamental frequency of a periodic signal $e^{j\omega_0 n}$ is ___________________

7. Odd signal satisfies ___________________ condition

8. The smallest value of T, for which $x(t) = x(t + T)$ is known as ___________________

9. EEG signal is an example of ___________________.

10. An energy signal has ___________________ power

11. A power signal has ___________________ energy

12. If the signal $x(t) = e^{at}$ represent a decay then a is ___________________

13. An impulse function $\delta(n)$ is defined by ___________________

14. $\delta(t - 1)$ exists at ___________________

15. $u(n + 1) = 1$ for all ___________________

16. $u(n) - u(n - 1) =$ ___________________

17. Integral of step function is ___________________

18. Fundamental frequency of below signal is ___________________

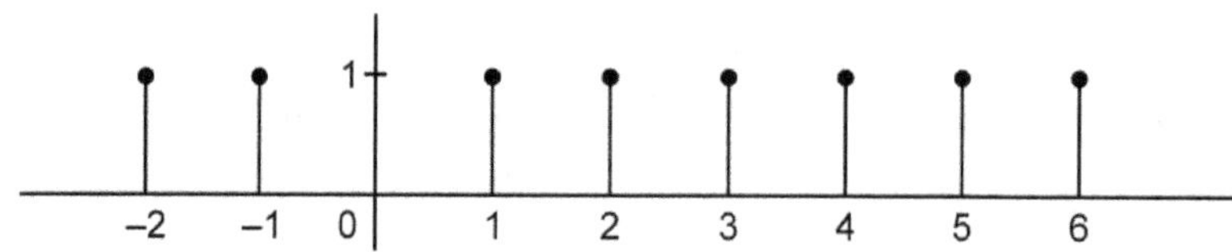

19. $x(n)$ is periodic / aperiodic pulse is ___________________

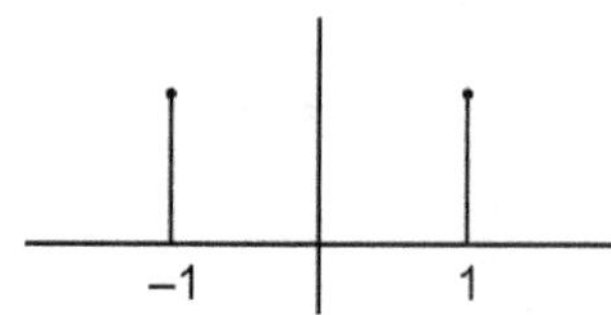

20. Total energy of rectangular pulse is ________________

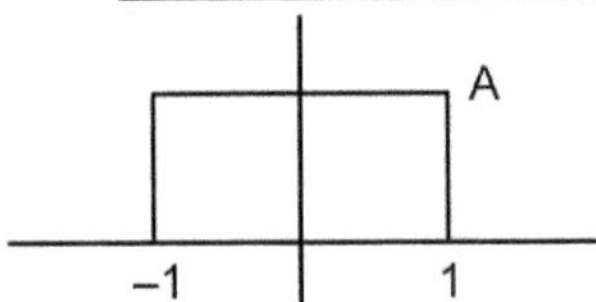

21. Total energy of x(n) shown is ________________

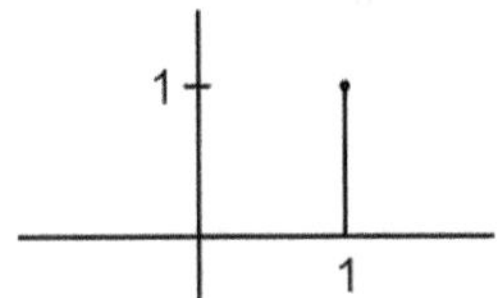

22. If x(t) is

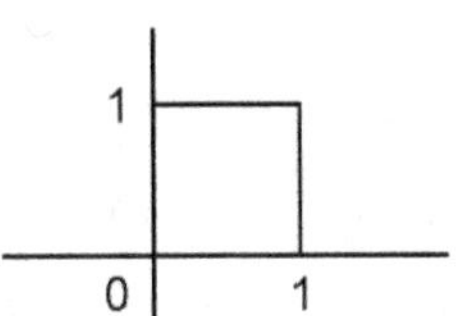

    then x(t – 1) is ________________
23. $x(n) = 5\sin 2n$, is x(n) periodic ________________
24. Unit impulse is odd/even function
25. Express following in terms of u(t)
    (a) (r(t) = ________________   (b) $\delta(t)$ = ________________
    (c) Signum (t) = ________________   (d) $\pi(t)$ = ________________
26. Periodic signal is energy or power signal?
27. What is the period of $2\cos\left(\dfrac{t}{4}\right)$ ________________

28. Find which of the signals are causal

    (a) $\left(\dfrac{1}{2}\right)^{n} u(n+3)$ (b) $e^{-2t} u(t-2)$ (c) $u(n+2) - u(n-2)$

29. Find which of the signals are energy or power signals

    (a) $e^{-3t} u(t)$ (b) $\left(\dfrac{1}{3}\right)^{n} u(n)$ (c) $\cos\left(\dfrac{\pi}{4}n\right)$

    (d) $u(t) - u(t-1)$ (e) $r(t) - r(t-2)$
30. Find the fundamental period of following signals
    (a) $2\sin(3t+1) + 3\sin(4t-1)$

(b) $\sin\left(\dfrac{7\pi}{3}\right)t$

(c) $e^{j\frac{7\pi}{3}n}$

SUMMARY OF CHAPTER

1. Signals can be classified into continuous and discrete in time domain

2. Continuous signals satisfy condition $x(T) = x(t + T)$ i.e., they are periodic with fundamental period T

3. The signal $x(t)$ is energy signal if $0 < \bar{t} < \alpha$

4. The signal $x(t)$ is power signal if $0 < P < \alpha$

5. The signal $x(-t)$ is obtained by reflecting $x(t)$ about $t = 0$

6. The signal $x(t)$ is even symmetric if $x(t) = x(-t)$

7. The signal $x(t)$ is odd symmetric if $x(t) = -x(-t)$

8. Unit impulse function $\delta(t)$ is characterizes by unit area and is connected at a single instant $t = 0$

9. Unit impulse is given by $u(t) = \int_{-\alpha}^{t} \delta(T)dT$; unit ramp is given by $r(t) = \int_{-\alpha}^{t} u(T)dT$

10. A signal can be either energy or a power signal.

11. A signal whose complete physical description is known in mathematical form, then it is deterministic

12. Two signals $x(T_1)$ and $x(T_2)$ are periodic, then $x(t) = x(T_1) + x(T_2)$ is periodic if T_1/T_2 is a rational number

13. Sampling function $S_a(t) = \dfrac{\sin t}{t}$

14. A signal is said to be causal if $x(t) = 0$, $t < 0$

15. Any signal that doesn't contain any singularities (delta function or its donatives) at $t = 0$ can be written as sum of causal part and anti-causal $x(t) = x^+(t) + x^-(t)$

16. A signal is non causal if it starts before $t = 0$

17. The area under curve $\int_{-\alpha}^{\alpha} \delta(t)dt$ is unity

18. Singularity is a point at which function doesn't possess a derivative

19. Singularity function are also defined as generalized function

20. A generalized function is defined by its effect on the other functions instead of by its value at every instant of time.

CHAPTER 2

Analogy between Signal and Vectors

Objectives of the Chapter

In this chapter, basics of vector, orthogonal functions, orthogonal vector space, approximating signals using orthogonal functions, orthogonality in signals were discussed with examples. Orthogonality in complex functions is also discussed.

2.1 VECTOR

A vector is a geometric object that has both magnitude, direction and sense. A vector is represented by a ray with a definite direction.

2.1.1 BASIC PROPERTIES OF VECTORS

1. **Vector equality:** Two vectors said to be equal if their magnitude and directions are same or if they have equal coordinates.

2. **Addition and Subtraction of vectors:** The addition of two vectors is performed according to parallelogram rule. If a and b are both sides of a parallelogram, $a + b$ is diagonal.

   Subtraction of two vectors: Place the ends of a and b at same point and measure the distance between other ends of a and b then it is equal to a – b.

3. **Dot product:** It is also called as inner product and is denoted by

   $$a \cdot b = |a|\,|b|\cos\theta$$

   where θ is angle between two vectors. The result is scalar.

   $$a \cdot b = a_x\, b_x + a_y\, b_y + a_z\, b_z$$

   a_x, a_y, a_z and b_x, b_y, b_z are x, y, z components of a, b vectors.

4. **Cross product:** It is also known as outer product and is denoted by

   $$a \times b = |a|\,|b|\,\sin\theta$$

   where n is unit vector perpendicular to both a and b

   Its resultant is also a vector

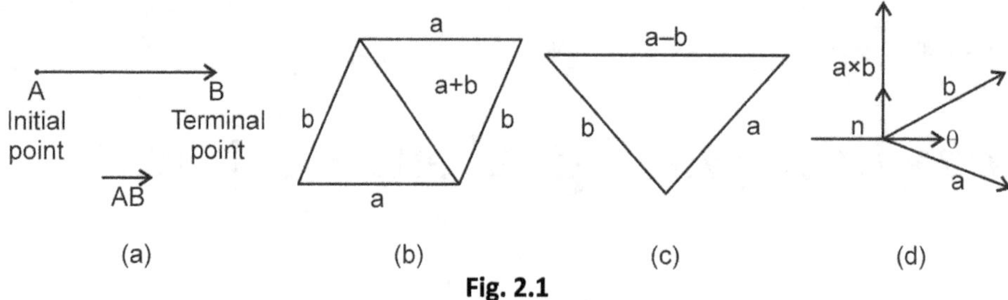

Fig. 2.1

2.2   PHASOR

A phasor is a representation of AC currents and voltages (sinusoidals) in the form of magnitude and angle $A\,\lfloor\theta$. It is also known as sinor.

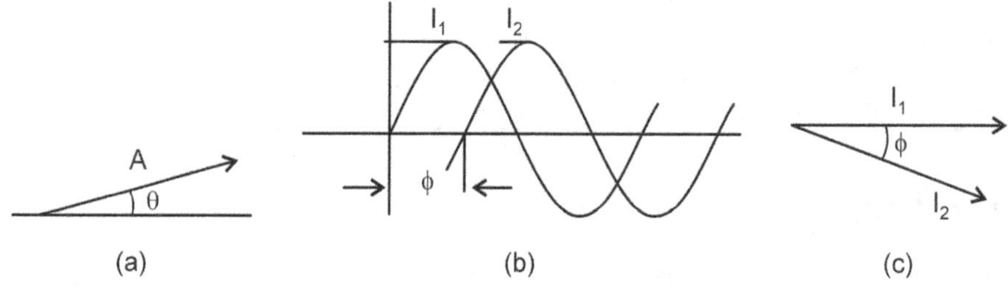

Fig. 2.2

AC current is a continuous signal and represented by

$$i_1 = I_1 \sin \omega t$$

$$i_2 = I_2 \sin (\omega t + \phi)$$

where ϕ is phase difference. They are represented by phasors as shown in figure

As phasor and vector are same, we can indicate a time varying signal using a vector. When a signal is changing from instant to instant in a manner which cannot be defined, we can represent it by a row vector.

2.3   ORTHOGONAL FUNCTIONS

If two functions are said to be orthogonal, if when multiplied together and integrated over the domain of interest, the integral must become zero.

The orthogonality property can be stated mathematically as

$$\langle g_m g_n \rangle = \int_a^b g_m(x)\, g_n(x)\, dx = \begin{cases} 0 & m \neq n \\ \|gm\|^2 & m = n \end{cases}$$

If $\|gm\| = 1$, then $g_m(x)$ is said to be orthonormal function.

Orthogonal property is mainly used to expand arbitrary functions with an infinite series expansion in terms of the given basis functions. Orthogonal means oriented at 90° to each other.

Using a set of orthogonal functions, we can represent a signal vector as approximately as possible so that error is very less.

Sine and Cosine functions belong to orthogonal functions.

$$\text{the definite integrals} \quad \left.\begin{array}{l} \int\limits_{-\pi}^{\pi} \cos mx \sin nx \, dx = 0 \\[3mm] \int\limits_{-\pi}^{\pi} \sin mx \cos nx \, dx = 0 \end{array}\right\} \quad \text{for } m \neq n$$

2.3.1 ORTHOGONAL POLYNOMIALS

A given signal can be approximated by a function

$$f(t) = C_o + C_1 P_1(t) + C_2 P_2(t) + \ldots\ldots$$

where $P_1(t)$ and $P_2(t)$ are orthogonal polynomials such that

$$\sum_{t=0}^{n} P_1(t) P_2(t) = 0$$

The least square error is given by

$$\sum \left\{ f(t) - C_0 - C_1 P_1 - C_2 P_2 \ldots\ldots \right\}^2 \text{ is minimum}$$

The necessary condition that partial derivatives which represent to C_0, C_1, C_2 be zero becomes

$$\sum -2 P_j(t) \left[f(t) - C_0 - C_1 P_1 - \ldots \right] = 0$$

Using the orthogonality relation, jth term product sum does not vanish

$$\sum P_j(t)\, y_j + \sum C_j\, P_j^2(t) = 0$$

$$\therefore \ C_j = \frac{\Sigma y_j p_j(t)}{\Sigma p_j^2(t)}$$

It can be shown that

$$P_1(t) = 1 - 2\frac{t}{n}$$

$$P_2(t) = 1 - \frac{6t}{n} + \frac{6t(t-1)}{n(n-1)}$$

$$P_3(t) = 1 - 12\frac{t}{n} + 30\frac{t(t-1)}{n(n-1)} - 20\frac{t(t-1)\,(t-2)}{n(n-1)\,(n-2)}$$

One of the main advantage of orthogonal functions is the coefficients are independent and if a fit is made with lower degree polynomial and if we want to go for higher degree, only additional coefficients are required to be calculated.

2.3.2   ORTHOGONAL VECTOR SPACE

The components of vector A along x, y, z axis can be expressed in terms of unit vectors a_x, a_y, a_z in respective directions as

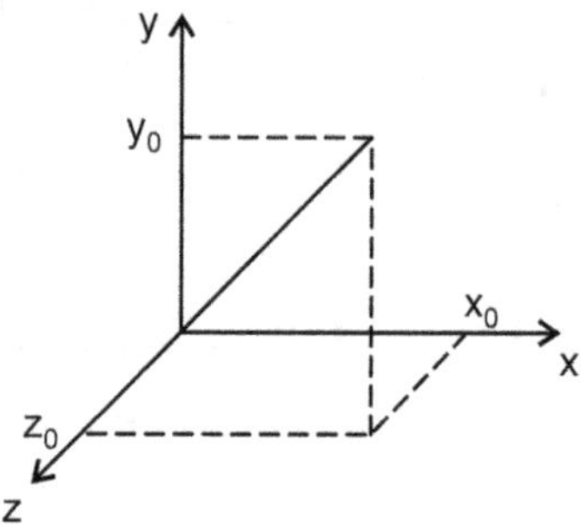

Fig. 2.3

$$x - \text{axis} = A \cdot a_x$$

$$y - \text{axis} = A \cdot a_y$$

$$z - \text{axis} = A \cdot a_z$$

or we can also express A as

$$A = x_0\, a_x + y_0\, a_y + z_0\, a_z$$

Q these three vectors are perpendicular to each other

$$a_x \cdot a_y = a_y \cdot a_z = a_z \cdot a_x = 0$$

$$a_x \cdot a_x = a_y \cdot a_y = a_z \cdot a_z = 1$$

Similarly if there are 'n'$\perp$ coordinates and 'n' unit vectors a_1, a_2, a_3 ..... a_n along these coordinates

$$A = A_1\, a_1 + A_2\, a_2 + \ldots\ldots$$

where A_1, A_2,....... are components of A in 'n' dimensional space

$$\therefore\ A \cdot a_r = A_1\, a_1 \cdot a_r + A_2\, a_2 \cdot a_r + \ldots\ldots\ldots A_r\, a_r \cdot a_r$$

$$= A_r$$

$$A_r = A \cdot a_r$$

since $a_i \cdot a_j = 0 \qquad i \neq j$

$\qquad a_j \cdot a_j = 1 \qquad i = j$

if a_1, a_2 .... are not unit vectors

$$\therefore\ a_i \cdot a_j = k \qquad \text{where } k \text{ is constant}$$

$$\therefore\ A \cdot a_r = A_r\, k$$

$$A_r = \frac{1}{k}\, A \cdot a_r$$

A vector is analogous to a orthogonal. Thus a signal can also be expressed as any function of 't' as sum of its components which are mutually orthogonal.

2.3.3   APPROXIMATION OF A SIGNAL USING ORTHOGONAL FUNCTIONS

Let us consider

$$g_1(t), g_2(t), \ldots\ldots g_n(t)$$

n functions which are orthogonal so that

$$\int_{t_1}^{t_2} g_i(t)g_j(t)dt = 0 \qquad i \neq j$$

$$\int_{t_1}^{t_2} g_i^2(t)dt = 1 \qquad i = j$$

Let an arbitrary function

$$f(t) = A_1 g_1(t) + A_2 g_2(t) + \ldots\ldots$$

$$= \sum_{r=1}^{n} A_r g_r(t)$$

2.3.4   ORTHOGONAL VECTORS

A vector has a magnitude and direction

if A_1 is approximated in terms of A_2

$$A_1 = a_{12} A_2 + A_c$$

A_c is error vector

Dot product of two vector

$$A . B = A B \cos \theta$$

$$\therefore \text{ Component A along B} = A \cos \theta = \frac{A.B}{B}$$

Similarly component A_1 long $A_2 = \dfrac{A_1 . A_2}{A_2}$

but $\dfrac{A_1 . A_2}{A_2} = a_{12} A_2;\ \ a_{12} = \dfrac{A_1 . A_2}{A_2^2}$

if A_1 and A_2 are orthogonal i.e., perpendicular to each other

$$A_1 \cdot A_2 = A_1 A_2 \cos \theta = A_1 A_2 \cos 90 = 0$$

$$\therefore a_{12} = 0$$

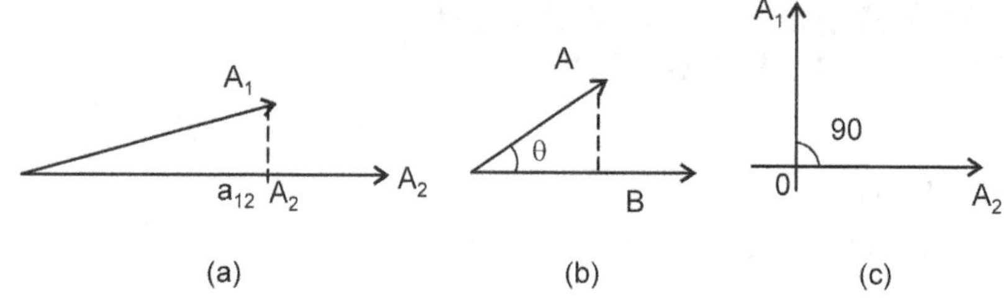

(a)   (b)   (c)

Fig. 2.4

2.3.5   ORTHOGONALITY IN SIGNALS

Let us consider two signals

$$f_1(t) \text{ and } f_2(t)$$

if f_1 is approximated in terms of f_2

$$f_1(t) = a_{12} f_2(t) + f_e(t)$$

$$f_e(t) = f_1(t) - a_{12} f_2(t)$$

average error within a time period

$$\frac{1}{(t_2 - t_1)} \int_{t_1}^{t_2} \{ f_1(t) - a_{12} f_2(t) \} dt$$

mean square error

$$e_r = f_c^2(t) = \frac{1}{t_2 - t_1} \int_{t_1}^{t_2} \{ f_1(t) - a_{12} f_2(t) \}^2 dt$$

a_{12} having minimum error

$$\frac{de_r}{da_{12}} = \frac{d}{da_{12}} \left\{ \frac{1}{t_2 - t_1} \int_{t_1}^{t_2} \sum f_1(t) - a_{12} f_2(t) \right\}^2 dt = 0$$

$$-2 \int_{t_1}^{t_2} f_1(t) f_2(t) dt + 2a_{12} \int_{t_1}^{t_2} f_2^2(t) dt = 0$$

$$a_{12} = \frac{\displaystyle\int_{t_1}^{t_2} f_1(t) f_2(t) dt}{\displaystyle\int_{t_1}^{t_2} f_2^2(t) dt}$$

if $\qquad a_{12} = 0$

i.e., $\qquad \displaystyle\int_{t_1}^{t_2} f_1(t)\,f_2(t)\,dt = 0$

then the two signals are said to be orthogonal, over the interval (t_1, t_2).

$\therefore \qquad e_r = \dfrac{1}{t_2 - t_1} \displaystyle\int_{t_1}^{t_2} \{f(t) - \Sigma A_r\,g_r(t)\}^2\,dt$

to minimize the error

$$\frac{\partial e_r}{\partial A_j} = 0$$

$$\frac{\partial}{\partial A_j}\left\{\frac{1}{t_2 - t_1}\int_{t_1}^{t_2}\left\{f(t) - \sum_{r=1}^{n} A_r g_r(t)\right\}^2 dt\right\} = 0$$

$$\int -2f(t)g_j(t) + 2A_j g_j^2(t)\,dt = 0$$

$$A_j = \frac{\displaystyle\int_{t_1}^{t_2} f(t)g_j(t)\,dt}{\displaystyle\int_{t_1}^{t_2} g_j^2(t)\,dt}$$

$$\boxed{A_j = \frac{1}{k_j}\int_{t_1}^{t_2} f(t)\,g_j(t)}$$

$\therefore \qquad f(t) \;=\; \dfrac{1}{k_1}g_1(t) + \dfrac{1}{k_2}g_2(t) + \ldots\ldots$

error when one term is considered

$$e_1 = \frac{1}{t_2 - t_1}\int_{t_1}^{t_2}\left\{f(t) - \frac{1}{k_1}g_1(t)\right\}^2 dt$$

SOLVED PROBLEMS

EXERCISE 1

1. **Show that the below functions are orthogonal**

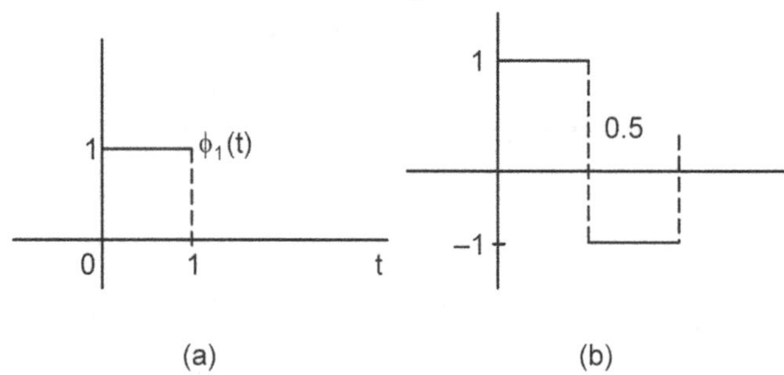

(a)   (b)

Fig. 2.5

(a) $\displaystyle\int_0^1 \phi_1(t)\,\phi_2(t) = \int_0^{0.5} 1.1.dt + \int_{0.5}^1 1.-1\,dt = 0$ hence they are orthogonal

2. **Show that the functions sin nw_0 t and sin mw_0t are orthogonal**

$$\int_{t_0}^{t_0+2\pi/w_0} \sin nw_0 t \times \sin mw_0\, dt =$$

$$= \int_{t_0}^{t_0+2\pi/w_0} \frac{1}{2}\{\cos(n-m)w_0 t - \cos(n+m)w_0 t\}\,dt$$

$$= \frac{1}{2w_0}\left[\frac{1}{(n-m)}\sin(n-m)w_0 t - \frac{1}{(n+m)}\sin(n+m)w_0 t\right]_{t_0}^{t_0+\frac{2\pi}{w_0}}$$

If n and m are integers, the above integral is 0, so they are orthogonal

let $t_0 = 0$

$$\therefore \quad = \frac{1}{2w_0}\left\{\frac{1}{(n-m)}\left(\sin(n-m)2\pi - \sin 0\right) - \frac{1}{(n+m)}\left(\sin(n+m)2\pi - \sin 0\right)\right\} = 0$$

3. **Split the signal x (t) = e^t into even and odd parts and show that they are orthogonal over an interval (– a, a).**

$$x(t) = e^t$$

$$x_e(t) = \frac{x(t)+x(-t)}{2} = \frac{e^t + e^{-t}}{2} = \cosh t \text{ and } x_0(t) = \sinh(t)$$

$$\therefore \quad \int_{-a}^{a} \cosh t \sinh t = \frac{1}{2}\int_{-a}^{a} \sinh(2t)\,dt$$

$$= \left. \frac{1}{2}\frac{\cosh 2t}{2} \right/_{-a}^{a} = 0$$

4. Approximate the rectangular function using a finite series of sinusoidal functions.

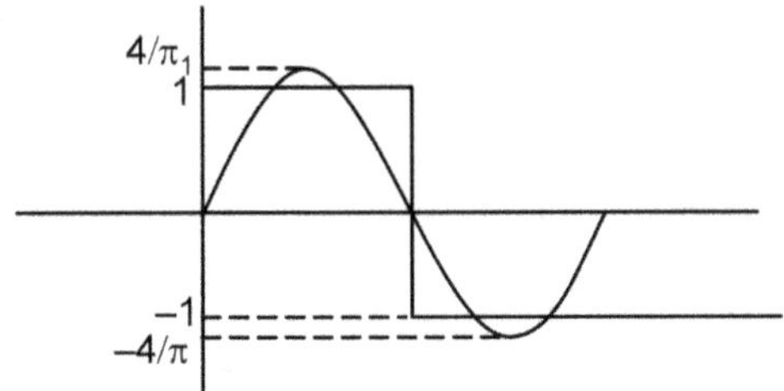

Fig. 2.6

Let

$$f(t) \quad = \quad A_1 \sin t + A_2 \sin 2\,t + \ldots\ldots\ldots\ldots A_n \sin nt$$

Constant

$$Ar \quad = \quad \frac{\displaystyle\int_{0}^{2\pi} f(t)\sin rt\, fdt}{\displaystyle\int_{r}^{2\pi} \sin^2 rt\, dt} = \frac{1}{\pi}\left\{\int_{0}^{\pi} \sin rt\,dt - \int_{\pi}^{2\pi} \sin rt\,dt\right\}$$

$$= \quad \frac{4}{\pi r} \quad \text{if } r \text{ is odd}$$

$$= \quad 0 \quad \text{if } r \text{ is even}$$

$$\therefore \quad f(t) = \frac{4}{\pi}\sin t + \frac{4}{3\pi}\sin 3t + \ldots.$$

5. Approximate the below function using exponential fourier series

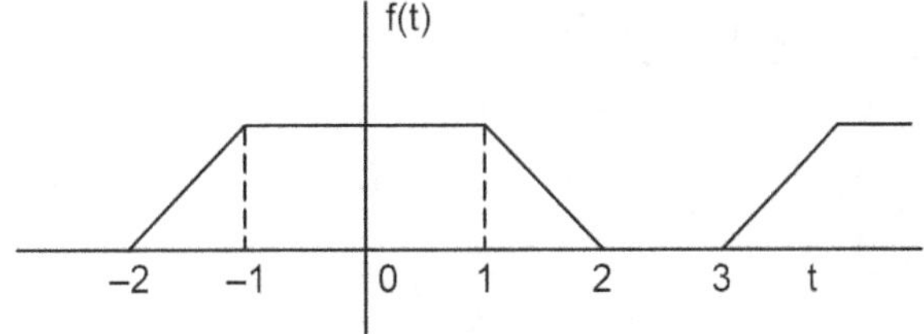

Fig. 2.7

$$T = 5$$

$$w_0 = \frac{2\pi}{5}$$

$$x(t) = t + 2 \qquad -2 \le t \le -1$$

$$= 1 \qquad\qquad -1 \le t \le 1$$

$$= 2 - t \qquad\quad 1 \le t \le 2$$

$$F_0 = \frac{1}{5}\left\{ \int_{-2}^{-1}(t+2)\,dt + \int_{-1}^{1} dt + \int_{1}^{2}(2-t)\,dt \right\} = \frac{3}{5}$$

$$F_n = \frac{1}{5}\left\{ \int_{-2}^{-1}(t+2)e^{-jnw_0}t\,dt + \int_{-1}^{+1} e^{-jnw_0}t\,dt + \int_{1}^{2}(2-t)e^{-jnw_0 t}\,dt \right\}$$

(A)   **(B)**   **(C)**

$$A = \int_{-2}^{-1}(t+2)e^{-jnw_0 t}\,dt = \frac{-t}{jnw_0}\left\{e^{-jnw_0 t}\right\}\int_{-2}^{-1} + \frac{1}{n^2 w_0^2}e^{-jnw_0 t}\int_{-2}^{-1} - \frac{2}{jnw_0}e^{-jnw_0 t}\int_{-2}^{-1}$$

$$= \frac{1}{jnw_0}\left\{e^{jnw_0} - 2e^{j2nw_0}\right\} + \frac{1}{n^2 w_0^2}\left\{e^{jnw_0} - e^{j2nw_0}\right\}^{-2} - \frac{2}{jnw_0}e^{jnw_0 t} + \frac{2}{jnw_0}e^{j2nw_0 t}$$

$$B = \int_{-1}^{1} e^{-jnw_0 t}\,dt = \frac{e^{jnw_0} - e^{-jnw_0}}{jnw_0}$$

$$C = \int_{1}^{2}(2-t)e^{-jnw_0}t\,dt = \frac{-2}{jnw_0}\left\{e^{-j2nw_0} - e^{-jnw_0}\right\} + \frac{2}{jnw_0}e^{-j2nw_0} - \frac{1}{jnw_0}e^{-jnw_0} - \frac{1}{n^2 w_0^2}e^{-j2nw_0} + \frac{1}{n^2 w_0^\alpha}e^{-jnw_0}$$

$$C_n = \frac{1}{5}\left\{ \frac{1}{n^2 w_0^2}\left(e^{jnw_0} - e^{j2nw_0}\right) - \frac{1}{n^2 w_0^2}\left(e^{-j2nw_0} - e^{-jnw_0}\right) \right\}$$

$$= \frac{5}{2r^2 \pi^>}\left\{ \cos\frac{2\pi n}{5} - \cos\frac{4\pi n}{5} \right\}$$

$$f(t) = \frac{3}{5} + \sum_{n=-\alpha}^{\alpha}\frac{5}{2n^2 \pi} = \left\{ \cos\frac{2\pi n}{5} - \frac{\cos 4\pi n}{5} \right\}e^{jnw_0 t}$$

EXERCISE 2

1. **Approximate the below function using Legendre Fourier series**

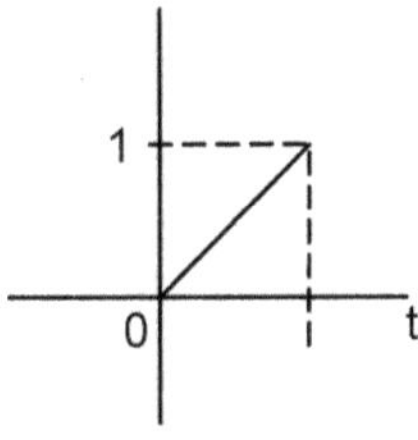

Fig. 2.8

$$f(t) = t$$

$$f(t) = C_0 \, P_0(t) + C_1 \, P_1(t) + \ldots\ldots\ldots C_n \, P_n(t)$$

$$C_0 = \frac{1}{2}\int_0^1 t \, dt = \frac{1}{2}\frac{1^2}{2} \Big/_0^1 = \frac{1}{4}$$

$$C_1 = \frac{3}{2}\int_0^1 t.t \, dt = \frac{3}{2}\int_0^1 t^2 \, dt = \frac{3}{2}\frac{t^3}{3} \Big/_0^1 = \frac{1}{2}$$

$$C_2 = \frac{5}{2}\int_0^1 t.\left(\frac{3}{2}t^2 - \frac{1}{2}\right) dt = \frac{5}{2}\left(\frac{3}{2}\frac{t^4}{4} - \frac{1}{2}\frac{t^2}{2}\right)\Big/_0^1 = \frac{5}{2}\left(\frac{3}{8} - \frac{1}{4}\right) = \frac{5}{2}\frac{3-2}{8} = \frac{5}{2},\frac{1}{8} = \frac{5}{16}$$

$$\therefore f(t) = \frac{1}{4} + \frac{1}{2}t + \frac{5}{16}\left(\frac{3}{2}t^2 - \frac{1}{2}\right)$$

2. **Find the trigonometric fourier series for x (t) shown in fig.**

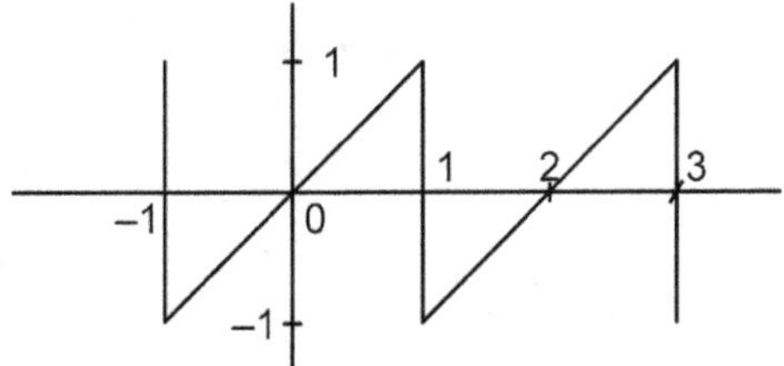

Fig. 2.9

$$a_0 = \frac{1}{T}\int_{t_0}^{t_0+T} f(t) \, dt \qquad\qquad w_0 \frac{2\pi}{2} = \pi$$

$$= \frac{1}{2}\int_{-1}^{1} t \, dt = 0$$

$$a_n = \frac{2}{T}\int_{-1}^{1} t\cos nw_0 t\ dt = \left\{\frac{t}{n\pi}\sin mw_0 t + \frac{1}{n^2\pi^2}\cos mw_0 t\right\}_{-1}^{+1} = 0$$

$$w_0\ \frac{2\pi}{2} = \pi$$

$$b_n = \frac{2}{T}\int_{t_0}^{t_0+T} t\sin n\pi t\ dt$$

$$= \left\{\frac{-t}{n\pi}\cos n\pi t + \frac{1}{n^2\pi^2}\cos n\pi t\right\}_{-1}^{1}$$

$$= \frac{-2}{n\pi}\cos n\pi = \frac{2}{\pi}\left\{\frac{-(-1)^n}{n}\right\}$$

$$x(t) = \sum_{n=1}^{\alpha}\frac{2}{\pi}\left\{\frac{-(-1)^n}{n}\right\}\sin nw_0 t$$

3. Find trigonometric Fourier series of x (t)

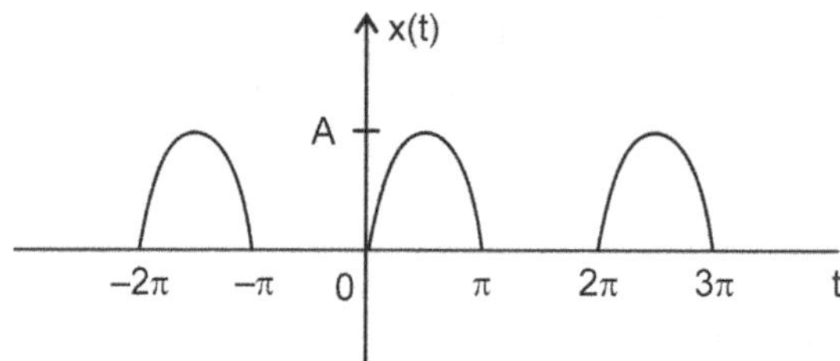

Fig. 2.10

$$w_0 = \frac{2\pi}{2\pi} = 1$$

$$a_0 = \frac{1}{2\pi}\int_{0}^{\pi} A\sin t\ dt = \frac{-A}{2\pi}\cos\Big|_{0}^{\pi} = \frac{A}{\pi}$$

$$a_n = \frac{2}{2\pi}\int_{0}^{\pi} A\sin t\cos nt\ dt = \frac{2A}{\pi(1-n^2)}$$

$$b_n = \frac{2}{2\pi}\int_{0}^{\pi} A\sin t\sin nt\ dt = 0$$

$$x(t) = \frac{A}{\pi} + \sum_{n=1}^{\alpha}\frac{2A}{\pi(1-n^2)}\cos nt$$

4. Find the trigonometric fourier series for x (t)

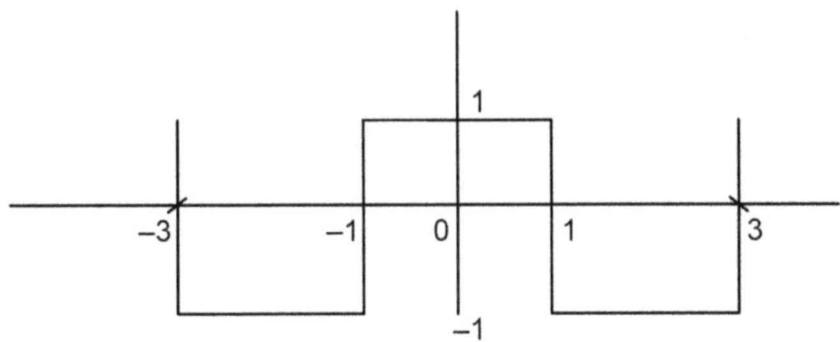

Fig. 2.11

$$w_0 = \frac{2\pi}{4} = \frac{\pi}{2}$$

$$a_0 = 0$$

$$a_n = \frac{4}{n\pi} \sin\frac{n\pi}{2}$$

$$b_n = 0$$

2.4   ORTHOGONALITY IN COMPLEX FUNCTIONS

Let f_1 (t) and f_2 (t) are complex functions of real variable t, then

$$f_1(t) \cong A_{12} f_2(t)$$

$$A_{12} = \frac{\displaystyle\int_{t_1}^{t_2} f_1(t) f_2^*(t)\, dt}{\displaystyle\int_{t_1}^{t_2} f_2(t) f_2^*(t)\, dt}$$

where f_2^* (t) is a complex conjugate of f_2 (t). The two complex functions f_1 (t) and f_2 (t) are orthogonal over (t_1, t_2) if

$$\int_{t_1}^{t_2} f_1(t) f_2^*(t)\, dt = \int_{t_1}^{t_2} f_1^*(t) f_2(t)\, dt = 0$$

for a set of complex functions

$$\int_{t_1}^{t_2} g_m(t) g_n^*(t) = 0 \qquad \text{if } m \neq n$$
$$\qquad\qquad\qquad km \quad \text{if } m = n$$

$$f(t) \quad = \quad A_1\, g_1(t) + A_2\, g_2(t) \ldots\ldots\ldots Argr(t)$$

$$\text{where} \quad Ar \quad = \quad \frac{1}{k}\int f(t) g(t)\, dt$$

2.4.1   EXAMPLES OF ORTHOGONAL FUNCTIONS

1. Legender Fourier Series

Legender polynomial

$$p_n(x) = \frac{1}{2^n \, n! \, dx^n} \frac{d^n}{}(x^2 - 1)^n \qquad n = 0, 1, 2 \ldots\ldots\ldots\ldots$$

$$\therefore \; P_0(x) = 1$$
$$P_1(x) = x$$

$$P_2(x) = \frac{3}{2}x^2 - \frac{1}{2} \text{ and } P_3(x) = (\frac{5}{2}x^3 - \frac{3}{2}x)$$

Orthogonality of these polynomials

$$\int P_m(t)P_n(t)\,dt = \begin{cases} 0 & m \neq n \\ \dfrac{2}{2m+1} & m = n \end{cases}$$

We can express any $f(t)$ is terms of Legendre polynomial series

$$f(t) = C_0 \, P_0(t) + C_1 \, P_1(t) + \ldots\ldots\ldots\ldots$$

$$C_r = \frac{\displaystyle\int_{-1}^{1} f(t)P_r(t)\,dt}{\displaystyle\int_{-1}^{1} P_r^2(t)\,dt}$$

$$= \frac{2r+1}{2}\int_{-1}^{1} f(t)\,P_r(t)\,dt$$

Ex:

$$f(t) = C_0 \, P_0(t) + C_1 \, P_1(t) + \ldots\ldots$$

$$f(t) = \begin{cases} 1 & -1 < t < 0 \\ -1 & 0 < t < 1 \end{cases}$$

$$C_0 = \frac{1}{2}\int_{-1}^{1} f(t)\,dt = 0$$

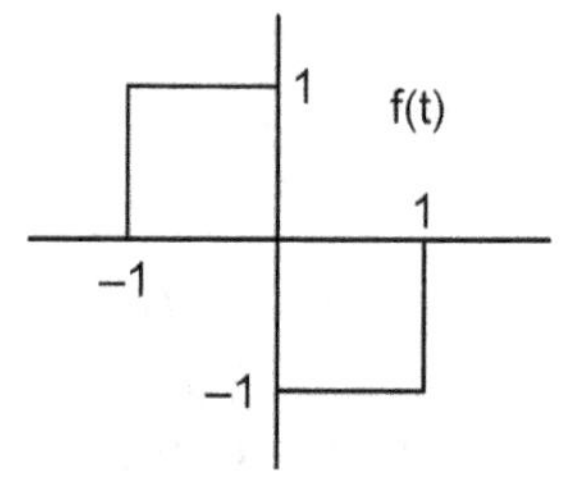

Fig. 2.12

$$C_1 = \frac{3}{2}\int_{-1}^{1} t\,f(t)\,dt = \frac{-3}{2}$$

$$C_2 = \frac{5}{2}\int_{-1}^{1} f(t)\left(\frac{3}{2}t^2 - \frac{1}{2}\right) dt = 0$$

$$C_r \quad \text{is} \ 0 \ \text{for} \ r \ \text{even}$$

$$C_3 \quad = \ 7/8$$

$$\therefore \quad f(t) \ = \ -\frac{3}{2}t + \frac{7}{8}\left(\frac{5}{2}t^3 - \frac{3}{2}t\right) + \dots\dots\dots$$

2. Trigonometric Fourier Series

The set of functions

$1, \ \cos w_0 t + \cos 2w_0 t + \dots\dots\cos nw_0 t$ and

$\sin w_0 t + \sin 2w_0 t + \dots\dots\dots\sin nw_0 t$ over $t_o, \ t_o + 2\pi/w_0$ are orthogonal functions

$\therefore$ any function $f(t) = a_0 + a_1 \cos w_0 t + a_2 \cos 2w_0 t + \dots \ b_1 \sin w_0 t + b_2 \sin 2w_0 t + \dots$

$$f(t) = a_0 + \sum_{n=1}^{\alpha}(a_n \cos nw_0 t + b_n \sin w_0 t)$$

$$\text{where} \quad a_n \ = \ \frac{\displaystyle\int_{t_0}^{t_0+T} f(t)\cos nw_0 t \ dt}{\displaystyle\int_{t_0}^{t_0+T} \cos^2 nw_0 t \ dt}$$

$$b_n \ = \ \frac{\displaystyle\int_{t_0}^{t_0+T} f(t)\sin nw_0 t \ dt}{\displaystyle\int_{t_0}^{t_0+T} \sin^2 nw_0 t \ dt}$$

$$a_n \ = \ \frac{2}{T}\int_{t_0}^{t_0+T} f(t) \cos nw_0 t \ dt$$

$$b_n \ = \ \frac{2}{T}\int_{t_0}^{t_0+T} f(t) \sin nw_0 t \ dt$$

Ex: Expand the function using trigonometric series

$$f(t) \ = \ a_0 + a_1 \cos 2pt + \dots\dots\dots$$

$$b_1 \sin 2pt + \dots\dots\dots\dots$$

$$a_0 \ = \ \frac{1}{T}\int_{0}^{T} f(t) \ dt = A/2$$

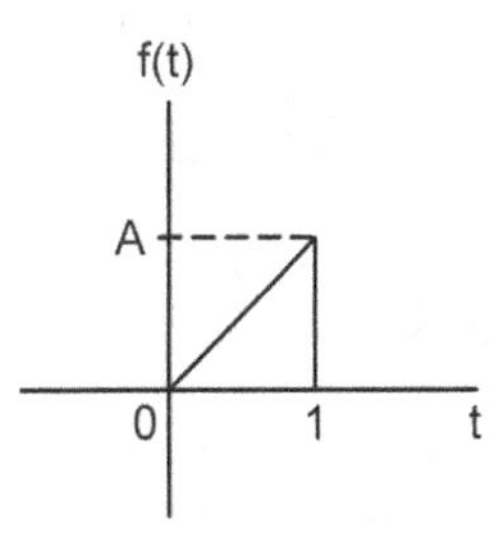

Fig. 2.13

$$a_n = \frac{2}{1}\int_0^1 At\cos 2\pi nt\, dt = 0$$

$$b_n = \frac{2}{1}\int_0^1 At\sin 2\pi nt\, dt = 0$$

$$= \frac{A}{2\pi^2 n^2}\{\sin 2\pi nt - 2\pi nt\cos 2\pi nt\}_0^1 = -A/\pi n$$

$$f(t) = \frac{A}{2} - \frac{A}{\pi}\sin 2\pi t - \frac{A}{2\pi}\sin 4\pi t + \ldots\ldots$$

3. Exponential Fourier Series

Set of exponential functions $e^{jnw_0 t}$ is orthogonal over an interval $(t_0, t_0 + 2\pi/w_0)$

$$I = \int_{t_0}^{t_0+2\pi/w_0} e^{jnw_0 t}\,(e^{jmw_0 t})\, dt$$

$$= \int_{t_0}^{t_0+2\pi/w_0} e^{jnw_0 t}\, e^{-jmw_0 t}\, dt$$

If $n = m$

$$I = \int_{t_0}^{t_0+2\pi/w_0} dt = \frac{2\pi}{w_0}$$

If $n \neq m$

$$I = \frac{1}{j(n-m)w_0} e^{j(n-m)w_0 t}\Big/_{t_0}^{t_0+2\pi/w_0}$$

$$= \frac{1}{j(n-m)w_0} e^{j(n-m)w_0 t_0}\left[e^{j2\pi(n-m)} - 1\right]$$

since n & m are integers

$$e^{j2\pi(n-m)} = 1 \qquad \therefore I = 0$$

$$\text{thus}\quad \int_{t_0}^{t_0+\frac{2\pi}{w_0}} e^{jnw_0 t}\left(e^{jmw_0 t}\right) dt = \frac{2\pi}{w_0} \qquad n = m$$

$$0 \quad n \neq m$$

$\therefore$ any function $f(t)$ can be expressed as

$$f(t) = F_0 + F_1 e^{jw_0 t} + F_2 e^{j2w_0 t} + \ldots\ldots$$

$$+ F_{-1} e^{-jw_0 t} + F_{-2} e^{-j2w_0 t} + \ldots\ldots$$

$$= \sum_{n=-\alpha}^{\alpha} F_n e^{jnw_0 t}$$

$$\text{where } F_n = \frac{\displaystyle\int_{t_0}^{t_0+T} f(t) e^{-jnw_0 t}\, dt}{\displaystyle\int_{t_0}^{t_0+T} e^{jw_0 nt}.e^{-jnw_0 t}\, dt}$$

$$= \frac{1}{T} \int_{t_0}^{t_0+T} f(t) e^{-jnw_0 t}\, dt$$

$$F_0 = \frac{1}{T} \int_{0}^{T} x(t)\, dt$$

Conversion of trigonometric series to Fourier series

$$a_0 = F_o$$

$$a_n = F_n + F_{-n}$$

$$b_n = j(F_n - F_{-n})$$

$$F_n = \frac{1}{2}(a_n - jb_n)$$

2.5 REPRESENTATION OF A FUNCTION BY CLOSED OR COMPLETE SET OF MUTUALLY ORTHOGONAL FUNCTIONS

If a function $f(t)$ is approximated by large number of orthogonal functions

$$f(t) = C_1\, g_1(t) + C_2\, g_2(t) + \ldots\ldots\ldots$$

such that right hand side infinite series converges to make error zero

$$\text{thus} \quad \int_{t_1}^{t_2} f^2(t)\, dt = \sum_{r=1}^{\alpha} A_r^2\, Kr$$

This representation is known as generalized fourier series representation.

A pair of real valued energy signals are orthogonal if

$$\int_{-\alpha}^{\alpha} x_1(t)\, x_2(t)\, dt = 0$$

A pair of real valued power signals are orthogonal if

$$\underset{T \to \alpha}{Lt}\ \frac{1}{2T}\int_{-T}^{T} x_1(t)\, x_2(t)\, dt = 0$$

Impulses are also orthogonal to one another

$$\int_{-a}^{a} \delta(t - t_1)\, \delta(t - t_2)\, dt = 0 \qquad \text{for } t_1 \neq \frac{1}{2}$$

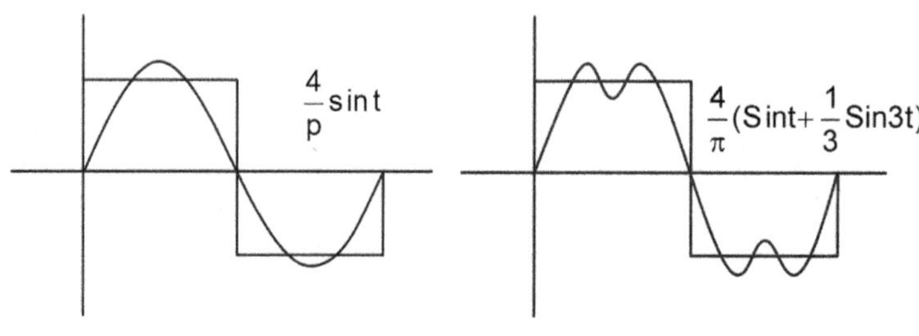

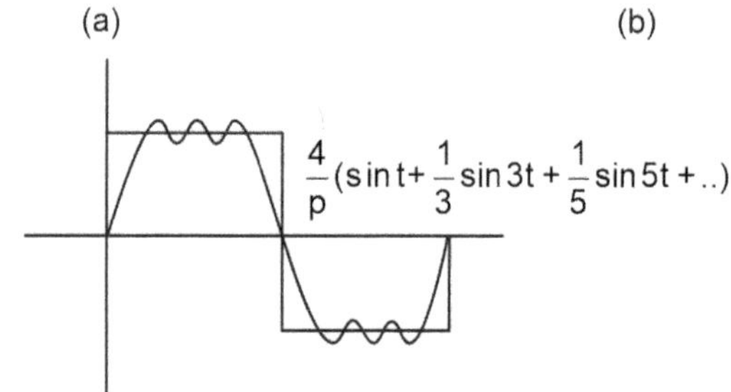

Fig. 2.14

Mean square error

$$e_r = \frac{1}{(t_2 - t_1)}\int_{t_1}^{t_2} \{f_1(t) - A_r\, f_2(t)\}^2\, dt$$

for one term

$$e = \frac{1}{2\pi}\int_{0}^{\pi}(1 - \frac{4}{\pi}\sin t)^2\, dt + \int_{\pi}^{2\pi}(1 - \frac{4}{\pi}\sin t)^2\, dt = 0.19$$

MATLAB PROGRAMS

1. Compute and plot of the fourier coefficients for the periodic signal shown.

Ans: For the signal $W_0 = 2$ (refer example 8)

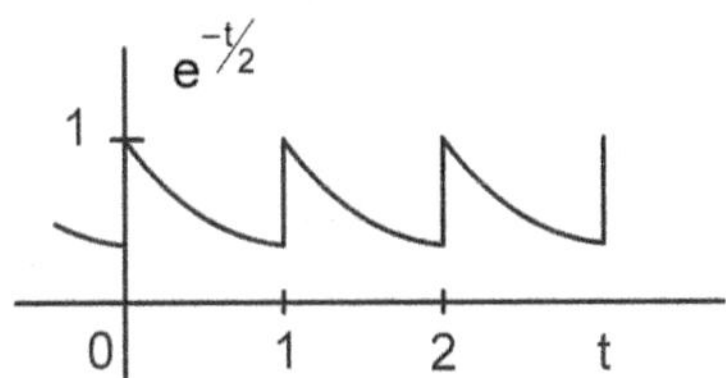

➢ N = 1:10, an(1) = 0.504, an(n+1) = 0.504 * 2/(1+16 * n^2);
➢ b_n(1) = 0; b_n(n+1) = 0.504*8*n/(1+16 × n^2);
➢ C_n(1) = an(1); C_n (n+1) = 8qrt (a_n(n+1).^2 + b_n (n+1),^2);
➢ The tan(1) = 0; the tan(n+1) = atan2(–b_n(n+1), a_n(n+1));
➢ n = [0, n]
➢ clf: subplot (2, 2, 1); stem (n, a, n', k');
➢ ylabel (an'); xlabel ('n');
➢ subplot (2, 2, 2); stem (n, b_n, 'k'); ylabel ('b_n'); xlabel ('n');
➢ subplot (2, 2, 3); stem (cn, 'k'); ylabel ('c_n'), xlabel ('n');
➢ subplot (2, 2, 4); stem (n, theta n, 'k'); ylabel ('\theta n [rad]'/); xlabel ('n').

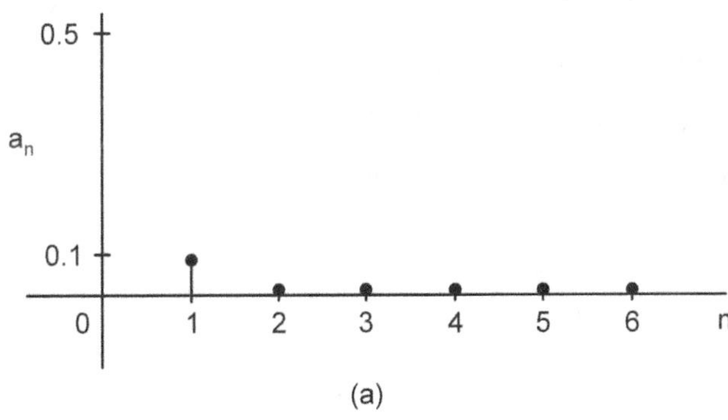

(a)

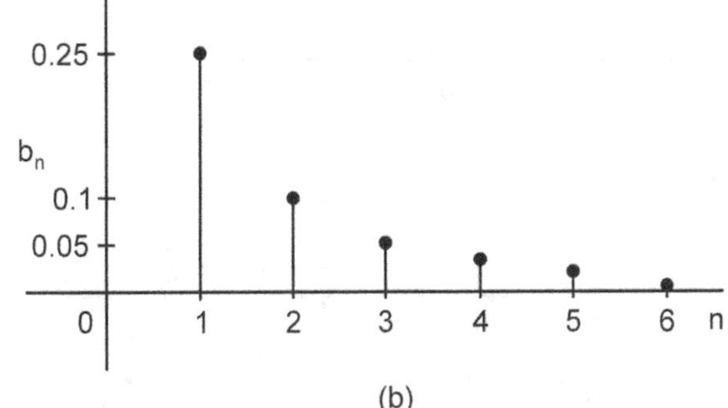

(b)

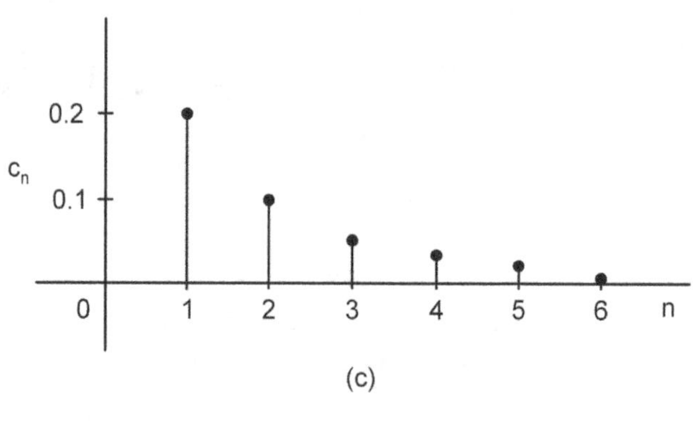

(c)

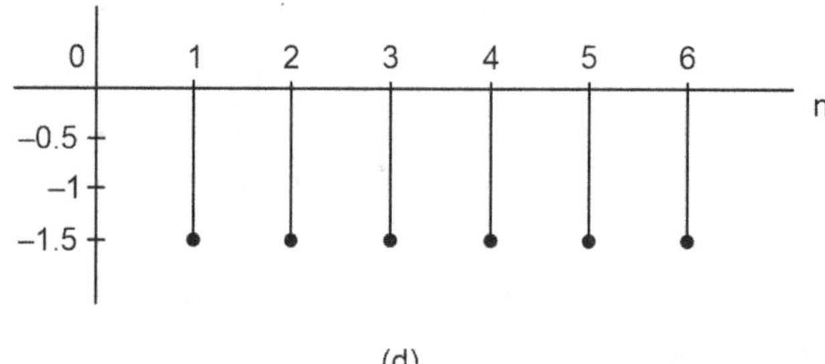

(d)

2. Plot the exponential fourier spectra for the signal in problem-1

> $n = (-10:10)$; $D_n = 0.504/(1 + j* 4 \times n)$;

> clf; subplot (2, 1, 1); stem (n, abs (D_n), ('k'));

> xlabel ('n'); y label ('$|D|$');

> subplot (2, 1, 2); stem (n, angle (D_n), ('k'));

> x label ('n'); y label (' angle D_n [rad]');

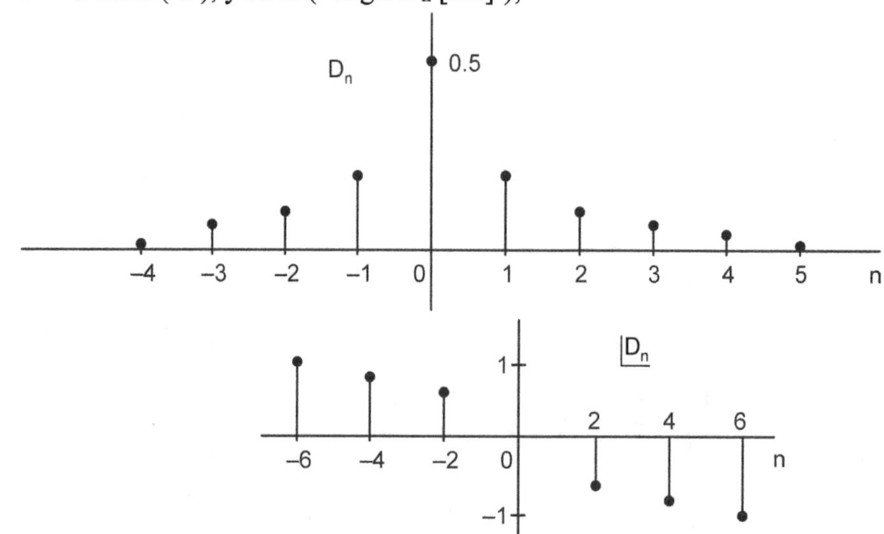

OBJECTIVE QUESTIONS

1. A vector is specified by _________ and _________
2. If θ is angle between two vectors then dot product is _________
3. If θ is angle between two vectors then cross product is _________
4. Dot product of two orthogonal vectors is _________
5. Functions $\sin mw_0t$ and $\sin nw_0t$ are _________
6. Functions $\cos mw_0t$ and $\cos nw_0t$ are _________
7. Vectors which are orthonormal has a magnitude _________

SUMMARY OF CHAPTER

1. Two functions $\phi_i(t)$ and $\phi_j(t)$ are said to be orthogonal over an interval $[0, T]$ if $\int_0^T \phi_i(t)\phi_j(t)\,dt = \begin{matrix} E & i=j \\ 0 & i \neq j \end{matrix}$ and are orthornormal over an interval $[0, T]$ if $E_i = 1$ for all i

2. Any arbitrary signal $x(t)$ can be expanded over an interval $[0, T]$ in terms of orthogonal basis function $\{\phi_n(t)\}$ as $x(t) = \sum_{n=-\alpha}^{\alpha} X_n\phi_n(t)dt$ where $X_n = \dfrac{1}{E_n}\int_0^T x(t)\phi_n(t)dt$

3. The complex exponentials $\phi_n(t) = e^{in\,wo1}$, $w_0 = \dfrac{2\pi}{T}$ are orthogonal over the interval $[0,T]$

4. The energy of sum of two orthogonal signals is equal to the sum of the energies of the two signals

5. A signal can be transformed to space which is composed of orthogonal vectors. The most common orthogonal vector space the in the set or site and cosine waves

6. If the vectors of an orthogonal basis normalized the resulting basis is orthonormal basis

7. Orthogonal basis for an inner product space V is a basis for V whose vectors are mutually orthogonal

8. A set of vectors S is orthonormal if every vector in S has magnitude 1 and the set of vectors are mutually orthogonal

9. If the eigen values of two eigen functions are the same, then the functions are said to be degenerate

10. An eigen basis is a basis in which every vector is an eigen vector

CHAPTER 3

Fourier Series

Objectives of the Chapter

This chapter describes need of Fourier series, Fourier series of periodic signals, properties of fourier series, Fourier frequency spectrum, power spectrum, continuous Fourier series, Parsevals theorem, Gibbs phenomena.

3.1 INTRODUCTION

Fourier: Jean Baptiste Joseph Fourier (March 21, 1768 – May 16, 1830) was a French mathematician. His investigation was on heat flow. According to him any function of a variable, whether continuous or discontinuous, can be expanded in a series of sines of multiples of the variable. Due to lack of precise notion of function and integral in the early 19th century, Fourier series became popular.

If $f(x)$ denotes a periodic function of real variable x, of period 2π, such that $f(x + 2\pi) = f(x)$. Fourier attempted to write such function as the sum of series of simple 2π periodic functions. Fourier series does not always converge (sum is equal to original function).

Example:

Consider a saw tooth wave

$$f(x) = x \quad \text{for} -\pi < x < \pi$$

In this case fourier coefficients are given by

$$a_n = \frac{1}{\pi} \int_{-\pi}^{\pi} x \cos(nx)\, dx = 0$$

$$b_n = \frac{1}{\pi} \int_{-\pi}^{\pi} x \sin(nx) = \frac{2(-1)^{n+1}}{n}$$

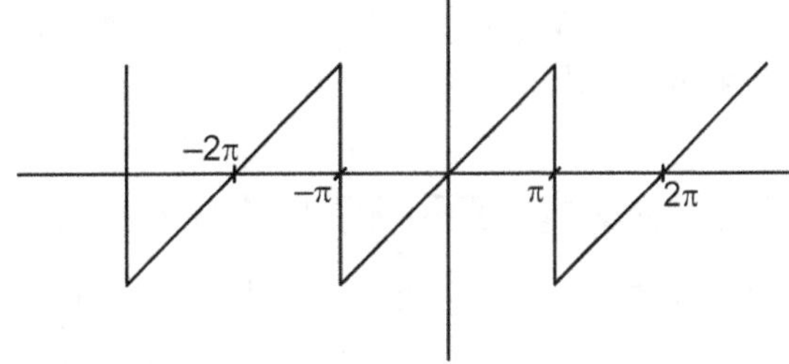

Fig. 3.1

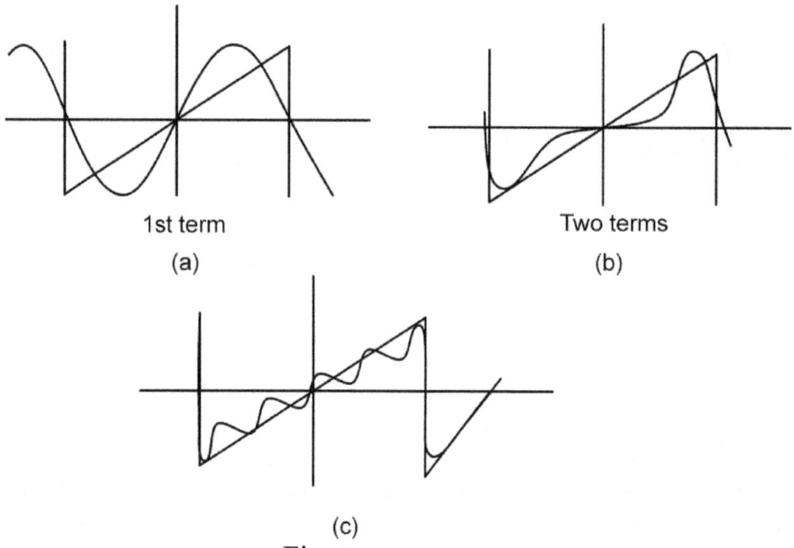

1st term

(a)

Two terms

(b)

(c)

Five terms

Fig. 3.2

$$\therefore \quad f(x) = 2\sum_{n=1}^{\alpha} \frac{(-1)^{n+1}}{n} \sin nx$$

$$x(t+T) = a_0 + \sum_{n=1}^{k} a_n \cos n\omega_0(t+T) + b_n \sin n\omega_0(t+T)$$

$$= a_0 + \sum_{n=1}^{k} a_n \cos(n\omega_0 t + 2n\pi) + b_n \sin(n\omega_0 t + 2n\pi)$$

$$= a_0 + \sum_{n=1}^{k} a_n \cos n\omega_0 t + b_n \sin n\omega_0 t$$

$$= x(t)$$

Hence the summation of sine and cosine functions of frequencies o, ω_0, $2\omega_0$, ------$k\omega_0$ is a periodic signal with period T.

3.2 FOURIER SERIES OF A PERIODIC FUNCTION

A periodic function $f(t)$ is represented by fourier series over a finite interval $(t_0, t_0 + T)$.

Outside this interval $f(t)$ its fourier series may not be equal.

If $f(t)$ is periodic, we can show that equality holds over $(-\alpha < t < \alpha)$.

Let $f(t)$ be represented by exponential fourier series

$$f(t) = \sum_{n=-\alpha}^{\alpha} F_n e^{jn\omega_0 t} \quad t_0 < t < t_0 + T \text{ and } \omega_0 = \frac{2\pi}{T}$$

If f (t) is periodic

$$f(t+T) = \sum_{n=-\alpha}^{\alpha} F_n \, e^{jn\omega_0 (t=T)}$$

$$= \sum_{n=-\alpha}^{\alpha} F_n \, e^{jn\omega_0 t} \cdot e^{jn\omega_0 T}$$

$$\sum_{n-\alpha}^{\alpha} F_n \, e^{jn\omega_0 t} \qquad\qquad Q \; e^{jn2\pi} = 1$$

The smallest value of T which for all n satisfies this condition is called as period.

3.2.1 FOURIER THEOREM

If f (t) is a periodic function over a period T then it can be represented by fourier series over $(-\alpha < t < \alpha)$

$$\therefore \quad f(t) = \sum_{n=-\alpha}^{\alpha} F_n \, e^{jn\omega_0 t} \qquad -\alpha < t < \alpha$$

$$F_n = \frac{1}{T} \int_{t_0}^{t_0+T} f(t) e^{-jn\omega_0 t} dt$$

or

$$F_n = \frac{1}{T} \int_{0}^{T} f(t) e^{-jn\omega_0 t} dt$$

3.3 PROPERTIES OF FOURIER SERIES

3.3.1 DIRICHLET CONDITIONS

Any periodic signal to be represented by FS must satisfy

1. f (t) must have only a finite number of maxima and minima

2. f (t) must possess finite number of discontinuities

3. f (t) is absolutely integrable over one period

$$\int_{0}^{T} |f(t)| dt < \alpha$$

3.3.2 PROPERTIES

If f(t) is a periodic signal with period T and $\omega_0 = \dfrac{2\pi}{T}$, then fourier coefficients of f(t) are denoted by C_n

(a) Linearity:

$\quad\quad$ If $FS\{f_1(t)\} = C_n$ $\quad\quad\quad$ then $FS\{f_1(t) + f_2(t)\} = AC_n + BD_n$

$\quad\quad\quad FS\{f_2(t)\} = D_n$

(b) Time shifting:

$$FS\{f(t - t_0)\} = e^{-j\omega_0 t_0} C_n$$

(c) Time reversal

$$FS\{f(-t)\} = C_{-n}$$

(d) Time Scaling

$$FS\{f(ta)\} = C_n \quad\quad \omega_n = a\omega_0$$

Property	Signal	Fourier Series Coefficients
	$x(t)$	a_n
	$y(t)$	b_n
Linearity	$Ax(t) + By(t)$	$Aa_n + Bb_n$
Time Shifting	$x(t - t_0)$	$e^{-j\omega_0 t_0}\, a_n$
Frequency Shifting	$e^{jw_0 t}\, x(t)$	a_{n-1}
Time reverse	$x(-t)$	a_{-n}
Time Scaling	$x(at)$	a_n
Conjugation	$x^*(t)$	$a^*{}_{-n}$
Multiplication	$x(t)\, y(t)$	$\displaystyle\sum_{k=-\alpha}^{\alpha} a\, b_{n-k}$
Differentiation	$\dfrac{d}{dt} x(t)$	$Jn\omega_0 a_n$
Integration	$\displaystyle\int_{-\alpha}^{\alpha} x(t)\, dt$	$\dfrac{1}{jn\omega_0} a_n$

3.3.3 FOURIER FREQUENCY SPECTRUM

Fourier series expansion of a periodic function is equivalent to resolving the function in terms of its frequency components $\omega_0,\ 2\omega_0 \ldots\ldots n\omega_0$; where $\omega_0 = \dfrac{2\pi}{T}$

If spectrum is known, we can find f (t). The spectrum exists at discrete values of ω, it is called line spectrum.

Exponential series is best to represent spectrum. Both $e^{j\omega t}$ and $e^{-j\omega t}$ can be used. We need two spectra: magnitude spectrum and phase spectrum.

$$f(t) = F_0 + F_1 e^{j\omega_0 t} + F_2 e^{j2\omega_0 t} + \ldots\ldots$$

$$+ F_1 e^{-j\omega_0 t} + F_2 e^{-j2\omega_0 t} + \ldots\ldots$$

3.3.4 POWER SPECTRUM

Power delivered by f (t) is

$$\frac{1}{T} \int_{-T/2}^{T/2} f^2(t)dt$$

$$= \frac{1}{T} \int f(t) \sum F_n \, e^{jn\omega_0 t} \, dt$$

$$= \frac{1}{T} \sum F_n \int f(t) e^{jn\omega_0 t} \, dt$$

$$= \sum F_n \, F_{-n} = \sum_{n=-\alpha}^{\alpha} |F_n|^2$$

$$= F_0^2 + F_1^2 + \ldots\ldots F_{-1}^2 + F_{-2}^2 + \ldots\ldots$$

3.3.5 PROPERTIES OF CONTINUOUS FOURIER SERIES

Let us denote $\Leftrightarrow$ as a transformation from periodic signal to fourier series and vice versa.

Let $x_1(t) \Leftrightarrow C_n$ and $x_2(t) \Leftrightarrow d_n$

1. **Linearity:** Let $x_1(t)$ and $x_2(t)$ are two periodic signals with period T and having fourier series coefficients C_n and d_n.

   $\therefore$ The fourier coefficients of linear combination of $x_1(t)$ and $x_2(t)$ are given by

   $$FS\{Ax_1(t) + Bx_2(t)\} \Leftrightarrow Ac_n + Bd_n$$

   Proof:

   $$\frac{1}{T} \int_{T_0} \{Ax_1(t) + Bx_2(t)\} e^{-j2\pi n f_0 t} \, dt$$

   $$= \frac{A}{T_0} \int_{T_0} x_1(t) e^{-j2\pi n f_0 t} \, dt + \frac{B}{T_0} \int_{T_0} x_2(t) e^{-j2\pi n f_0 t} \, dt = Ac_n + Bd_n$$

2. Multiplication:

$$x_1(t)\,x_2(t) \Leftrightarrow \sum_{k=-\alpha}^{\alpha} C_k D_{n-k}$$

Proof:

By multiplying x_1 (t) and x_2 (t)

$$x_3\ (t)\ =\ \sum_{k=-\alpha}^{\alpha} C_k\, e^{j2\pi k f_0 t}\ \sum_{m=-\alpha}^{\alpha} d_m e^{j2\pi m f_0 t}$$

$$=\ \sum_{k=-\alpha}^{\alpha}\sum_{m=-\alpha}^{\alpha} C_k d_m\, e^{j2\pi(k+m)f_0 t}$$

Let $n = m + k$ $\therefore\ m = n - k$

$$\sum_{k=-\alpha}^{\alpha}\sum_{n=-\alpha}^{\alpha} C_k d_{n-k}\, e^{j2\pi n f_0 T} = \sum\left(\sum C_k\, d_{n-k}\right) e^{j2\pi n f_0 T}$$

3. Time reversal:

Let $x\,(-t) \Leftrightarrow C_{-n}$

Proof:

$$x\ (-t)\ =\ \sum_{n=-\alpha}^{\alpha} C_n\, e^{j2\pi n f_0 (t)} = \sum C_n e^{j2\pi(-n)f_0 t}$$

$$=\ \sum_{m=\alpha}^{-\alpha} C_{-m}\, e^{j2\pi n f_0 t}$$

4. Time shifting:

$$x(t-T) \Leftrightarrow e^{-j2\pi n f_0 T} C_n$$

Proof:

$$x(t-T) = \sum_{n=-\alpha}^{\alpha} C_n\, e^{j2\pi n f_0 (t-T)}$$

$$=\ \sum_{n=-\alpha}^{\alpha} \left(e^{-j2\pi n f_0 T} C_n \right) e^{j2\pi n f_0 t}$$

5. Time Scaling:

$$x(at) = \sum_{n=-\alpha}^{\alpha} C_n\, e^{j2\pi n f_0 (at)}$$

Proof: Fourier series coefficients are same, but exponential basis functions have changed.

6. Conjugation:

$$x^*(t) \Leftrightarrow C^*_{-n}$$

Proof:

$$x^*(t) = \left(\sum_{n=-\alpha}^{\alpha} C_n e^{j2\pi n f_0 t} \right)^* = \sum_{m=\alpha}^{-\alpha} C^*_{-m} e^{j2\pi n f_0 t}$$

Let $m = -n$

3.4   PARSEVALS THEOREM

It states that the power of a periodic signal is equal to the sum of the powers of its fourier components.

Consider two periodic signals $x_1(t)$ and $x_2(t)$ with equal period T.

$$\frac{1}{T} \int_{t_0}^{t_0+T} x_1(t) x_2(t) = \frac{1}{T} \int_{t_0}^{t_0+T} \sum_{n=-\alpha}^{\alpha} C_n e^{jn\omega_0 t} \left\{ \sum_{n=-\alpha}^{\alpha} d_n e^{jn\omega_0 t} \right\} dt$$

$$= \frac{1}{T} \sum \sum C_n d_m \int_{t_0}^{t_0+T} e^{j(n-m)\omega_0 t} dt$$

$$= 0 \text{ for } \quad n \neq m$$

$$= \sum C_n d_n \quad n = m$$

If $x_1(t) = x_2(t) = x(t)$ then above equation can be written as

$$\frac{1}{T} \int_{t_0}^{t_0+T} |x(t)|^2 = \sum_{n=-\alpha}^{\alpha} |C_n|^2$$

In trigonometric form a periodic signal

$$x(t) = C_0 + \sum_{n=1}^{\alpha} C_n \cos(n w_0 t + \theta_n)$$

Every term in right hand side of the above equation is a power signal.

The sum of the powers of all the sinusoidal components

$$P_x = C_0^2 + \frac{1}{2} \sum_{n=1}^{\alpha} C_n^2$$

3.5 GIBBS PHENOMENON

The fourier series of a piece wise continuously differentiable periodic function behaves at a jump discontinuity: the nth partial sum of the fourier series has large oscillations near the jump, which might increase the maximum of the partial sum above that of the function itself. The overshoot does not die out as the frequency increases, but becomes finite.

When a square wave is approximated using sine wave

$$f(t) = \sin t = \frac{1}{3}\sin 3t + \frac{1}{5}\sin 5t + \dots$$

As the number of terms rises, the error of the approximation reduced in width and energy, but converges to a fixed height

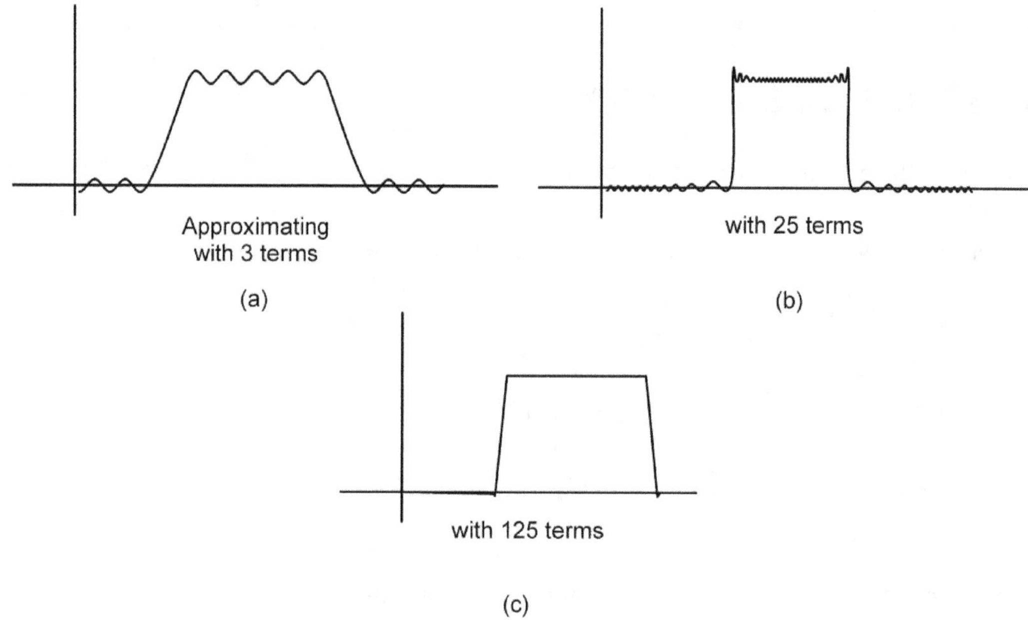

Fig. 3.3

At any jump point of a piece wise continuously differentiable function with a jump of 'a', the nth partial fourier series will overshoot this jump by approximately 'a' at one end and undershoot it by the same amount at the other end.

Gibbs over shoot is a manifestation of the subtle non uniform convergence of fourier series.

3.6 IMPORTANT POINTS REGARDING FOURIER COEFFICIENTS

1. The coefficients of exponential fourier series of an even periodic waveform on t will be real and its trigonometric fourier series will contain only cosine terms.

2. The coefficients of exponential fourier series of an odd periodic waveform on t will be imaginary and its trigonometric fourier series will contain only sine terms.

3. A periodic waveform with half wave symmetry doesn't have any average value and doesn't contain any even harmonics.

4. If a periodic waveform is even on t and is half wave symmetry, its fourier series expansion will contain only cosine functions at odd harmonics.

5.   If a periodic waveform is odd on t and is half wave symmetry, its fourier series expansion will contain only sine functions at odd harmonics.

3.7 COMPACT FORM OF FOURIER SERIES

If $x(t)$ is real and a_n, b_n are real for all n, trigonometric fourier series can be expressed as

$$x(t) = C_0 + \sum_{n=1}^{\alpha} C_n \cos(nw_0 t + \theta_n) \text{ where}$$

$$C_0 = a_0$$

$$C_n = \sqrt{a_n^2 + b_n^2} \quad \text{and} \quad \theta = \tan^{-1}\left(\frac{-b_n}{a_n}\right) \quad \theta = \tan^{-1}\left(\frac{-b_n}{a_n}\right)$$

3.8 EXPONENTIAL FOURIER SERIES

According to Euler's equality, cos $nw_0 t$ and sin $nw_0 t$ can be expressed in terms of $e^{jnw_0 t}$ and $e^{-jnw_0 t}$.

$$x(t) = \sum_{n=-\alpha}^{\alpha} D_n e^{jnw_0 t}$$

Multiplying both sides by $e^{-jmw_0 t}$

$$\int_{T_0} x(t) e^{-jmw_0 t} dt = \sum_{n=-\alpha}^{\alpha} D_n \int_{T_n} e^{j(n-m)w_0 t} dt$$

We have

$$\int_{T_0} e^{jnw_0 t} e^{-jmw_0 t} dt = 0 \qquad m \neq n$$

$$= T_0 \qquad m = n$$

$$\therefore \qquad D_m T_0 = \int_{T_0} x(t) e^{-jmw_0 t} dt$$

$$\therefore \qquad D_m = \frac{1}{T_0} \int_{T_0} x(t) e^{-jmw_0 t} dt$$

Exponential fourier series can be expressed as

$$x(t) = \sum_{n=-\alpha}^{\alpha} D_n e^{jnw_0 t}$$

$$\text{where} \quad D_n = \frac{1}{T_0} \int_{T_0} x(t) e^{-jnw_0 t} dt$$

Comparing trigonometric and exponential series

$$D_0 = a_0$$

$$D_n = \frac{1}{2}(a_n - jb_n)$$

$$D_{-n} = \frac{1}{2}(a_n + jb_n)$$

also $a_n - jb_n = C_n \, e^{j\theta n}$ where $C_n = \sqrt{a_n^2 + b_n^2}$

$$\theta_n = \tan^{-1}\left(\frac{-b_n}{a_n}\right)$$

Hence $D_0 = a_0 = C_0$

$$D_n = \frac{1}{2}C_n e^{j\theta n} \qquad D_{-n} = \frac{1}{2}C_n e^{-j\theta n}$$

3.9 LIMITATIONS OF FOURIER SERIES METHOD

(a) The fourier series can be used only for periodic inputs which starts from $t = -\alpha$

(b) The fourier methods can be applied readily to asymptotically stable systems. It cannot handle unstable or marginally stable systems.

The first limitation can be overcome by representing a periodic signals in terms of exponentials. This representation can be achieved through fourier integral.

The second limitation can be overcome by using exponentials e^{st}, where s is not restricted to imaginary, but takes complex value. The generalization leads to Laplace integral.

SOLVED PROBLEMS

EXERCISE – 1

For the signal shown

1. $$x(t) = 2 + \cos\left(\frac{2\pi}{3}t\right) + 4\sin\left(\frac{5\pi}{3}t\right)$$

Find fundamental frequency ω_0 and the fourier series coefficients C_n such

$$x(t) = \sum C_n e^{jn\omega_0 t}$$

Solution: Time period of signal $\cos\dfrac{2\pi t}{3}$

$$T_1 = \frac{2\pi}{\frac{2\pi}{3}} = 3$$

Time period of signal $\sin \dfrac{5\pi t}{3}$

$$T_2 = \frac{2\pi}{\frac{5\pi}{3}} = \frac{6}{5}$$

$$\therefore \quad \frac{T_1}{T_2} = \frac{5}{2}$$

$$2T_1 = 5T_2$$

$$T = 6 \text{ sec}$$

$$\omega_0 = \frac{2\pi}{T} = \frac{\pi}{3}$$

$$x(t) = 2 + 0.5\left\{ e^{\frac{j2\pi t}{3}} + e^{\frac{-j2\pi t}{3}} \right\} + \frac{2}{j}\left\{ e^{\frac{j5\pi t}{3}} - e^{\frac{-j5\pi t}{5}} \right\}$$

$$= -2je^{\frac{-j5\pi t}{5}} + 0.5e^{\frac{j2\pi t}{3}} + 2 + 0.5e^{\frac{-j2\pi t}{3}} + 2je^{\frac{-j5\pi t}{3}}$$

$$= -2je^{-j(-5)\left(\frac{\pi}{3}\right)t} + 0.5e^{-j(-2)\left(\frac{\pi}{3}\right)t} + 2 + 0.5e^{-j2\left(\frac{\pi}{3}\right)t} + 2je^{-j5\left(\frac{\pi}{3}\right)t}$$

$$= C_{-5}e^{-j(-5)\omega_0 t} + C_{-2}e^{-j(-2)\omega_0 t} + C_0 + C_2 e^{-j2\omega_0 t} + C_5 e^{-j5\omega_0 t}$$

$$C_{-5} = -2j; \quad C_{-2} = 0.5; \quad C_0 = 2; \quad C_2 = 0.5; \quad C_5 = 2j$$

2. Find fourier series coefficients for the signal shown below

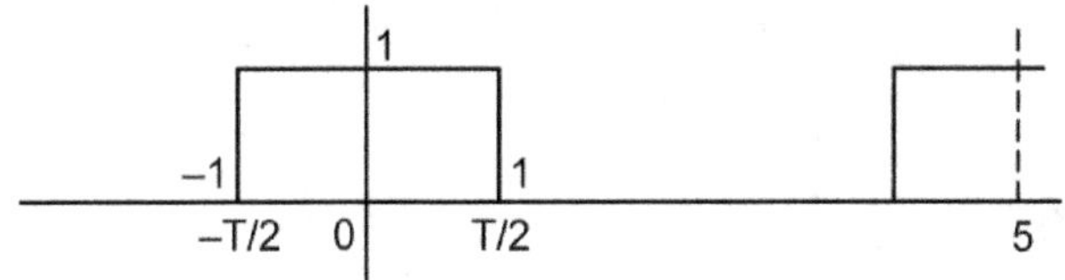

Fig. 3.4

$$C_n = \frac{1}{T}\int_{t_0}^{t_0+T} x(t)e^{-jn\omega_0 t}\, dt$$

$$= \frac{1}{T}\cdot\frac{1}{-jn\omega_0}e^{-jn\omega_0 t}\Big|_{-T/2}^{T/2}$$

$$= \frac{1}{n} \left\{ \frac{e^{jn\omega_0 T/2} - e^{-jn\omega_0 T/2}}{j\omega_0 T} \right\}$$

$$= \frac{2}{T} \frac{\sin(n\omega_0 T)}{n\omega_0}$$

$$T = 5, \ \frac{T}{2} = 1$$

$$\omega_0 = \frac{2\pi}{5}$$

$$C_n = \frac{2}{5} \frac{\sin\left(n\frac{2\pi}{5}\right)}{n\frac{2\pi}{5}} = \frac{\sin\left(\frac{2\pi n}{5}\right)}{n\pi}$$

$$\underset{n \to o}{Lt} \frac{\sin\theta}{\theta} = 1$$

$$C_0 = \underset{n \to o}{Lt} \frac{\sin\left(\frac{2\pi n}{5}\right)}{n\pi} = \frac{2}{5} \lim_{n \to o} \frac{\sin\left(\frac{2\pi n}{5}\right)}{\frac{2\pi n}{5}} = \frac{2}{5}$$

$$C_1 = \frac{\sin\left(\frac{2\pi}{5}\right)}{\pi} = 0.3027$$

$$C_2 = \frac{\sin\left(\frac{4\pi}{5}\right)}{2\pi} = 0.09$$

3. Sketch the signal $x(t) = t^2$ using trigonometric series over $(-1, 1)$

$$\omega_0 = \frac{2\pi}{2} = \pi$$

$$a_0 = \frac{1}{2} \int_{-1}^{1} t^2 \, dt = \frac{1}{2} \frac{t^3}{5} \Big/ \Big._{-1}^{1} = \frac{1}{3}$$

$$a_n = \frac{2}{T} \int_{t_0}^{t_0+T} x(t) \cos n\omega_0 t = \int_{-1}^{1} t^2 \cos n\pi t \, dt = \frac{4}{n^2\pi^2}(-1)^n$$

$$b_n = 0$$

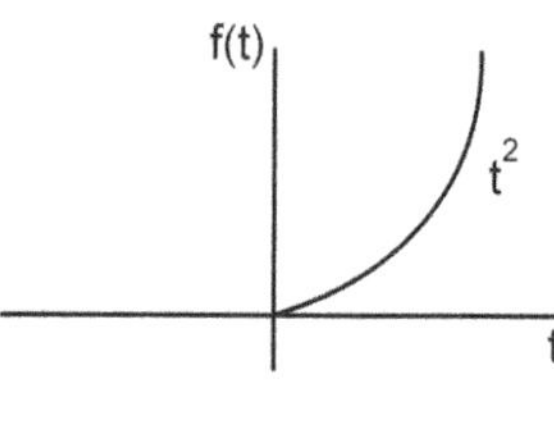

Fig. 3.5

4. Expand the periodic function by the exponential fourier series.

$$f(t) = V_0, \qquad \frac{T}{2} < t < \frac{T}{2}$$

$$F_n = \frac{1}{T} \int_{-T/2}^{T/2} V_0 e^{-jn\omega_0 t}\, dt$$

$$\omega_0 = \frac{2\pi}{2T} = \frac{\pi}{T}$$

$$= \frac{V_0}{T} \left. \frac{e^{-jnw_0 t}}{-jnw_0} \right|_{-T/2}^{T/2}$$

$$= \frac{V_0}{\pi} \frac{e^{-jn\,2\pi\,T/2} - e^{jn\,2\pi\,T/2T}}{-jn\frac{\pi}{\pi}}$$

$$= \frac{V_0}{-jn\pi} \left(e^{-jn\pi\,T/2T} - e^{jn\pi\,T/2T} \right)$$

$$= \frac{2V_0}{n\pi} \left\{ \frac{e^{jn\pi\,T/2T} - e^{jn\pi\,T/2T}}{2j} \right\}$$

$$= \frac{2V_0}{n\pi} \sin\left(n\pi\frac{T}{2T} \right)$$

$$= \frac{TV_0}{T} \frac{\sin n\pi\frac{T}{2T}}{n\pi\,T/2T} = \frac{V_0 T}{T} \frac{\sin \alpha}{\alpha}$$

$$\alpha = \frac{n\pi T}{2T}$$

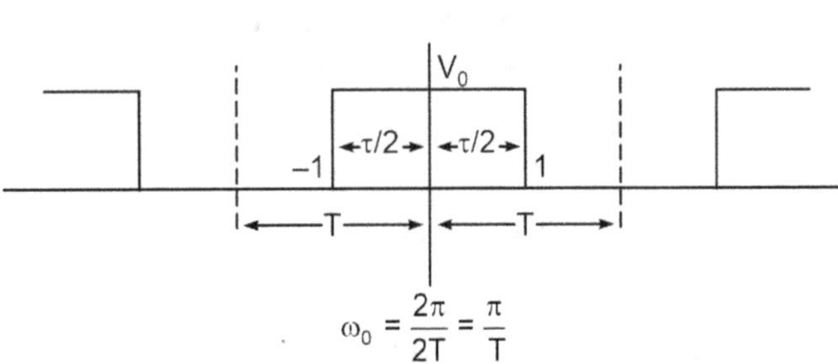

$$\omega_0 = \frac{2\pi}{2T} = \frac{\pi}{T}$$

Fig. 3.6

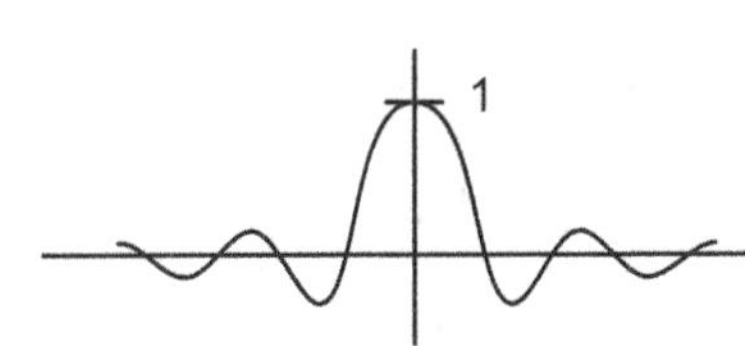

Fig. 3.7

5. Find the fourier series for $\cos\omega_0 t + \sin^2 \omega_0 t$

$$\textit{Ans:}\, x\,(t) = \cos\omega_0 t + \sin^2 \omega_0 t$$

$$= \frac{1}{2}\left\{ e^{jw_0 t} + e^{-jw_0 t} \right\} + \frac{1}{2j}\left\{ e^{jw_0 t} - e^{-jw_0 t} \right\}^2$$

$$= \frac{1}{2} e^{jw_0 t} + e^{-jw_0 t} - \frac{1}{4}\left\{ e^{jw_0 t} - 2 + e^{-2jw_0 t} \right\}$$

$$= \frac{1}{2} + \frac{1}{2} e^{jw_0 t} + \frac{1}{2} jw_0 t - \frac{1}{4} e^{2jw_0 t} - \frac{1}{4} e^{-2jw_0 t} + \ldots\ldots$$

$$C_0 = \frac{1}{2},\ C_1 = \frac{1}{2},\ C_{-1} = \frac{-1}{2},\ C_2 = \frac{-1}{4}$$

6. Find the exponential fourier series for square wave

   Average value is 0, and the wave is odd

   $$f(t) = F_0 + F_1 e^{j\omega_0 t} + F_2 e^{j2\omega_0 t} + \ldots\ldots\ldots$$

   $$F_{-1} e^{-j\omega_0 t} + F_{-2} e^{-j2\omega_0 t} + \ldots\ldots\ldots$$

   $$F_n = \frac{1}{2\pi} \int_{-\pi}^{0} -Ae^{-jn\omega t}\, d\omega t + \int_{0}^{\pi} Ae^{-jn\omega t}\, d\omega t$$

   $$= \frac{\Delta}{2\pi} - \frac{1}{-jn} e^{-jn\omega t} \Big/_{-\pi}^{0} + \frac{1}{-jn} e^{-jn\omega t} \Big/_{0}^{\pi}$$

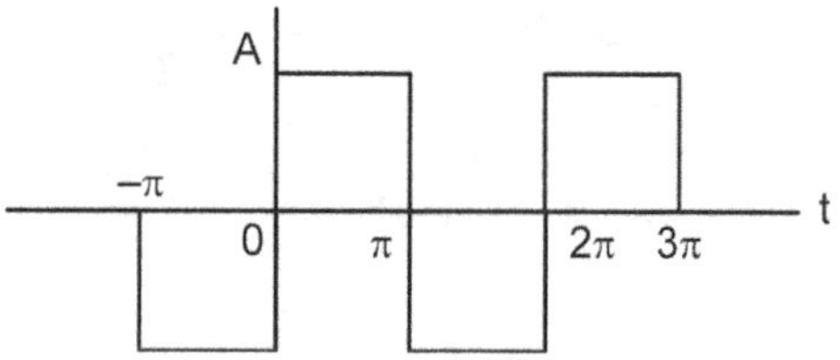

   for n even $e^{jn\pi} = 1, \quad F_n = 0$

   n odd $e^{jn\pi} = 1, \quad F_n = -j\left(\dfrac{2A}{n\pi}\right)$

   $$f(t) = \ldots + j\frac{2A}{3\pi} e^{-j3\omega t} + j\frac{2A}{\pi} e^{j\omega t} - j\frac{2A}{\pi} e^{j\omega t} - j\frac{2A}{3\pi} e^{j3\omega t} - \ldots$$

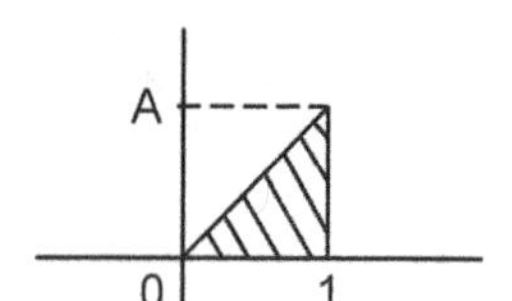

Fig. 3.8

7. Approximate the function

   $$f(t) = At \qquad 0 < t < 1$$

   $$f(t) = \sum_{n=-\alpha}^{\alpha} F_n e^{j2\pi nt}$$

   $$F_n = \frac{1}{T} \int_{t_0}^{t_0+T} f(x) e^{j2n\pi t}$$

Fig. 3.9

8. Find the exponential fourier series for

   $$F_n = \frac{1}{2\pi} \int_{0}^{\pi} A\sin\omega t\, e^{-jn\omega t}\, d\omega t$$

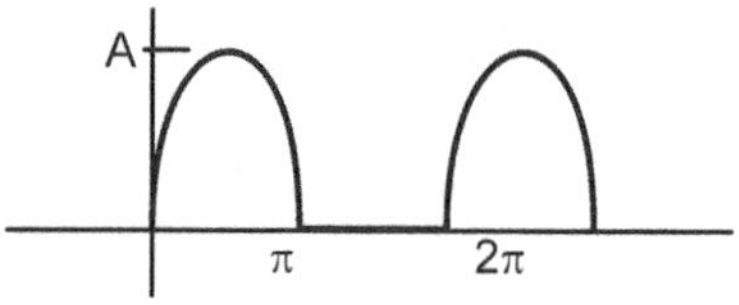

   $$= \frac{1}{2\pi} \int_{0}^{\pi} A\sin\theta e^{-jn\theta}\, d\theta$$

   $$= \frac{A}{2\pi}\left\{ \frac{e^{-jn\theta}}{(1-n^2)}(-jn\sin\theta - \cos\theta) \right\} \Big/_{0}^{\pi}$$

   $$= \frac{A}{2\pi(1-n^2)}\left(e^{-jn\pi} + 1\right)$$

   for n   odd $e^{-jn\pi} = -1 \quad F_n = 0$

   n   even $F_n = \dfrac{A}{\pi(1-n^2)}$

Fig. 3.10

PROBLEMS FROM PREVIOUS PAPERS

1. Find the exponential fourier series and plot magnitude and phase spectrum for the signal shown in figure.

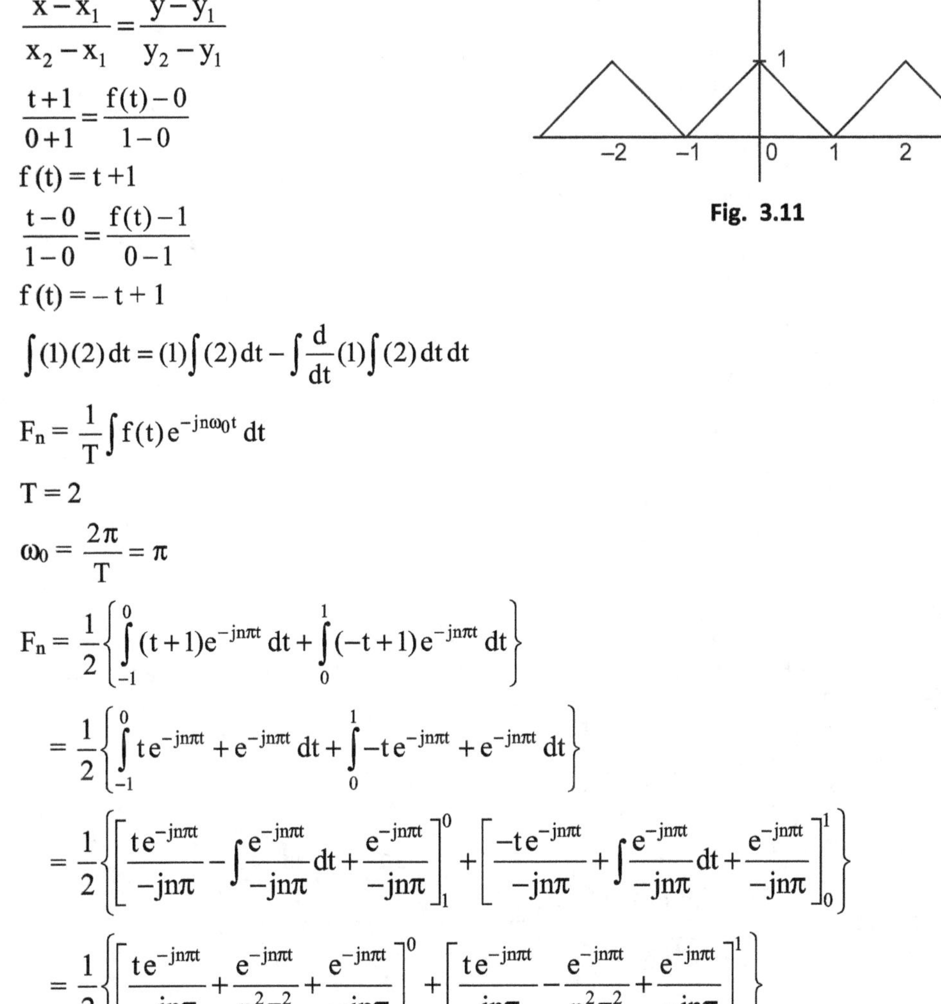

Fig. 3.11

$$\frac{x - x_1}{x_2 - x_1} = \frac{y - y_1}{y_2 - y_1}$$

$$\frac{t + 1}{0 + 1} = \frac{f(t) - 0}{1 - 0}$$

$$f(t) = t + 1$$

$$\frac{t - 0}{1 - 0} = \frac{f(t) - 1}{0 - 1}$$

$$f(t) = -t + 1$$

$$\int (1)(2)\,dt = (1)\int (2)\,dt - \int \frac{d}{dt}(1)\int (2)\,dt\,dt$$

$$F_n = \frac{1}{T}\int f(t)\,e^{-jn\omega_0 t}\,dt$$

$$T = 2$$

$$\omega_0 = \frac{2\pi}{T} = \pi$$

$$F_n = \frac{1}{2}\left\{\int_{-1}^{0}(t+1)e^{-jn\pi t}\,dt + \int_{0}^{1}(-t+1)e^{-jn\pi t}\,dt\right\}$$

$$= \frac{1}{2}\left\{\int_{-1}^{0} t\,e^{-jn\pi t} + e^{-jn\pi t}\,dt + \int_{0}^{1} -t\,e^{-jn\pi t} + e^{-jn\pi t}\,dt\right\}$$

$$= \frac{1}{2}\left\{\left[\frac{t\,e^{-jn\pi t}}{-jn\pi} - \int \frac{e^{-jn\pi t}}{-jn\pi}\,dt + \frac{e^{-jn\pi t}}{-jn\pi}\right]_{-1}^{0} + \left[\frac{-t\,e^{-jn\pi t}}{-jn\pi} + \int \frac{e^{-jn\pi t}}{-jn\pi}\,dt + \frac{e^{-jn\pi t}}{-jn\pi}\right]_{0}^{1}\right\}$$

$$= \frac{1}{2}\left\{\left[\frac{t\,e^{-jn\pi t}}{-jn\pi} + \frac{e^{-jn\pi t}}{n^2\pi^2} + \frac{e^{-jn\pi t}}{-jn\pi}\right]_{-1}^{0} + \left[\frac{t\,e^{-jn\pi t}}{jn\pi} - \frac{e^{-jn\pi t}}{n^2\pi^2} + \frac{e^{-jn\pi t}}{-jn\pi}\right]_{0}^{1}\right\}$$

$$= \frac{1}{2}\left\{\frac{1}{n^2\pi^2} - \frac{1}{jn\pi} - \frac{e^{+jn\pi}}{jn\pi} - \frac{e^{jn\pi}}{n^2\pi^2} + \frac{e^{jn\pi}}{jn\pi} + \frac{e^{-jn\pi}}{jn\pi} - \frac{e^{-jn\pi}}{n^2\pi^2} - \frac{e^{-jn\pi}}{+jn\pi} + \frac{1}{n^2\pi^2} + \frac{1}{jn\pi}\right\}$$

$$= \frac{1}{2}\left\{\frac{2}{n^2\pi^2} - \frac{e^{jn\pi}}{n^2\pi^2} - \frac{e^{-jn\pi}}{n^2\pi^2}\right\}$$

$$= \frac{1}{2n^2\pi^2}\left\{2 - e^{jn\pi} - e^{-jn\pi}\right\}$$

2. Find trigonometric fourier series for periodic square waveform shown is figure

Nov $-$ 2008

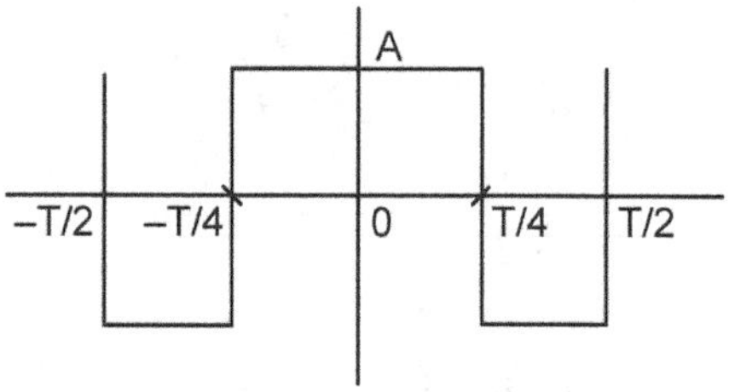

$$T = \frac{3T}{4}$$

$$\omega_0 = \frac{2\pi}{3T/4} = \frac{3\pi}{3T}$$

$$f(t) = a_0 + \sum_{n=1}^{\alpha} a_n \cos \omega_0 t + \sum_{n=1}^{\alpha} b_n \sin n\omega_0 t$$

$$a_0 = \frac{1}{T} \int_0^T f(t)\,dt$$

$$= \frac{4}{3T} \left\{ \int_{-T/4}^{T/4} A\,dt + \int_{T/4}^{T/2} -A\,dt \right\}$$

$$= \frac{4A}{3T} \left\{ \frac{T}{4} + \frac{T}{4} + -\frac{T}{2} + \frac{T}{4} \right\}$$

$$= \frac{A}{3}$$

$$a_n = \frac{2}{T} \int f(t) \cos n\omega_0\,dt$$

$$= \frac{8}{3T} \left\{ \int_{-T/4}^{T/4} A \cos n\omega_0 dt + \int_{T/4}^{T/2} A \cos n\omega_0 t\,dt \right\}$$

$$\text{First term} = \frac{8A}{3T} \left. \frac{\sin n\omega_0 t}{n\omega_0 t} \right|_{-T/4}^{T/4} = \frac{8A}{3T} \frac{\sin n \frac{3\pi}{3T} \cdot \frac{T}{4} - \frac{\sin ns\pi}{3T - T/4}}{n\frac{8\pi}{3T}}$$

$$= \frac{2A}{n\pi} \sin \frac{8\pi}{12} = \frac{2A}{n\pi} \sin \frac{n2\pi}{3}$$

$$\text{Second term} = \frac{8A}{3T} \left(\sin \frac{n8\pi}{3T} T/2 - \sin \frac{8\pi}{3T} T/4 \right) \Big/ \frac{n8\pi}{3T}$$

Fig. 3.12

$$b_n = \frac{2}{T}\int f(t)\sin n\omega_0 t\,dt$$

$$= \frac{8}{3T}\int A\sin n\omega_0\,dt = \frac{8}{3T}\left\{-\frac{\cos n\omega_0 t}{n}\Bigg|_{-T/4}^{T/4} + \frac{-\cos n\omega_0 t}{n\omega_0}\Bigg|_{T/4}^{T/2}\right\}$$

$$= \frac{-8A}{3T}\left[\left\{\frac{\cos n\dfrac{8\pi}{3T}\dfrac{T}{4} - \cos\dfrac{8\pi}{3T}\dfrac{T}{4}}{n\dfrac{8\pi}{3T}}\right\} + \left\{\frac{\cos n\dfrac{8\pi}{3T}\cdot\dfrac{T}{2} - \cos n\dfrac{8\pi}{3T}\dfrac{T}{4}}{n\dfrac{8\pi}{3T}}\right\}\right]$$

$$= \frac{-A}{n\pi}\left\{\cos n\frac{4\pi}{3} - \cos n\frac{2\pi}{3}\right\}$$

$$a_n = \frac{2A}{n\pi}\sin n\frac{2\pi}{3} + \frac{A}{n\pi}\left\{\sin n\frac{4\pi}{3} - \sin n\frac{2\pi}{3}\right\} = \frac{A}{n\pi}\left\{\sin n\frac{2\pi}{3} + \sin n\frac{4\pi}{3}\right\}$$

$$\therefore\quad f(t) = \frac{A}{3} + \frac{A}{n\pi}\sin\frac{2\pi}{3} + \frac{A}{n\pi}\sin n\frac{4\pi}{3} + \frac{A}{n\pi}\cos n\frac{2\pi}{3} - \frac{A}{n\pi}\cos n\frac{4\pi}{3}$$

3. Find trigonometric fourier expansion of periodic waveform

Feb 2008, May 2005

$$T = T$$

$$\omega_0 = \frac{2\pi}{T}$$

$$a_0 = \frac{1}{T}\int f(t)\,dt$$

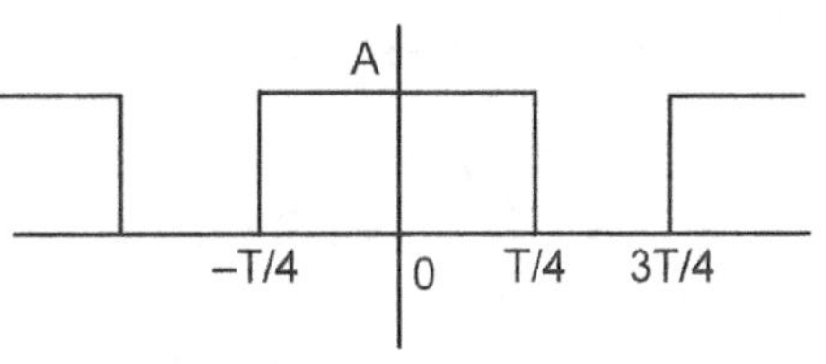

Fig. 3.13

$$= \frac{1}{T}\int_{-T/4}^{T/4} A\,dt = \frac{1}{T}\left\{\frac{T}{4} + \frac{T}{4}\right\} = \frac{A}{2}$$

$$a_n = \frac{2}{T}\int_{-T/4}^{T/4} A\cos n\omega_0 t\,dt = \frac{2A}{n\pi}\sin\frac{n\pi}{2}$$

$$b_n = \frac{2}{T}\int_{-T/4}^{T/4} A\sin n\omega_0 t\,dt = 0$$

4. Obtain trigonometric forurier series for waveform shown in fig.   Sep 2007, Nov 2009

$$a_0 = \frac{1}{T}\int_0^{T/2} A\sin\omega t\,dt \qquad = \frac{A}{T}(-\cos\omega t)\Bigg|_0^{T/2} = \frac{A}{T}\{1 - \cos\omega T/2\}$$

if $T = T$

$$= \frac{2A}{2\pi}\{2\} = \frac{A}{\pi}$$

$$a_n = \frac{2}{T}\int_0^{T/2} A\sin\omega t\cos n\,\omega t\,dt$$

if $T = 2\pi$

$\omega = 1$

$$= \frac{2}{2\pi}\int_0^{T/2} A\sin t\cos nt\,dt$$

$$= \frac{2A}{2\pi}\int_0^{\pi} \sin(1+n)t + \sin(1-n)t\,dt$$

$$= \frac{A}{\pi}\left\{-\frac{\cos(1+n)t}{1+n}\Big|_0^{\pi} - \frac{\cos(1-n)t}{1-n}\Big|_0^{\pi}\right\}$$

$$= \frac{A}{\pi}\left(\frac{2}{1+n} + \frac{2}{1-n}\right) = \frac{2A}{\pi(1-n^2)}$$

$$b_n = \frac{2}{T}\int_0^{T/2} A\sin\omega t\sin n\,\omega t\,dt$$

$$= \frac{2}{2\pi}\int_0^{\pi} A\sin t\sin nt\,dt = 0$$

$$f(t) = \frac{A}{\pi} + \sum_{n=2}^{\alpha}\frac{2A}{\pi(1-n^2)}\cos nt$$

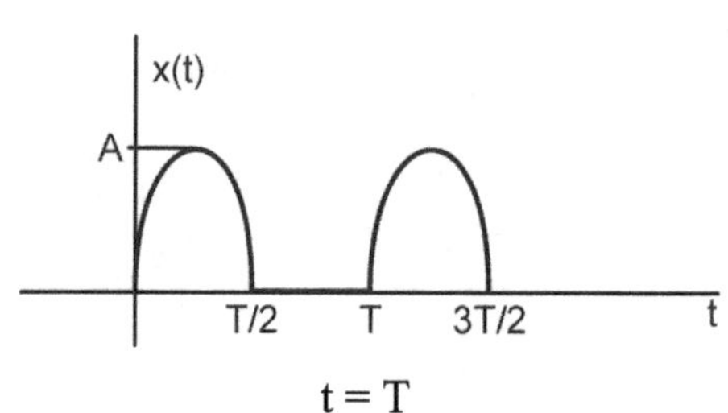

$t = T$

Fig. 3.14

5. Find the exponential fourier series for the impulse train.   Sep 2007

$$\delta_T(t) = \sum_{n=-\alpha}^{\alpha}\delta(t - nT_0)$$

$$F_n = \frac{1}{T}\int f(t)e^{-jn\omega_0 t}\,dt$$

$$= \frac{1}{T_0}\int_{-T_0/2}^{T_0/2}\delta_T(t)e^{-jn\omega_0 t}\,dt = \frac{1}{T_0}$$

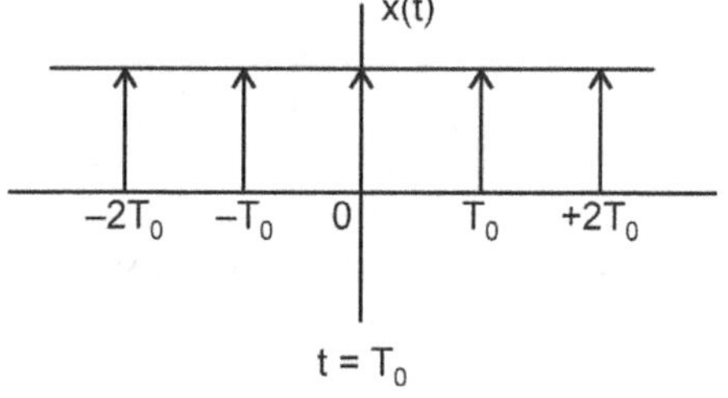

$t = T_0$

Fig. 3.15

6. Explain the exponential fourier series for the waveform shown fig.   Sep 2007

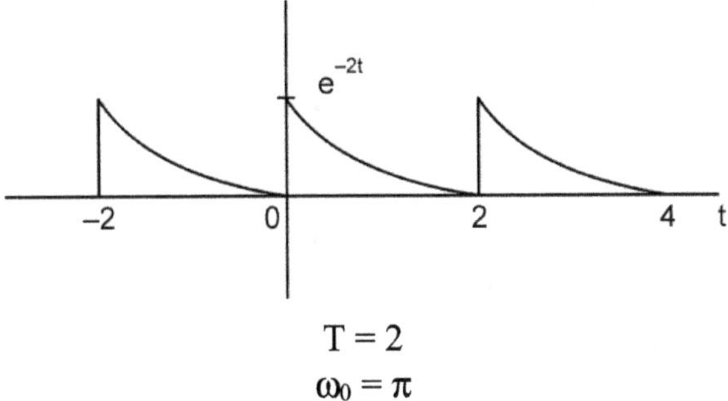

$$T = 2$$
$$\omega_0 = \pi$$

Fig . 3.16

$$F_n = \frac{1}{T}\int_0^T f(t)e^{-jn\omega_0 t}\,dt$$

$$= \frac{1}{2}\int_0^2 e^{-2t}\,e^{-jn\pi t}\,dt$$

$$= \frac{1}{2}\int_0^2 e^{-(2+jn\pi)t}\,dt = \frac{1}{2}\left.\frac{e^{-(2+jn\pi)t}}{-(2+jn\pi)}\right|_0^2 = \frac{1}{2}\left\{\frac{e^{-2(2+jn\pi)}}{-(2+jn\pi)}+\frac{1}{(2+jn\pi)}\right\}$$

$$= \frac{1}{2}\left\{\frac{1}{(2+jn\pi)}-\frac{e^{-2(2+jn\pi)}}{(2+jn\pi)}\right\}$$

$$= \frac{1-e^{-(2+jn\pi)}}{2(2+jn\pi)}$$

$$f(t) = \frac{1-e^{-4}}{4}+\frac{1-e^{-(2+j\pi)}}{2(2+j\pi)}$$

7. Find the exponential fourier series for saw tooth   Sep 2007

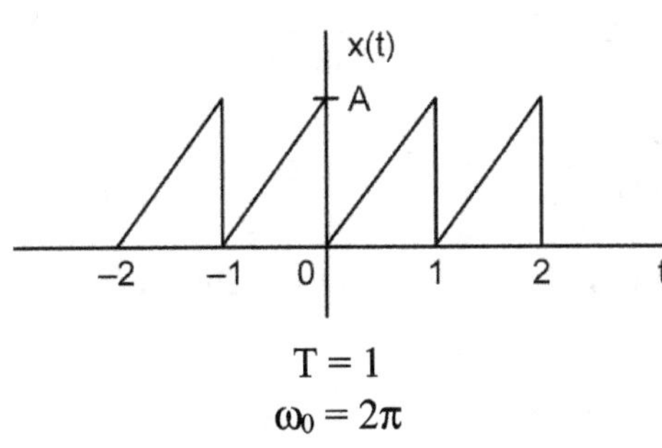

$$T = 1$$
$$\omega_0 = 2\pi$$

Fig. 3.17

$$f(t) = At$$

$$F_n = \frac{1}{T}\int_0^T f(t)e^{-jn\omega_0 t}\,dt$$

$$= \frac{1}{T}\int_0^1 At\cdot e^{-jn2\pi t}\,dt$$

$$= A\left\{\frac{te^{-jn2\pi t}}{-jn2\pi} - \int\frac{e^{-jn2\pi t}}{-jn2\pi}\,dt\right\}_0^1$$

$$= A\left\{\frac{te^{-jn2\pi t}}{-jn2\pi} - \frac{e^{-jn2\pi t}}{n^2 4\pi^2}\right\}_0^1$$

$$= A\left\{\frac{e^{-jn2\pi t}}{-jn2\pi} - \frac{e^{-jn2\pi t}}{n^2 4\pi^2}\right\}$$

$$= A\cdot e^{-jn2\pi}\left\{\frac{1}{-jn2\pi} - \frac{1}{4n^2\pi^2}\right\}$$

8. Find the compact trigonometric fourier series for the periodic signal x(t) shown below

For the signal $T_0 = \pi$, $W_0 = \dfrac{2\pi}{T_0} = 2\,\text{rad}/s$

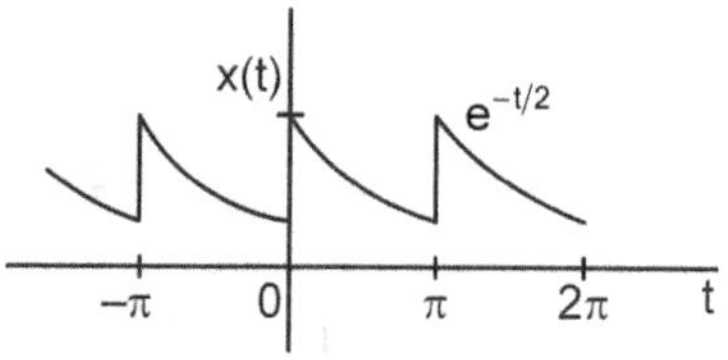

$$\therefore\ x(t) = a_0 + \sum_{n=1}^{\alpha} a_n\cos2nt + b_n\sin2nt$$

where

Fig. 3.18

$$a_0 = \frac{1}{\pi}\int_{T_0} x(t)\,dt = \frac{1}{\pi}\int_0^\pi e^{-t/2}\,dt = 0.504$$

$$a_n = \frac{2}{\pi}\int_0^\pi e^{-t/2}\cos2nt\,dt = 0.504\left(\frac{2}{1+16n^2}\right)$$

$$b_n = \frac{2}{\pi}\int_0^\pi e^{-t/2}\sin2nt\,dt = 0.504\left(\frac{8_n}{1+16n^2}\right)$$

$$\therefore\ x(t) = 0.504\left\{1 + \sum_{n=1}^{\alpha}\frac{2}{1+16n^2}\left(\cos2nt + 4\pi\sin2nt\right)\right\}$$

9. Find the compact trigonometric fourier series for signal shown

For the signal

$$T_0 = 2$$

$$W_0 = \frac{2\pi}{2} = \pi$$

$$x(t) = a_0 \sum_{n=1}^{\alpha} a_n \cos n\pi t + b_n \sin n\pi t$$

Where $x(t) = 2AT$ $|t| < \dfrac{1}{2}$

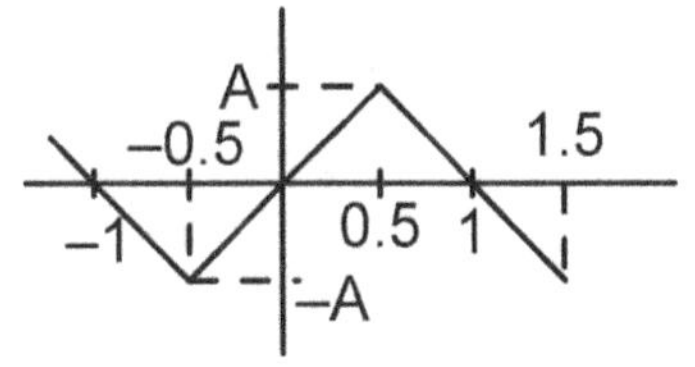

Fig. 3.19

$$2A(1-t)\frac{1}{2} < t < \frac{3}{2}$$

DC value or average value is 0 for the signal

$$\therefore \quad a_0 = 0$$

and $\quad a_n = \dfrac{2}{2} \displaystyle\int_{-1/2}^{3/2} x(t)\cos n\pi t\, dt$

$$= \int_{-1/2}^{1/2} 2At \cos n\pi t\, dt + \int_{1/2}^{3/2} 2At\,(1-t)\cos n\pi t\, dt$$

$$= 0$$

$$b_n = \int_{-1/2}^{1/2} 2At \sin n\pi t\, dt + \int_{1/2}^{3/2} 2A(1-t)\sin n\pi t\, dt$$

$$b_n = \frac{8A}{n^2\pi^2}\sin\left(\frac{n\pi}{2}\right)$$

$$= 0 \text{ for n even}$$

$$= \frac{8A}{n^2\pi^2} \qquad n = 1, 5, 9, 13\ldots..$$

$$= \frac{-8A}{n^2\pi^2} \qquad n = 3, 7, 11, 15\ldots..$$

$$x(t) = \frac{8A}{\pi^2}\left\{ \sin\pi t - \frac{1}{9}\sin 3\pi t + \frac{1}{25}\sin 5\pi t - \frac{1}{49}\sin 7\pi t + \ldots.. \right\}$$

10.   Find the compact fourier series for signal shown

$$T_0 = 2\pi$$

$$W_0 = \frac{2\pi}{T_0} = 1$$

$$x(t) = a_0 + \sum_{n=1}^{\alpha} a_n \cos nt + b_n \sin nt$$

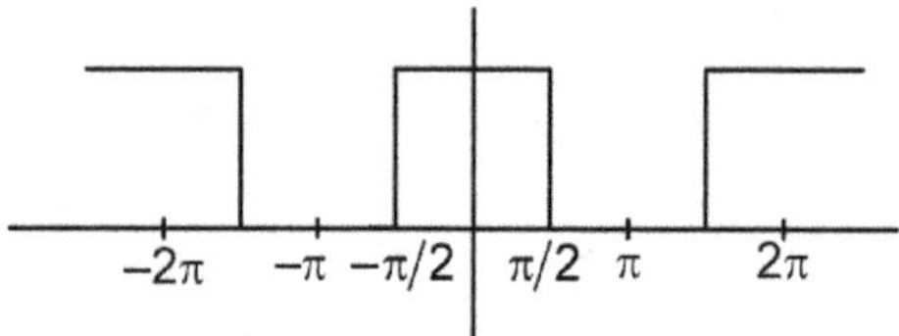

Fig. 3.20

Where

$$a_0 = \frac{1}{T_0}\int_{T_0} x(t)\,dt$$

$$a_n = \frac{1}{\pi}\int_{-\pi/2}^{\pi/2}\cos nt\,dt = \frac{2}{n\pi}\sin\left(\frac{n\pi}{2}\right)$$

$$= \frac{1}{2\pi}\int_{-\pi/2}^{\pi/2} dt = \frac{1}{2}$$

$$= 0 \qquad n\ \text{even}$$

$$b_n = \frac{1}{\pi}\int_{-\pi/2}^{\pi/2}\sin nt\,dt = 0$$

$$= \frac{2}{n\pi}\qquad 1, 5, 9, 13$$

$$= \frac{-2}{n\pi}\qquad 3, 7, 11, 15$$

$$\therefore\quad x(t) = \frac{1}{2} + \frac{2}{\Pi}\left(\cos t - \frac{1}{3}\cos 3t + \frac{1}{3}\cos 5t - \frac{1}{7}\cos 7t + ...\right)$$

$\therefore$ in compact form

$$C_0 = \frac{1}{2}, \qquad C_n = 0\ n\ \text{even}$$

$$\frac{2}{n\pi}\ n\ \text{odd}$$

$$\theta_n = 0\ \text{ for all } n \neq 3, 7, 11$$

$$-\pi\quad n = 3, 7, 11, 15$$

11.   Find the fundamental frequency of signal shown

$$x(t) = 2 + 7\cos\left(\frac{1}{2}t + \theta_1\right) + 3\cos\left(\frac{2}{3}t + \theta_2\right) + 5\cos\left(\frac{7}{6}t + \theta_3\right)$$

Ans: The largest number of which all the frequencies are integer multiples is the fundamental frequency. The ratio of any two frequencies is a rational number then the frequencies are harmonically related.

The greatest number of $\dfrac{1}{2}, \dfrac{2}{3}, \dfrac{7}{6}$ is $\dfrac{1}{6}$

i.e., $3\left(\dfrac{1}{6}\right) = \dfrac{1}{2}$, $4\left(\dfrac{1}{6}\right) = \dfrac{2}{3}$ and $7\left(\dfrac{1}{6}\right) = \dfrac{7}{6}$

The ratios are 3:4:7 which are rational.

Hence fundamental frequency is 1/6

12. Determine $x(t) = \cos\left(\dfrac{2}{3}t + 30^0\right) + \sin\left(\dfrac{4}{5}t + 45^0\right)$ is periodic and what harmonics are present in it?

Ans: $W_0 = 2/15$ and $T_0 = 15\pi$

$\therefore$ In first term $\dfrac{\cancel{2}}{\cancel{3}} \times \dfrac{\cancel{15}}{\cancel{2}}$, 5th harmonic

and in second term $\dfrac{\cancel{4}}{\cancel{5}} \times \dfrac{\cancel{15}}{\cancel{2}}$, 6th harmonic are present.

OBJECTIVE QUESTIONS

1. Trigonometric fourier series of a periodic function has _____ terms
   (a) Sine   (b) Cosine   (c) Sine and cosine   (d) dc and cosine

2. In fourier series expansion a_n will be zero for _____ function and b_n will be zero for _____ function
   (a) Odd, even   (b) Odd, odd   (c) Even, odd   (d) Even, even

3. For the periodic signal $x(t) = 30 \sin 100t + 10 \cos 300t + 6 \sin (500t + \pi/4)$ the fundamental frequency is (radians)   (Gate 2013)
   (a) 100   (b) 300   (c) 500   (d) 1500

4. Consider the periodic square wave in figure find the ration of the power of 7th harmonic and 5th harmonic
   (Gate 2016)

   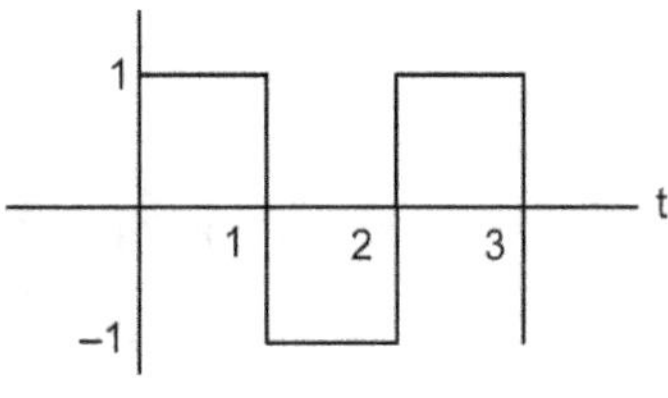

   $\therefore$ Power of 7th harmonic $= 1/7^2$

   Power of 5th harmonic $= 1/5^2$

   Ratio $= \dfrac{25}{49} = 0.5$

5. Fourier series of an odd periodic function has   (Gate 1994)
   (a) Odd harmonics   (b) Even harmonics
   (c) Cosine terms   (d) Sine terms

6. Find the average power of g(t) is

   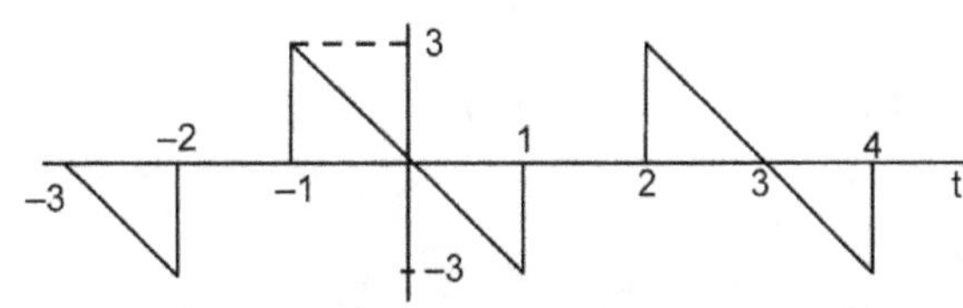

$$g(t) = x\left(\frac{t-1}{2}\right)$$

$$\therefore \quad x = -3t$$

Period of waveform is 3

$$\text{Power} = \frac{1}{T}\int_0^{} [x(t)]^2 \, dt$$

$$= \frac{1}{3}\left[\int_{-1}^{0}(-3t)^2 \, dt + \int_0^1 (-3t)^2 \, dt + \int_1^2 0^2 \, dt\right]$$

$$= 2$$

6. Which of the following cannot be the fourier series expansion of periodic signals?

GATE 2015

(a) $x(t) = 2 \cos t + 3 \cos 3t$ 　　　　(b) $x(t) = 2 \cos \pi t + 7\cos t$

(c) $x(t) = \cos t + 0.5$ 　　　　(d) $x(t) = 2\cos 1.5\pi t + \sin 3.5\pi t$

8. Choose the function for which a fouier series cannot be defined 　　　　GATE 2002

(a) $3\sin(25t)$ 　　　　(b) $4\cos(20t+3)+2\sin(710t)$

(c) $\exp(-1\,t1)\sin(25t)$ 　　　　(d) 1

9. Fourier series expansion of a real periodic signal with fundamental frequency f_0 is

$$g(t) = \sum_{n=-\alpha}^{\alpha} c_n e^{j2\pi n f_0 t} \quad \text{it is given that } C_3 = 3 + j5 \text{ then } C - 3 \text{ is}$$

(a) $5 + 3j$ 　　　(b) $-3 - j5$ 　　　(c) $-5 + 3j$ 　　　(d) $3 - j5$ 　　GATE 2003

10. If a signal $f(t)$ has energy E, the energy of the signal $f(2t)$ is equal to 　　　　GATE 2001

(a) E 　　　(b) 2E 　　　(c) E/2 　　　(d) 4E

11. A signal $2\cos\left(\frac{2\pi}{3}t\right) - \cos \pi t$ is the input of an LTI system with transfer function

$H(s) = e^s + e^{-s}$. If C_k is k^{th} coefficient in exponential fourier series, then C_3 is

GATE 2016

(a) 0 　　　(b) 1 　　　(c) 2 　　　(d) 3

SUMMARY OF CHAPTER

1. Periodic signal $x(t)$ of period T can be expanded in exponential Fourier series as
$$x(t) = \sum_{n=-\alpha}^{\alpha} X_n e^{jn\,w_o t}$$ where X_n are called Fourier series coefficients and are given by
$$X_n = \frac{1}{T}\int_0^T x(t)e^{-jnw_o t}dt$$

2. For periodic signals both magnitude and phase spectra are line spectra

3. For real valued signals magnitude spectrum has even symmetry and phase has odd symmetry

4. If signal $x(t)$ is real valued signal, then it can be expanded in trigonometric series of the form $x(t) = a_0 + \displaystyle\sum_{n=1}^{\alpha}\{a_n \cos(nw_o t) + b_n \sin(nw_o t)\}$

5. An alternative form of Fourier series (polar form) $x(-t) = c_0 + \displaystyle\sum_{n=1}^{\alpha} c_n \cos(nw_o t + \theta_n)$

   where $c_0 = a_0$, $c_n = \sqrt{a_n^2 + b_n^2}$, $\theta_n = \tan^{-1}\dfrac{b_n}{a_n}$ or $c_n = \begin{cases} 2|X_n| & n \geq 1 \\ X_o & n = 0 \end{cases}$ $\quad \angle X_n = \theta_n$

6. For the Fourier series to converge, the signal $x(t)$ must be absolute integrable, have only a finite number of maxima and minima and have a finite number of discontinuities over any period

7. If $x(t)$ satisfies Dirichlet conditions, then its Fourier series is guarantied to converge point wise at all points

8. A plot showing each of harmonic amplitude in the wave is called the line spectrum

9. When a continuous function $x(t)$ is synthesized by first N turn of Fourier series the synthesized function approaches $x(t)$ for all t as $N \to \alpha$ then no gibbs phenomenon appears

10. Parsevols relation states that the total average power in a periodic signal equals to the sum of the average powers in all of its harmonic components

CHAPTER 4

Fourier Transforms

Objectives of the Chapter

This chapter explains how to derive fourier transform from fourier series, fourier transform of arbitrary signal, fourier transform of standard signals, fourier transform of periodic signals, properties of fourier transforms. Fourier transforms involving impulse function and signum function, Introduction to hilbert transform.

4.1 INTRODUCTION

Fourier transform can be considered as a special case of Laplace transform with s = jw, but only difference lies in the nature of convergence of Laplace and fourier integrals.

Fourier transform is a systematic way to decompose generic functions into super position of symmetric functions. Fourier series are useful in approximating and analyzing periodic signals but not for aperiodic. Whereas fourier transform can be used for aperiodic functions.

Fourier transform is an operation that transform one complex valued function of a real variable into another. It also may be known as frequency domain representation of the original function. It describes which frequencies are present in the signal function.

Fourier transform is essentially inner product of the signal f(t) with the complex exponential $e^{j\alpha t}$, evaluated at different frequencies.

Continuous time signals that have finite duration in time will have non zero frequency content. Conversely continuous time signals that are band limited in frequency domain will have infinite duration.

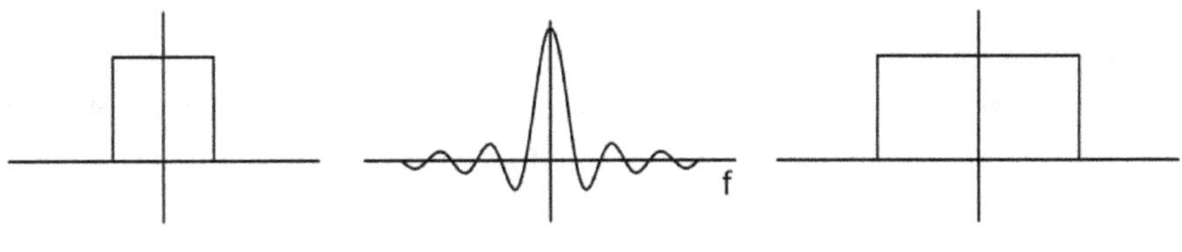

(a)

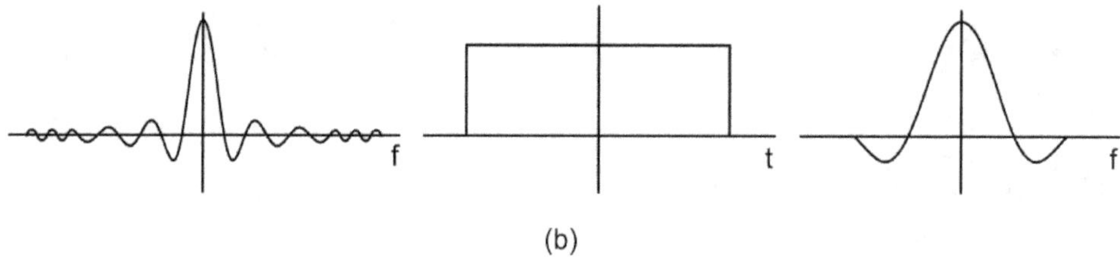

(b)

Fig. 4.1

As T is large, ω becomes small, that is as T tends to infinity the spectrum consists of infinite number of zero amplitude components, TF_n components are occurring at $n\omega_0$ frequencies. As T increases ω_0 decreases and TF_n becomes closer.

As $T \to \alpha$, TF_n becomes continuous

we have

$$F_n = \frac{1}{T}\int_{-T/2}^{T/2} f(t)e^{-jn\omega_0 t}\,dt$$

$$f(t) = \sum_n F_n\, e^{jn\omega_0 t}$$

$$T(n\omega_0) = TF_n = \int_{T/2}^{T/2} f(t)e^{-jn\omega_0 t}\,dt$$

$$f(t) = \frac{1}{T}\sum F(n\omega_0)e^{jn\omega_0 t}$$

$$= \frac{1}{2\pi}\sum F(n\omega_0)e^{jn\omega_0 t}\,\omega_0$$

as $T \to \alpha$, $\omega_0 \to 0$ and spectrum becomes continuous
and let $n\omega_0 = \omega$

$$\mathcal{F}^{-1}\{F(\omega)\} = f(t) = \frac{1}{2\pi}\int_{-\alpha}^{\alpha} F(\omega)e^{j\omega t}\,dt$$

$$\mathcal{F}\{f(t)\} = F(\omega) = \int_{-\alpha}^{\alpha} f(t)e^{-j\omega t}\,dt$$

Example

Let us consider a aperiodic signal $f(t) = A$ shown in figure. It is made equal to rectangular periodic signal whose $T \to \alpha$. As we can approximate this rectangular function by sin ωt, the aperiodic signal also can be estimated by using same function sin ωt whose $\omega \to 0$.

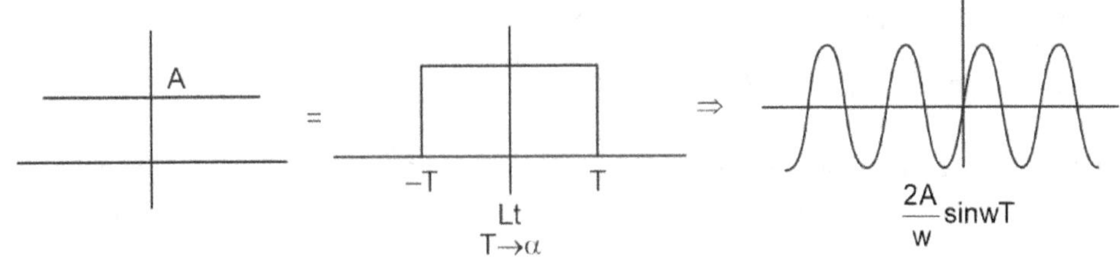

Fig. 4.2

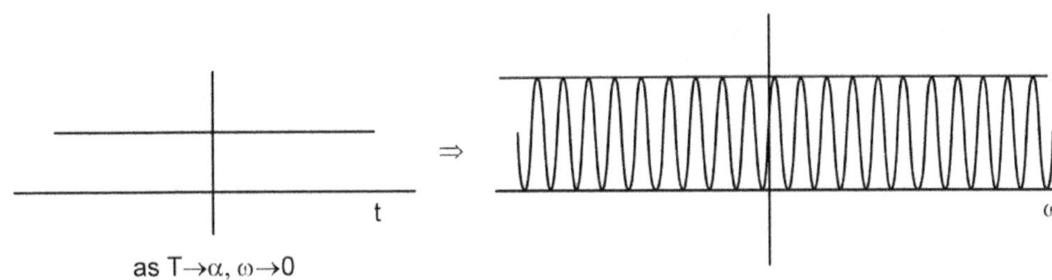

Fig. 4.3

4.2 FOURIER TRANSFORM OF NON PERIODIC FUNCTION

Fourier representation is a way of expressing a signal in terms of exponentials. Fourier spectrum of a signal consists different sinusoids having finite amplitudes and phases required to synthesize the signal. Frequency spectrum of periodic signal is easily visible, but the spectrum of aperiodic signal is not easy to visualize as it has continuous spectrum continuous means, the spectrum exists everywhere for 'ω' but amplitude of each component is zero. Hence we must measure spectral density per unit bandwidth.

Approach - 1

Let any function f(t) be represented in a finite interval $-\dfrac{T}{2} < t < \dfrac{T}{2}$, and then let T go to infinite.

Approach - 2

If a periodic functions f(t) is constructed and if its T is made α, it has only one cycle $-\alpha < t < \alpha$
.

These two are same. But later one is easier.

If T is made large, frequency becomes small and frequency spectrum becomes denser, amplitude becomes small. However shape is un altered.

Let us consider two functions f(t) and f_T (t), where $f_T(t)$ is periodic.

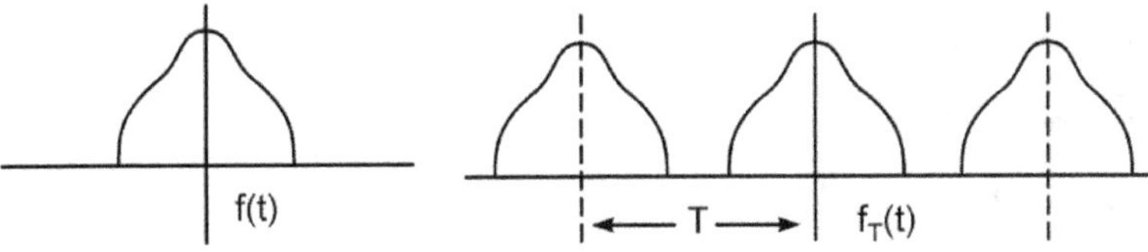

Fig. 4.4

Where f(t) repeats for every T secs it becomes $f_T(t)$.

If T is made large $\underset{T \to 0}{Lt}\ f_T(t) = f(t)$

Thus fourier series representing $f_T(t)$ over the entire interval will also represent f(t) over entire interval

if $T = \alpha$

$\therefore\ f_T(T)$ represented in terms of fourier series

$$f_T(t) = \sum_{n=-\alpha}^{\alpha} F_n e^{jn\omega_0 t} \qquad \omega_0 = \frac{2\pi}{T}$$

$$F_n = \frac{1}{T} \int_{-T/2}^{T/2} f_T(t) e^{-jn\omega_0 t}\, dt$$

F_n is the amplitude of frequency component $n\omega_0$.

As T is large, ω_0 is small and spectrum will be dense.

The spectrum exists for every value of ω, and no longer discrete, but it is continuous.

Let $n\omega_0 = \omega_n$ and let $T\ F_n(\omega_n) = F(\omega_n)$ where F_n or $F_n(\omega_n)$ both are same.

$$\therefore\ f_T(t) = \frac{1}{T} \sum_{n=-\alpha}^{\alpha} F(\omega_n) e^{-jn\omega_0 t}\, dt$$

$$F(\omega_n) = T\,F_n = \int_{-T/2}^{T/2} f_T(t) e^{-jn\omega_0 t}\, dt$$

$$T = 2\pi / \omega_0$$

$$f_T(t) = \frac{\omega_0}{2\pi} \sum_{n=-\alpha}^{\alpha} F(\omega_n) e^{-j\omega_n t}$$

As T is large, ω_0 becomes small.

Let $n\omega_0 = \omega_n$

$\therefore\ F_n$ is a function of ω_n

And denote $F_n(\omega_n) = F(\omega_n)$

Let $TF_n(\omega_n) = F(\omega_n)$

Then

$$f_T(t) = \frac{1}{T}\sum_{n=-\alpha}^{\alpha} F(\omega_n)e^{j\omega_n t}$$

$$F(\omega_n) = \int_{-T/2}^{T/2} f_T(t)e^{-j\omega_n t}$$

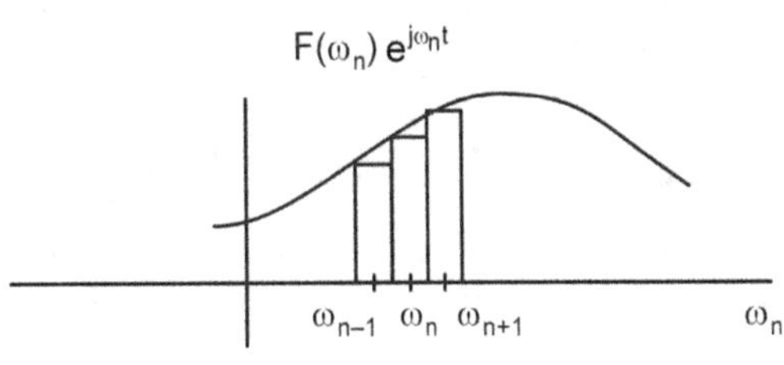

Fig. 4.5

$$T = \frac{2\pi}{\omega_0}$$

$$f_T(t) = \frac{1}{2\pi}\sum F(\omega_n)e^{j\omega_n t}\omega_0$$

The curve is a continuous function of ω and is given by $F(\omega)e^{j\omega t}$

$$f(t) = \frac{1}{2\pi}\int_{-\alpha}^{\alpha} F(\omega)e^{j\omega t}\,d\omega$$

$$F(\omega) = \int_{-\alpha}^{\alpha} f(t)e^{-j\omega t}\,dt$$

$$F(\omega) = \mathscr{F}\left[f(t)\right]$$

$$f(t) = \mathscr{F}^{-1}\left[F(\omega)\right]$$

$$\mathscr{F}\{f(t)\} = \int_{-\alpha}^{\alpha} f(t)e^{-j\omega t}\,dt$$

$$\mathscr{F}^{-1}\{f(t)\} = \frac{1}{2\pi}\int_{-\alpha}^{\alpha} F(\omega)e^{j\omega t}\,d\omega$$

4.3 EXISTENCE OF FOURIER TRANSFORM

If $\displaystyle\int_{-\alpha}^{\alpha} f(t)e^{-j\omega t}\,dt$ is finite then fourier transform exists.

Q magnitude of $e^{-j\omega t}$ is unity

$$\int_{-\alpha}^{\alpha} f(t)\,dt \text{ must be finite}$$

In the case of singularity functions this condition may not be true.

Thus for such functions fourier transform exists if T is finite. If T is made large and finite we can find fourier transform.

Absolute integrability within $(-\alpha, \alpha)$ is sufficient but not compulsory for a function f(t) to have fourier transform.

SINGULARITY FUNCTIONS

Singularity is a point at which a function does not possess a derivative. Each singularity function has a singular point at the origin and is zero elsewhere. These are also defined as generalized functions.

Step function u(t), impulse function $\delta(t)$ and their derivatives are known as singularity functions. These are defined by its effect on other functions instead of by its value at every instant of time.

Unit impulse function

$$\delta(t) = 0 \qquad t \neq 0$$

$$\int_{-\alpha}^{\alpha} \delta(t)\,dt = 1$$

or

$$\delta(t) = \underset{T \to 0}{Lt} \frac{1}{T}\{u(t) - u(t - T)\}$$

Fig. 4.6

As T goes to zero, the singularity function can be defined. Ideally there are no such true functions. But they are very useful in engineering. So they can be defined by un idealized step functions shown as above

4.4 FOURIER TRANSFORMS OF SOME SIGNALS

Most of the functions start at finite time at t = 0, to make this we can combine. Heaviside ( ) unit step ( ) function.

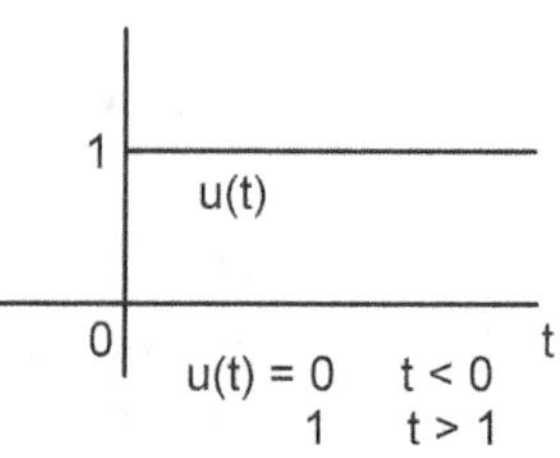

Fig. 4.7

1. Exponential function

$$F(\omega) = \mathcal{F}\{e^{-at}u(t)\}$$

$$= \int_{-\alpha}^{\alpha} e^{-at} e^{-j\omega t} \, dt$$

$$= \int_{0}^{\alpha} e^{-(a+j\omega)t} \, dt$$

$$= \frac{-1}{a+j\omega} e^{-(a+j\omega)t} \Big|_{0}^{\alpha}$$

$$= \frac{1}{a+j\omega}$$

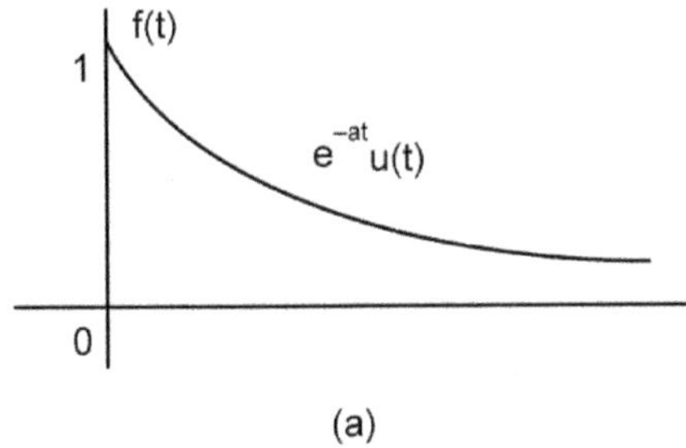
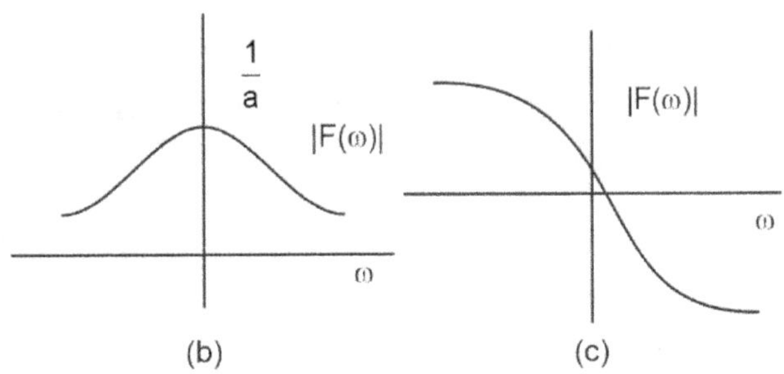

Fig. 4.8

Spectral density function of $e^{-at}u(t)$ is

$$\frac{1}{a+j\omega} = \frac{1}{\sqrt{a^2+\omega^2}} e^{-j\tan^{-1}\frac{\omega}{a}}$$

$$\text{Magnitude} = \frac{1}{\sqrt{a^2+\omega^2}}$$

$$\text{Phase} = \tan^{-1}\frac{\omega}{a}$$

2. Find the fourier transform for below functions

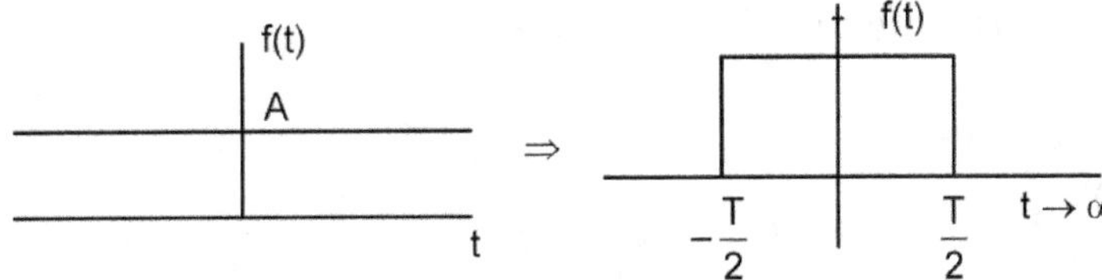

Fig. 4.9

$f(t) = A$

$f(t) = A = \underset{T \to \alpha}{Lt}\ \pi(t)$

$$\therefore\ \mathscr{F}\big(f(t)\big) = \int_{-\alpha}^{\alpha} A\, e^{-j\omega t}\, dt = \underset{T \to \alpha}{Lt} \int_{-T/2}^{T/2} A \cdot e^{-j\omega t}\, dt$$

$$= \underset{T \to \alpha}{Lt}\ A \cdot \frac{e^{-j\omega t}}{-j\omega} \Big|_{-T/2}^{T/2} = \underset{T \to \alpha}{Lt}\ A \left\{ \frac{e^{-j\omega t/2} - e^{j\omega T/2}}{-j\omega} \right\}$$

$$= \underset{T \to \alpha}{Lt}\ \frac{2A}{\omega} \left\{ \frac{e^{j\omega T/2} - e^{-j\omega T/2}}{2j} \right\}$$

$$= \underset{T \to \alpha}{Lt}\ \frac{2A}{\omega}\ \sin\left(\omega T/2\right)$$

$$= \underset{T \to \alpha}{Lt}\ \frac{AT}{\omega T/2}\ \sin\left(\omega T/2\right) = \underset{T \to \alpha}{Lt}\ AT \sin c\left(\frac{\omega T}{2}\right)$$

$$= A\ \underset{T \to \alpha}{Lt}\ T\delta a\left(\frac{\omega T}{2}\right)$$

$A2\pi$	AT
(a)	(b)

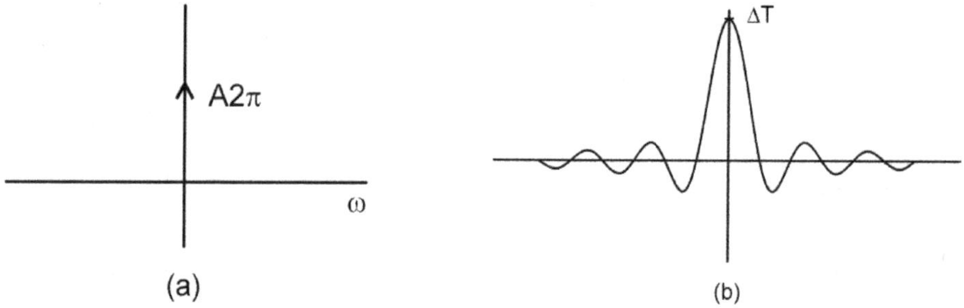

Fig. 4.10

3. Find the fourtier transform of $\delta(t)$

$$\mathscr{F}\ \{f(t)\} = \int_{-\alpha}^{\alpha} f(t)e^{-j\omega t}\, dt$$

$$= \int_{-\alpha}^{\alpha} \delta(t)e^{-j\omega t}\, dt = 1.e^{-j\omega 0} = 1$$

but $\ \delta(t) = 1\ $ at $\ \ t = 0$

$$\therefore\ F\,(\omega) = 1.$$

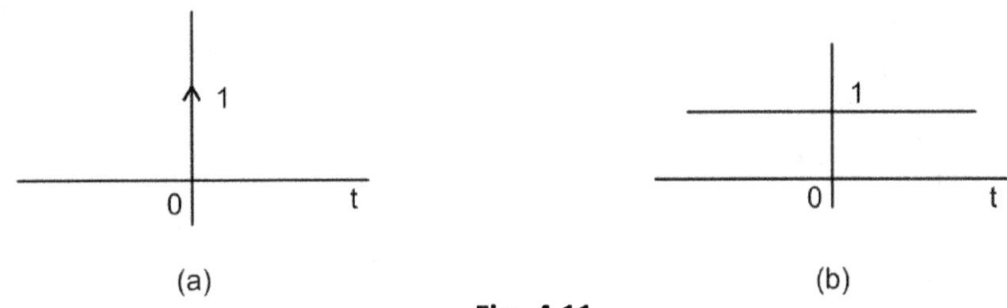

(a)   (b)

Fig. 4.11

4. Find the fourier transform of

$$f(t) = \cos\omega_0 t$$

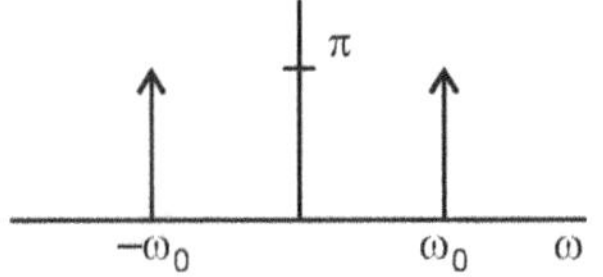

Fig. 4.12

$$\mathscr{F}\{f(t)\} = \int_{-\alpha}^{\alpha} \cos\omega_0 t \, e^{-j\omega t} \, dt$$

$$= \operatorname*{Lt}_{T\to\alpha} \int_{-T/2}^{T/2} \frac{e^{j\omega_0 t} + e^{-j\omega_0 t}}{2} e^{-j\omega t} \, dt$$

$$= \frac{1}{2} \operatorname*{Lt}_{T\to\alpha} \int_{-T/2}^{T/2} e^{-j(\omega-\omega_0)t} + e^{-j(\omega+\omega_0)t} \, dt$$

$$= \frac{1}{2} \operatorname*{Lt}_{T\to\alpha} \left\{ \frac{e^{-j(\omega-\omega_0)t}}{-j(\omega-\omega_0)} + \frac{e^{-j(\omega+\omega_0)t}}{-j(\omega+\omega_0)} \right\}\Bigg|_{-T/2}^{T/2}$$

$$= \operatorname*{Lt}_{T\to\alpha} \frac{\sin\left((\omega-\omega_0)\,T/2\right)}{(\omega-\omega_0)} + \frac{\sin\left((\omega+\omega_0)\,T/2\right)}{(\omega+\omega_0)}$$

$$= \operatorname*{Lt}_{T\to\alpha} \frac{T}{2}\left\{ \delta a(\omega-\omega_0)\tfrac{T}{2} + \delta a(\omega+\omega_0)\tfrac{T}{2} \right\}$$

$$= \pi\left\{ \delta(\omega-\omega_0) + \delta(\omega+\omega_0) \right\}$$

5. Find the fourier transform of signum function

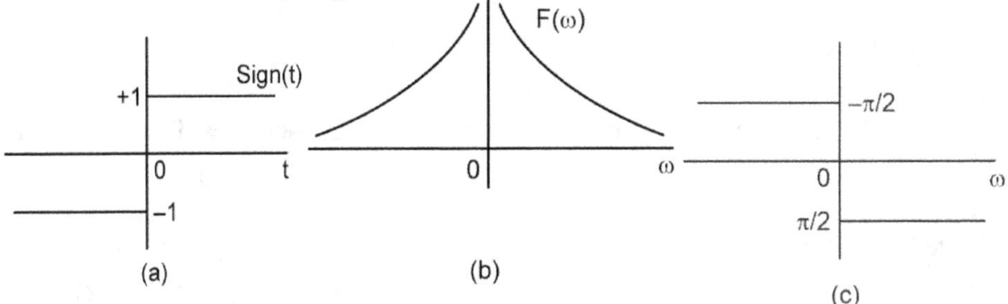

(a)   (b)   (c)

Fig. 4.13

$$\text{Sgn (t)} = +1 \quad t > 0$$
$$= -1 \quad t < 0$$

$$\mathscr{F}\{f(t)\} = \lim_{a \to 0} \int_{-\alpha}^{\alpha} e^{-a|t|} \text{Sgn}(t) e^{-j\omega t} \, dt$$

$$= \lim_{a \to 0} \int_{-\alpha}^{0} -e^{(a-j\omega)t} + \lim_{a \to 0} \int_{0}^{\alpha} e^{-a|t|} e^{-j\omega t}$$

$$= \lim_{a \to 0} -\frac{e^{(a-j\omega)t}}{(a-j\omega)} \Big|_{-\alpha}^{0} + \frac{e^{-(a+j\omega)t}}{-(a+j\omega)} \Big|_{0}^{\alpha}$$

$$= -\frac{1}{a-j\omega} + \frac{1}{a+j\omega} = \frac{2}{j\omega}$$

$$|F(j\omega)| = \frac{2}{\omega} \quad \underline{F(j\omega)} = \tan^{-1}\omega$$

6. Find $\mathscr{F}$T of unit step function

$$u(t) = 1 \quad t \geq 0$$
$$= 0 \quad t < 0$$

$$F(\omega) = \int_{-\alpha}^{\alpha} f(t) \, e^{-j\omega t} \, dt$$

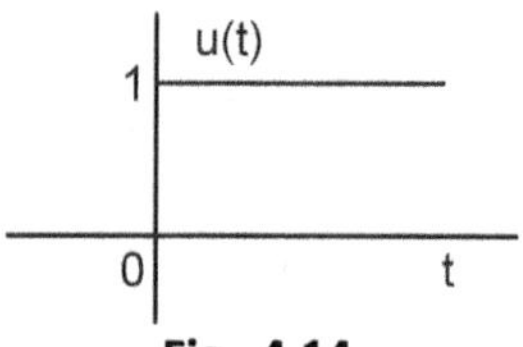

Fig. 4.14

$$= \int_{0}^{\alpha} 1 \cdot e^{-j\omega t} \, dt$$

$$= \frac{e^{-j\omega t}}{-j^{\omega}} \Big|_{0}^{\alpha} = \frac{1}{j^{\omega}}$$

7. Find Fourier transform of sampling function

$$f(t) = \frac{k}{\pi} \text{Sa}(kt)$$

$$\mathscr{F}\{f(t)\} = \int_{-\alpha}^{\alpha} \frac{k}{\pi} \text{Sa}(kt) \, dt = 1$$

As k is made large, amplitude becomes large. Function oscillates fast and decays very rapidly.

$$\lim_{k \to \alpha} \frac{k}{\pi} \text{Sa}(kt) = \delta(t)$$

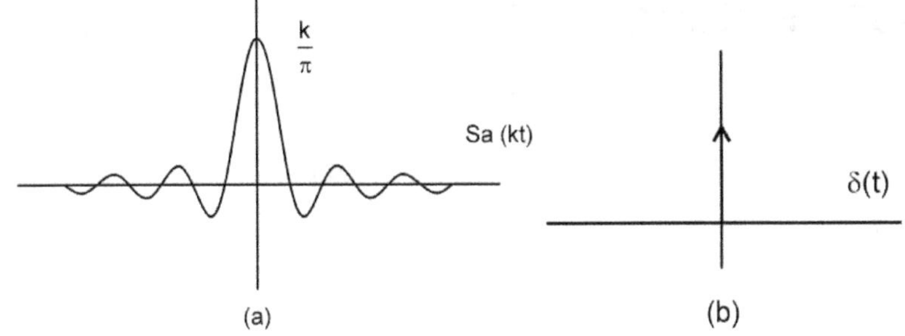

(a)   (b)

Fig. 4.15

8. Find the fourier transform of f(t) given below

$$f(t) = \Delta\left(\frac{t}{T}\right) = \begin{cases} 1 - \dfrac{2|t|}{T} & |t| < T \\ 0 & \text{otherwise} \end{cases}$$

$$F(\omega) = \int_{-\alpha}^{\alpha} \Delta\left(\frac{t}{T}\right) e^{-j\omega t}\, dt$$

$$= \int_{-T/2}^{0}\left(1 + \frac{2t}{T}\right) e^{-j\omega t}\, dt + \int_{0}^{T/2}\left(1 - \frac{2t}{T}\right) e^{-j\omega t}\, dt$$

$$= \frac{2}{T}\left[\left\{\frac{t e^{-j\omega t}}{-j\omega} - \frac{1 e^{-j\omega t}}{(j\omega)^2}\right\}\Bigg|_{-T/2}^{0} + \frac{e^{-j\omega t}}{-j\omega}\Bigg|_{-T/2}^{0}\right] + -\frac{2}{T}\left[\left\{\frac{t e^{-j\omega t}}{-j\omega} - \frac{1 e^{-j\omega t}}{(j\omega)^2}\right\}\Bigg|_{-T/2}^{0} + \frac{e^{-j\omega t}}{-j\omega}\Bigg|_{-T/2}^{0}\right]$$

$$= \frac{2}{T}\left\{\frac{e^{j\omega T/2}}{(j\omega)^2} + \frac{e^{-j\omega T/2}}{(j\omega)^2} - \frac{2}{(j\omega)^2}\right\}$$

$$= \frac{2}{T}\left\{\frac{e^{j\omega T/4} - e^{-j\omega T/4}}{j\omega}\right\}^2 = \frac{8}{\omega^2 T}\sin^2\frac{\omega T}{4}$$

$$= \frac{T}{2}\left(\frac{\sin \omega T/4}{\omega T/4}\right)^2 = \frac{T}{2}\sin c^2\left(\frac{\omega T}{4}\right)$$

(a)   (b)

Fig. 4.16

4.5 (a) FOURIER TRANSFORM

Table 4.1

f (t)	f (ω)

1.

$f(t) = e^{-at}\,u(t)$

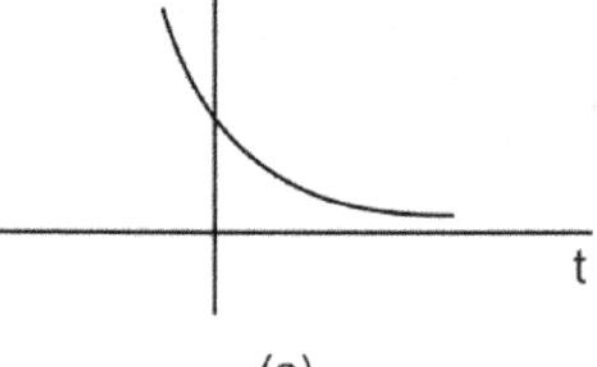

(a)

$F(w) = \dfrac{1}{a+j\omega}$

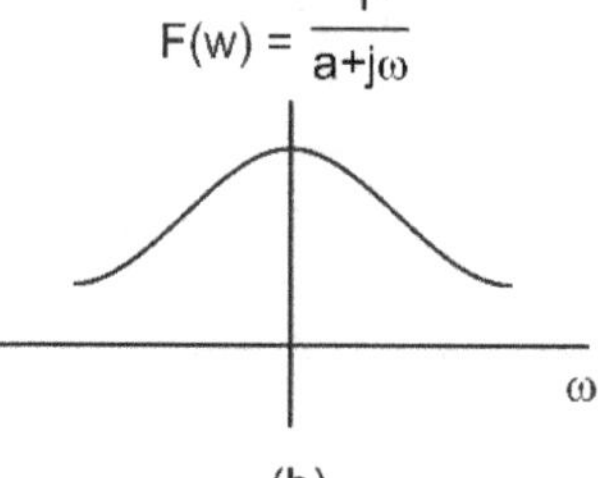

(b)

2.

$t\cdot e^{-at}\,u(t)$

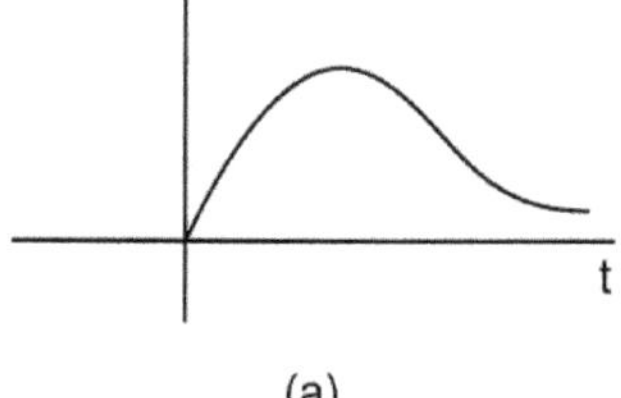

(a)

$\dfrac{1}{(a+j\omega)^2}$

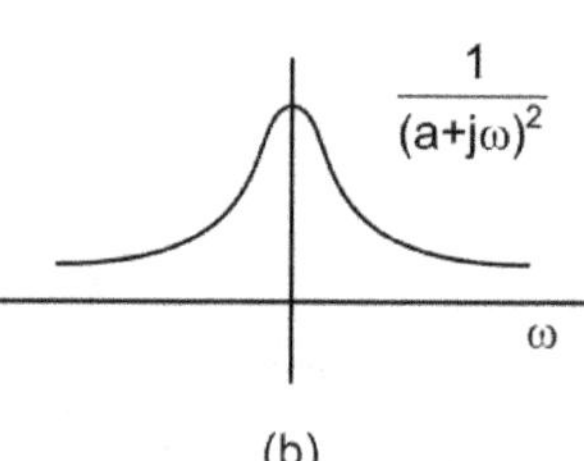

(b)

3.

$e^{-a|t|}$

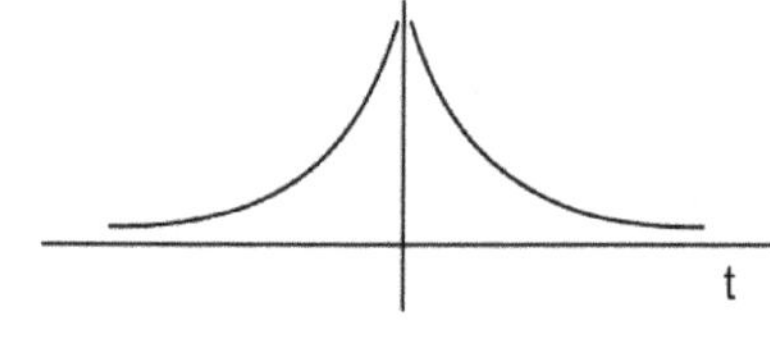

(a)

$\dfrac{2a}{a^2+\omega^2}$

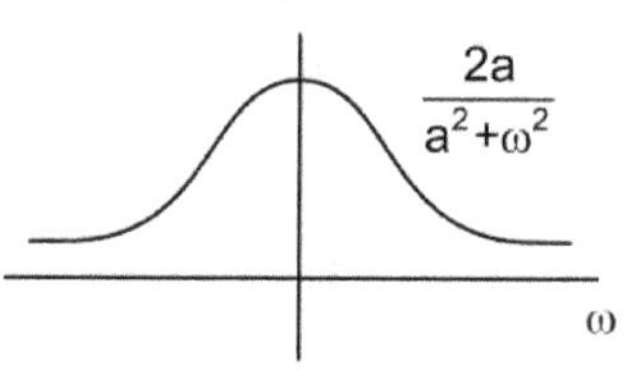

(b)

4.

$G_{2T}(t)$

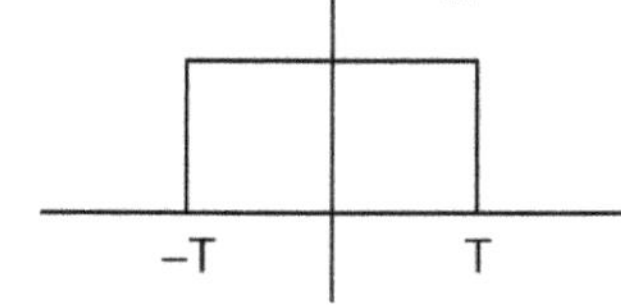

(a)

$T\,Sa\left(\dfrac{\omega T}{2}\right)$

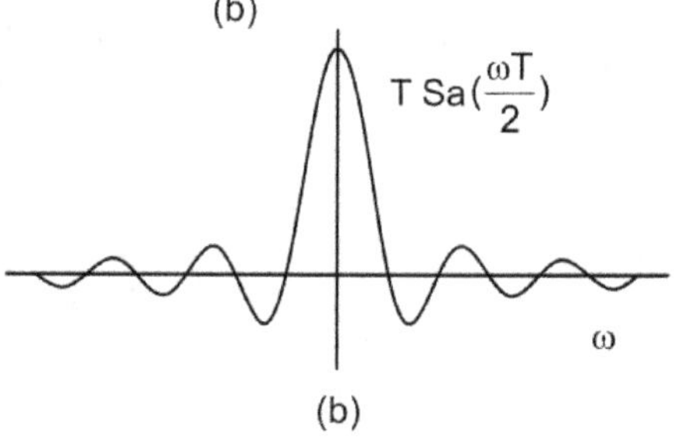

(b)

5.

$Sa\,(wt)$

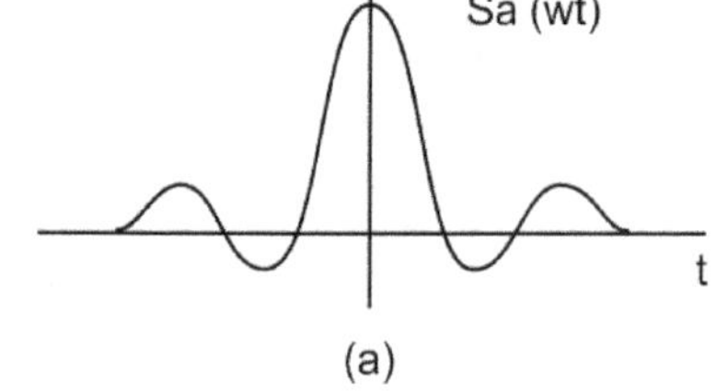

(a)

$\dfrac{\pi}{\omega}G(\omega)$

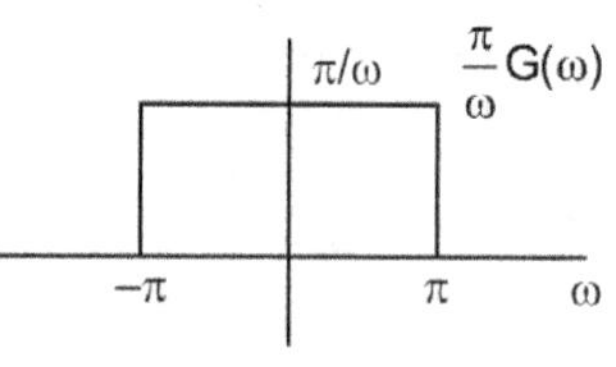

(b)

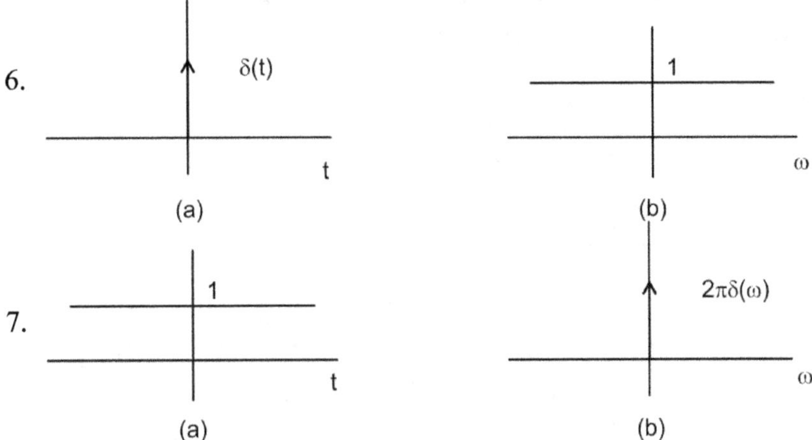

Fig. 4.17

4.5 (b) FOURIER TRANSFORMS OF SOME USEFUL FUNCTIONS

	x(t)	X(w)
1.	$e^{-at}u(t)$	$\dfrac{1}{a+j\omega}$ $a>0$
2.	$e^{at}u(-t)$	$\dfrac{1}{a-j\omega}$ $a>0$
3.	$e^{-a|t|}$	$\dfrac{2a}{a^2+\omega^2}$ $a>0$
4.	$te^{-at}u(t)$	$\dfrac{1}{(a+j\omega)^2}$ $a>0$
5.	$t^n e^{-at}u(t)$	$\dfrac{n!}{(a+j\omega)^{n+1}}$ $a>0$
6.	$\delta(t)$	1
7.	1	$2\pi\delta(\omega)$
8.	$e^{j\omega_o t}$	$2\pi\delta(\omega-\omega_o)$
9.	$\cos\omega_o t$	$\pi[\delta(\omega-\omega_o)+\delta(\omega+\omega_o)]$
10.	$\sin\omega_o t$	$j\pi[\delta(\omega+\omega_o)-\delta(\omega+\omega_o)]$

	$x(t)$	$X(\omega)$
11.	$u(t)$	$\pi\delta(\omega) + \dfrac{1}{j\omega}$
12.	$\text{sgn}(t)$	$\dfrac{2}{j\omega}$
13.	$\cos\omega_o t\, u(t)$	$\dfrac{\pi}{2}[\delta(\omega-\omega_o)+\delta(\omega+\omega_o)] + \dfrac{j\omega}{\omega_o^2-\omega^2}$
14.	$\sin\omega_o t\, u(t)$	$\dfrac{\pi}{2j}[\delta(\omega-\omega_o)-\delta(\omega+\omega_o)] + \dfrac{\omega_o}{\omega_o^2-\omega^2}$
15.	$e^{-at}\sin\omega_0 t\, u(t)$	$\dfrac{\omega_o}{(a+j\omega)^2+\omega_o^2}$
16.	$e^{-at}\cos\omega_o t\, u(t)$	$\dfrac{a+j\omega}{(a+j\omega)^2+\omega_o^2}$
17.	$\text{Rect}\left(\dfrac{t}{\tau}\right)$	$T\sin c\left(\dfrac{\omega\tau}{2}\right)$
18.	$\dfrac{\omega}{\pi}\sin(\omega t)$	$\text{Rect}\left(\dfrac{\omega}{2\omega}\right)$
19.	$\Delta\left(\dfrac{t}{\tau}\right)$	$\dfrac{T}{2}\sin c^2\left(\dfrac{\omega\tau}{4}\right)$
20.	$\dfrac{\omega}{2\pi}\sin c^2\left(\dfrac{\omega t}{2}\right)$	$\Delta\left(\dfrac{\omega}{2\omega}\right)$
21.	$\displaystyle\sum_{n=-\alpha}^{\alpha}\delta(t-nT)$	$\omega_o\displaystyle\sum_{n=-\alpha}^{\alpha}\delta(\omega-n\omega_o)$
22.	$e^{-t^2/2\sigma^2}$	$\sigma\sqrt{2\pi}\,e^{-\sigma^2\omega^2/2}$

4.6 PROPERTIES OF FOURIER TRANSFORMS

1. Symmetry Property:

If $f(t) \leftrightarrow F(\omega)$

$F(t) \leftrightarrow 2\pi f(-\omega)$

$$2\pi f(-t) = \int_{-\alpha}^{\alpha} F(\omega)e^{-j\omega t}\,d\omega$$

Let $\omega = x$

$$2\pi f(-t) = \int_{-\alpha}^{\alpha} F(x)e^{-jxt}\,dx$$

$$2\pi f(-\omega) = \int_{-\alpha}^{\alpha} F(x)e^{-jx\omega}\,dx$$

Let $x = t$

$$2\pi f(-\omega) = \int_{-\alpha}^{\alpha} F(t)e^{-j\omega t}\,dt$$

$$= \mathscr{F}\{F(t)\}$$

$$F(t) \leftrightarrow 2\pi f(-\omega)$$

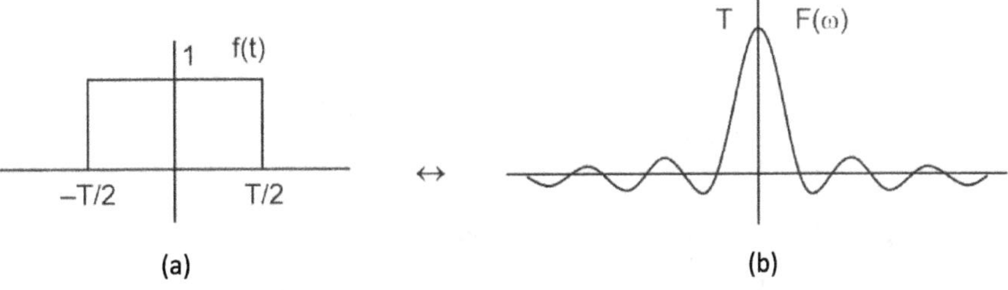

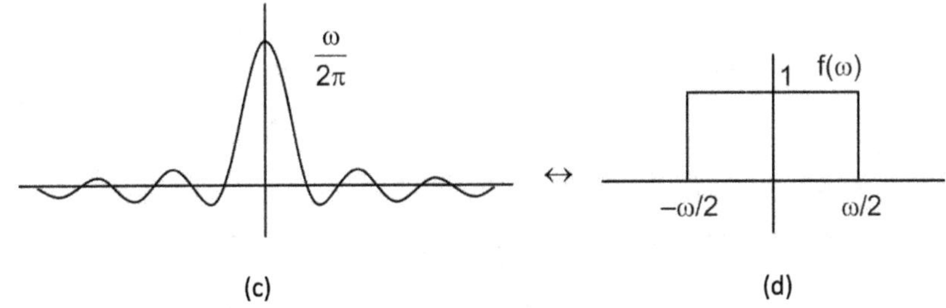

Fig. 4.18

2. Linearity Property:

If $f_1(t) \leftrightarrow F_1(\omega)$

     $f_2(t) \leftrightarrow F_2(\omega)$

for any a_1, a_2

$$a_1 f_1(t) + a_2 f_2(t) \leftrightarrow a_1 F_1(\omega) + a_2 F_2(\omega)$$

3. Scaling Property:

If $f(t) \leftrightarrow F(\omega)$

then for any constant a

$$f(at) \leftrightarrow \frac{1}{|a|} F\left(\frac{\omega}{a}\right)$$

Proof:

$$\mathcal{F}\{f(at)\} = \int_{-\alpha}^{\alpha} f(at)e^{-j\omega t}\, dt$$

Let $x = at$

$dx = adt$

$$= \frac{1}{a}\int_{-\alpha}^{\alpha} f(x)e^{-j\omega x/a}\, dx$$

$$= \frac{1}{a}F\left(\frac{\omega}{a}\right)$$

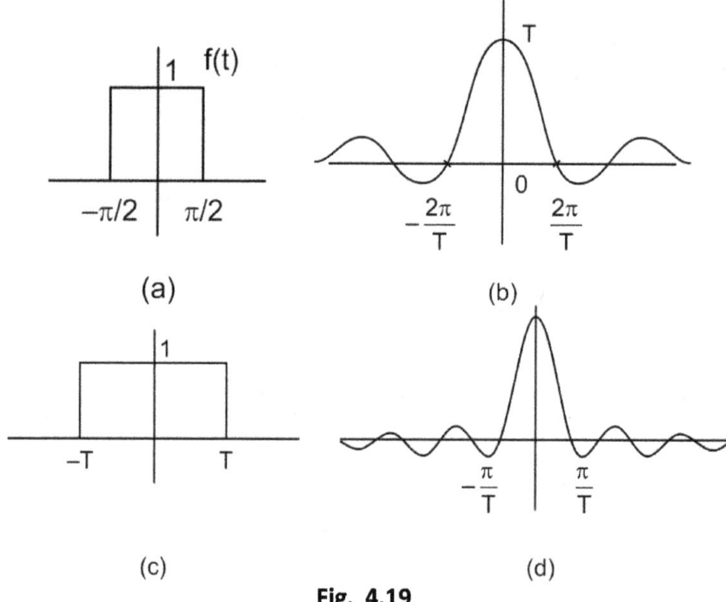

Fig. 4.19

Compression in time domain is equal to frequency expansion.

4. Time Shifting Property:

If $f(t) \leftrightarrow F(\omega)$

$$\mathcal{F}(t-t_0) \leftrightarrow F(\omega)e^{-j\omega t_0}$$

Proof:

$$\mathcal{F}\{f(t-t_0)\} = \int_{-\alpha}^{\alpha} f(t-t_0)e^{-j\omega(t-t_0)}\,dt$$

$$\text{Let } x \quad = \quad t-t_0$$

$$dx \quad = \quad dt$$

$$= \quad \int_{-\alpha}^{\alpha} f(x)e^{-j\omega(x+t_0)}\,dx$$

$$= \quad F(\omega)e^{-j\omega t_0}$$

When a function is shifted in time domain by to secs, its magnitude remains unchanged $F(\omega)$ but phase changes by an amount $-\omega t_0$.

5. Frequency Shifting Property:

$$\text{If} \quad f(t) \leftrightarrow F(\omega)$$

$$f(t)e^{j\omega_0 t} \leftrightarrow F(\omega - \omega_0)$$

Proof:

$$F(\omega) \quad = \quad \int_{-\alpha}^{\alpha} f(t)\,e^{-j\omega t}\,dt$$

$$\mathcal{F}\{f(t)e^{j\omega_0 t}\} = \quad \int_{-\alpha}^{\alpha} f(t)e^{j\omega_0 t}\,e^{-j\omega t}\,dt$$

$$= \quad \int_{-\alpha}^{\alpha} f(t)e^{-j(\omega - \omega_0)t}\,dt$$

$$= \quad F(\omega - \omega_0)$$

Frequency translation theorem:

A shift of ω_0 in frequency domain is equal to multiplication by $e^{j\omega_0 t}$ in time domain.

Ex: Modulation

$$f(t)\cos\omega_0 t \; = \; \frac{1}{2}\{f(t)e^{j\omega_0 t} + f(t)e^{-j\omega_0 t}\}$$

$$f(t) \leftrightarrow F(\omega)$$

$$f(t)\cos\omega_0 t \; \leftrightarrow \; \frac{1}{2}\{F(\omega + \omega_0) + F(\omega - \omega_0)\}$$

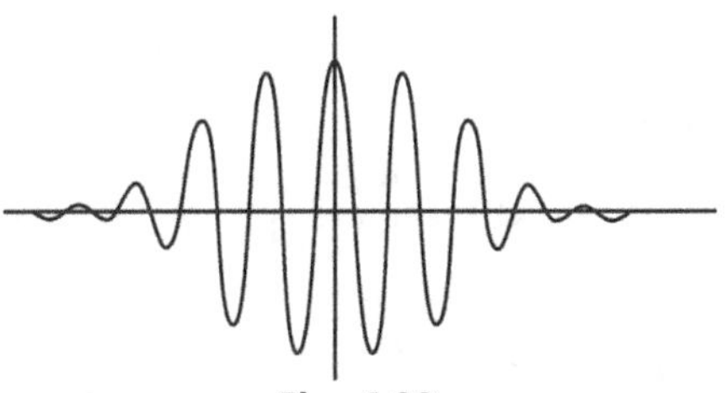

Fig. 4.20

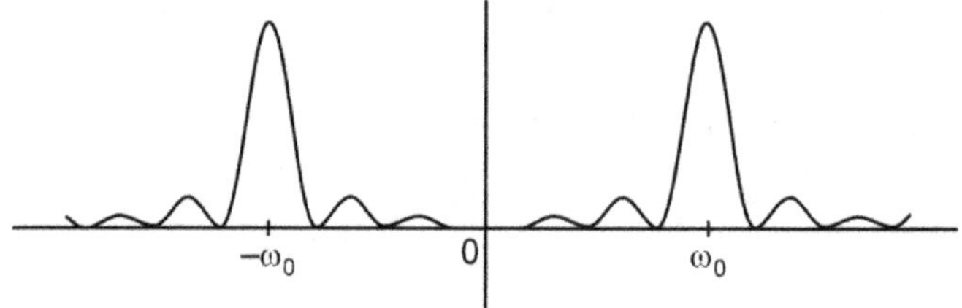

Fig. 4.21

$$f(t)\sin\omega_0 t \leftrightarrow \frac{1}{\Sigma}\{F(\omega+\omega_0)-F(\omega-\omega_0)\}$$

6. Time differentiation and Integration:

If $f(t) \leftrightarrow F(\omega)$

$$\frac{df}{dt} \leftrightarrow (j\omega)\,F(\omega)$$

and

$$\int_{-\alpha}^{T} f(t)\,dt \leftrightarrow \frac{1}{j\omega}F(\omega)$$

Proof:

$$f(t) = \frac{1}{2\pi}\int_{-\alpha}^{\alpha} F(\omega)e^{j\omega t}\,d\omega$$

$$\therefore \quad \frac{df}{dt} = \frac{1}{2\pi}\frac{d}{dt}\int_{-\alpha}^{\alpha} F(\omega)e^{j\omega t}\,d\omega$$

$$= \frac{1}{2\pi}\cdot j\omega\int_{-\alpha}^{\alpha} F(\omega)e^{j\omega t}\,d\omega$$

$$\frac{df}{dt} \leftrightarrow j\omega F(\omega)$$

Let

$$\phi(t) = \int_{-\alpha}^{T} f(t)\,dt$$

and

$$\frac{d}{dt}\phi(t) = f(t)\,20$$

$$f(t) \leftrightarrow j\omega\phi(\omega)$$

$$\phi(\omega) \leftrightarrow \frac{1}{j\omega}f(t)$$

$$\frac{1}{j\omega}F(\omega) = \phi(\omega)$$

$$\leftrightarrow \int_{-\alpha}^{T} f(t)\,dt$$

7. Frequency Differentiation:

If $\quad f(t) \leftrightarrow F(\omega)$

$$-jtf(t) \leftrightarrow \frac{dF}{d\omega}$$

Proof:

$$F(\omega) = \int_{-\alpha}^{\alpha} f(t)e^{-j\omega t}\,dt$$

$$\therefore \quad \frac{dF}{d\omega} = \frac{d}{d\omega}\int_{-\alpha}^{\alpha} f(t)e^{-j\omega t}\,dt$$

$$\frac{dF}{d\omega} = \int_{-\alpha}^{\alpha} -jtf(t)e^{-j\omega t}\,dt$$

$$-jtf(t) \leftrightarrow \frac{dF}{d(\omega)}$$

$$(-jt)^{n}f(t) = \frac{d^{n}f}{d\omega^{n}}$$

$$\frac{1}{-jt}f(t) \leftrightarrow \int F(\omega)\,d\omega$$

8. Time Reversal Property:

If $\quad f\{f(t)\} = F(\omega)$

then

$$\mathscr{F}\{f(-t)\} = F(-\omega)$$

$$\mathscr{F}\{f(-t)\} = \int_{-\alpha}^{\alpha} f(t)e^{-j\omega t}\,dt = \int_{-\alpha}^{\alpha} f(t)e^{j\omega t}\,dt$$

$$= \int f(t)e^{-(-j\omega t)}\,dt$$

$$= F(-\omega)$$

9. Conjugation:

If $\quad \mathscr{F}\{f(t)\} = F(\omega)$

then

$$\mathscr{F}\{f^*(t)\} = F(-\omega)$$

$$F(\omega) = \int_{-\alpha}^{\alpha} f(t)e^{-j\omega t}\, dt$$

$$F^*(\omega) = \int_{-\alpha}^{\alpha} f(t)e^{j\omega t}\, dt$$

$$F^*(-\omega) = \int_{-\alpha}^{\alpha} f(t)e^{-j\omega t}\, dt$$

$$\mathscr{F}\{f^*(t)\}$$

10. Multiplication Property:

If $\quad \mathscr{F}\{x_1(t)\} \Leftrightarrow F_1(\omega)$

$\quad \mathscr{F}\{x_2(t)\} \Leftrightarrow F_2(\omega)$

then

$$\mathscr{F}\{x_1(t)+x_2(t)\} \Leftrightarrow \frac{1}{2\pi}\{x_1(\omega)+x_2(\omega)\}$$

(a) Multiplication by $\sin\omega_0 t$

If $\quad f(t) \leftrightarrow F(\omega)$

$$f(t)\sin\omega_0 t \leftrightarrow \frac{j}{2}\{F(\omega+\omega_0)-F(\omega-\omega_0)\}$$

(b) Multiplication by $\cos\omega_0 t$

If $\quad f(t) \leftrightarrow F(\omega)$

$$f(t)\cos\omega_0 t \leftrightarrow \frac{1}{2}\{F(\omega+\omega_0)+F(\omega-\omega_0)\}$$

(c) Multiplication by complex exponential

If $\quad f(t) \leftrightarrow F(\omega)$

$$f(t)e^{j\omega_0 t} \leftrightarrow F(\omega-\omega_0)$$

4.7 INVERSE FOURIER TRANSFORM

As fourier transform is linear, if $x_1(t) \Leftrightarrow X_1(\omega)$ and $x_2(t) \Leftrightarrow X_2(\omega)$ then $a_1x_1(t) + a_2 x_2(t) \Leftrightarrow a_1X_1(\omega) + a_2X_2(\omega)$

Examples

(a) Find the inverse fourier transform of $\delta(\omega)$

$$f^{-1}\{\delta(\omega)\} = \frac{1}{2\pi}\int_{-\alpha}^{\alpha}\delta(\omega)e^{j\omega t}d\omega = \frac{1}{2\pi} \text{ or } 1 \Leftrightarrow 2\pi\delta(\omega)$$

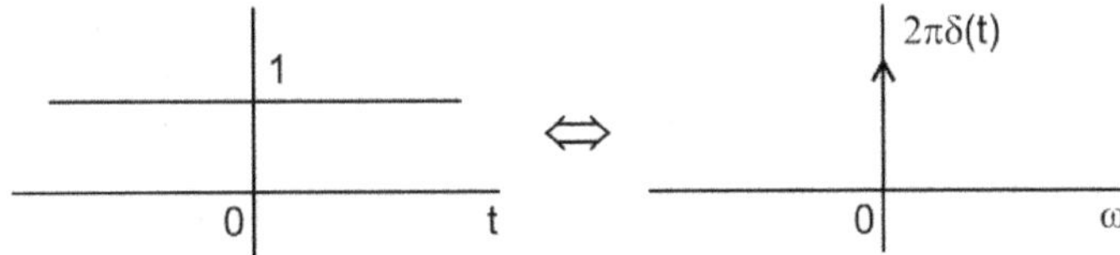

(b) Find the inverse transform of $\delta(\omega - \omega_o)$

Using sampling property of impulse function

$$f^{-1}\{\delta(\omega - \omega_o)\} = \frac{1}{2\pi}\int_{-\alpha}^{\alpha}\delta(\omega - \omega_o)e^{j\omega t}d\omega = \frac{1}{2\pi}e^{j\omega t} \text{ or } e^{j\omega_o t} \Leftrightarrow 2\pi\delta(\omega - \omega_o)$$

Similarly $e^{-j\omega_o t} \Leftrightarrow 2\pi\delta(\omega + \omega_o)$

$$\therefore \quad \cos\omega_o t = \frac{1}{2}(e^{j\omega_o t} + e^{-j\omega_o t}) = \pi\{\delta(\omega + \omega_o) + \delta(\omega - \omega_o)\}$$

4.8 TIME-FREQUENCY DUALITY OF FOURIER TRANSFORM

The inverse transform equation can be obtained from the direct transform equation by replacing $x(t)$ with $X(\omega)$, t with ω and ω with t. There are only two minor differences, the factor 2π appears only in inverse operate, and exponential indices are of opposite sign.

Example: If $x(t) \Leftrightarrow X(\omega)$, they $x(t - t_o) \Leftrightarrow X(\omega)_e^{-j\omega t o}$ duality of this property is $x(t)$ $x(\omega)_e^{j\omega_o t} \Leftrightarrow X(\omega - \omega_o)$.

SOLVED PROBLEMS

EXERCISE 2

1. Find the fourier transform of

$$f(t) = e^{-3t}\{u(t + 2) - u(t - 3)\}$$

$$\mathscr{F}\{f(t)\} = \int_{-2}^{3} e^{-3t}\,e^{-j\omega t}\,dt = \int_{-2}^{3} e^{-(j\omega+3)t}\,dt$$

$$= \frac{-1}{3+j\omega}\,e^{-(j\omega+3)t}\,\Big/_{-2}^{\;3}$$

$$= \frac{e^{-(j\omega+3)2} - e^{-(j\omega+3)3}}{j\omega+3}$$

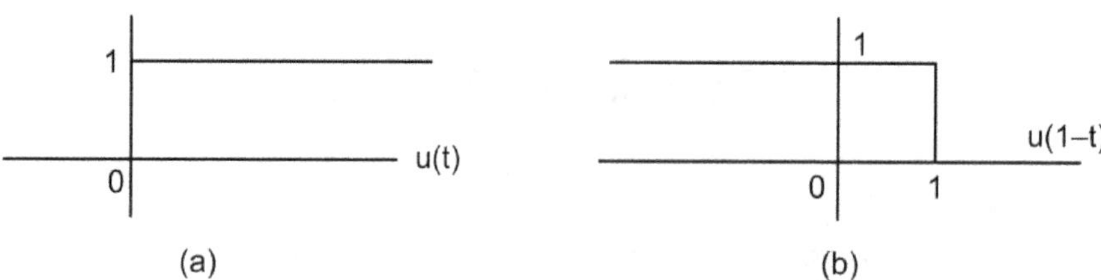

(a)   (b)   (c)

Fig. 4.22

2. Find the fourier transform of

$$f(t) = \frac{1}{a}\,e^{t/a}\,u(-t)$$

$$\mathscr{F}\{f(t)\} = \int_{-\alpha}^{0} \frac{1}{a}\,e^{t/a}\,e^{-j\omega t}\,dt$$

$$= \frac{1}{a}\int_{-\alpha}^{0} e^{\left(\frac{1}{a}-j\omega\right)t}\,dt$$

$$= \frac{1}{a}\,\frac{e^{\left(\frac{1}{a}-j\omega\right)t}}{\left(\frac{1}{a}-j\omega\right)}\,\Big/_{\alpha}^{\;0}$$

$$= \frac{1}{a}\,\frac{1}{\left(\frac{1}{a}-j\omega\right)} = \frac{1}{1-j\omega a}$$

3. Find the fourier transform of

$$f(t) = t^2 u(t)\,u(1-t)$$

(a)   (b)

Fig 4.23

$$\mathcal{F}\{f(t)\} = \int_0^1 t^2 e^{-j\omega t}\, dt$$

$$= t^2 \cdot \frac{e^{-j\omega t}}{-j\omega} - \int 2t \frac{e^{-j\omega t}}{-j\omega}\, dt$$

$$= \frac{t^2 \cdot e^{-j\omega t}}{-j\omega} + \frac{2}{j\omega} \left\{ \frac{t e^{-j\omega t}}{-j\omega} + \frac{1}{j\omega}\frac{e^{-j\omega t}}{-j\omega} \right\} \Bigg/ \int t e^{-j\omega t}\, dt \Bigg|_0^1 = \frac{t \cdot e^{-j\omega t}}{-j\omega t} - \int \frac{e^{-j\omega t}}{-j\omega}\, dt$$

$$= \frac{e^{-j\omega}}{-j\omega} + \frac{2}{j\omega}\left\{ \frac{e^{-j\omega}}{j\omega} + \frac{1}{j\omega}\frac{e^{-j\omega}}{-j\omega} - \frac{1}{j\omega} \cdot \frac{1}{-j\omega} \right\}$$

$$= -\frac{e^{-j\omega}}{j\omega} + \frac{2}{j\omega}\left\{ \frac{e^{-j\omega}}{j\omega} + \frac{1}{\omega^2}e^{-j\omega} - \frac{1}{\omega^2} \right\}$$

$$= -\frac{e^{-j\omega}}{j\omega} - \frac{2e^{-j\omega}}{\omega^2} + \frac{2}{j\omega^3}e^{-j\omega} - \frac{2}{j\omega^3}$$

4. Find fourier transform of

$$f(t) = \delta(t+1) + \delta(t+\tfrac{1}{2}) + \delta(t-\tfrac{1}{2}) + \delta(t-1)$$

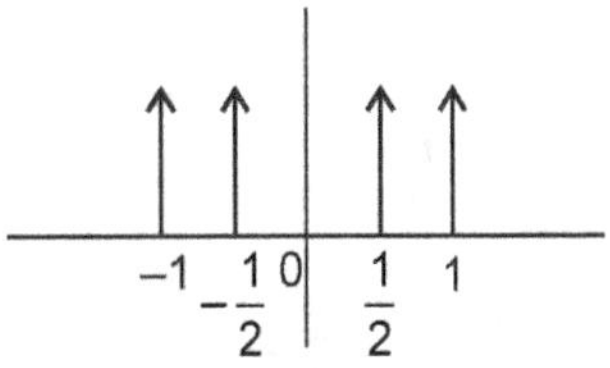

$$\mathcal{F}\{f(t)\} =$$

$$\int_{-\alpha}^{\alpha} \left\{ \delta(t+1) + \delta(t+\tfrac{1}{2}) + \delta(t-\tfrac{1}{2}) + \delta(t-1) \right\} e^{-j\omega t}\, dt$$

$$= e^{j\omega} + e^{-j\omega} + e^{-0.5j\omega} + e^{0.5j\omega}$$

$$= 2\cos j\omega + 2\cos 0.5 j\omega$$

Fig. 4.24

4. $f(t) = e^{-t}\operatorname{sgn}(t)$

$$= e^{-t}\{2u(t) - 1\}$$

$$= 2e^{-t}u(t) - e^{-t}$$

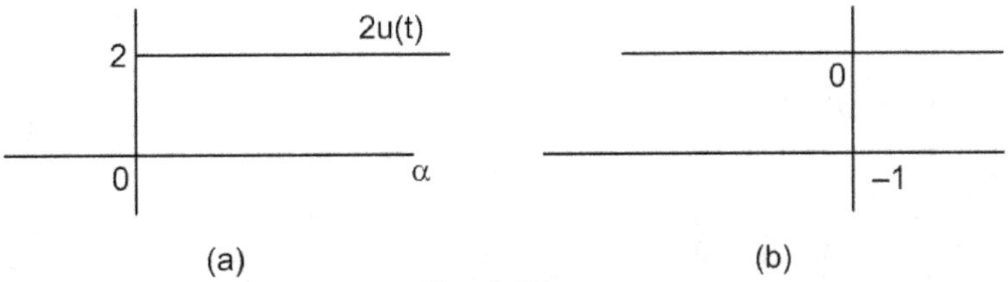

Fig. 4.25

$$\mathscr{F}\{f(t)\} = 2\int_0^\alpha e^{-t}e^{-j\omega t}\,dt - \int_{-\alpha}^\alpha e^{-t}\cdot e^{-j\omega t}\,dt$$

$$= 2\int_0^\alpha e^{-t(j\omega+1)}\,dt - \int_{-\alpha}^\alpha e^{-(j\omega+1)t}\,dt$$

5. $f(t) = \sin c(t)\,u(t)$

$$\mathscr{F}\{f(t)\} = \int_0^\alpha \sin c(t)\,e^{-j\omega t}\,dt$$

$$= \int_0^\alpha \frac{\sin t}{t}e^{-j\omega t}\,dt$$

$$= \frac{1}{2j}\int_0^\alpha \frac{e^t - e^{-t}}{t}\cdot e^{-j\omega t}\,dt$$

$$= \frac{1}{2j}\int_0^\alpha \frac{e^{-t(j\omega-1)}}{t} - \frac{e^{-(j\omega+1)t}}{t}\,dt$$

6. Find the fourier transform of $f(t) = \dfrac{1}{T}e^{t/T}u(-t)$

Ans:
$$f(t) = \frac{1}{T}e^{t/T}u(-1)$$

$$= \int_{-\alpha}^0 f(t)e^{-j\omega t}\,dt$$

$$= \frac{1}{T}\int_{-\alpha}^0 e^{t/T}e^{-j\omega t}\,dt$$

$$= \frac{1}{T}\int_{-\alpha}^0 e^{t\left(\frac{1}{T}-j\omega\right)}\,dt$$

$$= \frac{1}{T}\frac{e^{t\left(\frac{1}{T}-j\omega\right)}}{\left(\frac{1}{T}-j\omega\right)}\bigg|_{-\alpha}^0$$

$$= \frac{1}{T\left(\frac{1}{T}-j\omega\right)} = \frac{1}{1-j\omega T}$$

7. Find the fourier transform of $f(t) = \dfrac{1}{T}\left\{1 - \dfrac{|t|}{T}\right\}, \quad |t| < T$

Ans: $\quad f(t) = \dfrac{1}{T}\left\{1 + \dfrac{t}{T}\right\} \qquad (-T, 0)$

$$= \dfrac{1}{T}\left\{1 - \dfrac{t}{T}\right\} \qquad (0, T)$$

$$F(\omega) = \int_{-\alpha}^{0} f(t)e^{-j\omega t}\,dt + \int_{0}^{T} f(t)e^{-j\omega t}\,dt$$

$$= \int_{-T}^{0} \dfrac{1}{T}\left\{1 + \dfrac{t}{T}\right\}e^{-j\omega t}\,dt + \int_{0}^{T} \dfrac{1}{T}\left\{1 - \dfrac{t}{T}\right\}e^{-j\omega t}\,dt$$

$$= \dfrac{1}{T}\int_{-T}^{0}\left(\dfrac{t}{T}+1\right)e^{-j\omega t}\,dt + \int_{0}^{T}\left(1 - \dfrac{t}{T}\right)e^{-j\omega t}\,dt$$

$$= \dfrac{1}{T}\left[\left\{\left(1 + \dfrac{t}{T}\right)\dfrac{e^{-j\omega t}}{-j\omega} - \dfrac{1}{T}\dfrac{e^{-j\omega t}}{(-j\omega)^2}\right\} + \left\{\left(1 - \dfrac{t}{T}\right)\dfrac{e^{-j\omega t}}{-j\omega} - \left(-\dfrac{1}{T}\right)\dfrac{e^{-j\omega t}}{(j\omega)^2}\right\}_{0}^{T}\right]$$

$$= \dfrac{1}{T}\left\{-\dfrac{1}{j\omega} + \dfrac{1}{T\omega^2} - \dfrac{e^{j\omega T}}{T\omega^2} + \dfrac{1}{j\omega}\dfrac{e^{-j\omega T}}{T\omega^2} + \dfrac{1}{T\omega^2}\right\}$$

$$= \dfrac{1}{T^2\omega^2}\left\{2 - e^{j\omega T} - e^{-j\omega T}\right\}$$

$$= \dfrac{2 - 2\cos\omega T}{T^2\omega^2} = \dfrac{2\left\{1 - 2\sin^2\frac{\omega T}{2}\right\}}{T^2\omega^2}$$

$$= \dfrac{4\sin^2\frac{\omega T}{2}}{T^2\omega^2}$$

8. Show that fourier transform of $f(t)$ may also be expressed as

(a) $\quad F(j\omega) = \displaystyle\int_{-\alpha}^{\alpha} f(t)\cos\omega t\,dt - j\int_{-\alpha}^{\alpha} f(t)\sin\omega t\,dt$

(b) $\quad F(j\omega) = 2\displaystyle\int_{0}^{\alpha} f(t)\cos\omega t\,dt \qquad$ if $f(t)$ is even function

(c) $\quad F(j\omega) = -2j\displaystyle\int_{0}^{\alpha} f(t)\sin\omega t\,dt \qquad$ if $f(t)$ is odd function

Ans:

(a)
$$\mathcal{F}\{f(t)\} = F(j\omega) = \int_{-\alpha}^{\alpha} f(t)e^{-j\omega t}\,dt$$

$$= \int_{-\alpha}^{\alpha} f(t)[\cos\omega t - j\sin\omega t]\,dt$$

$$= \int_{-\alpha}^{\alpha} f(t)\cos\omega t\,dt - j\int_{-\alpha}^{\alpha} f(t)\sin\omega t\,dt........$$

$$\qquad\qquad\text{even}\qquad\qquad\text{odd}$$

(b)
$$F(j\omega) = \int_{-\alpha}^{\alpha}\text{even} - j\int_{-\alpha}^{\alpha}\text{odd}$$

$$= \text{odd}\int_{-\alpha}^{\alpha} - j\ \text{even}\ \Big|_{-\alpha}^{\alpha}$$

$$\text{odd}\int_{-\alpha}^{\alpha} -0 = 2\int_{0}^{\alpha} f(t)\cos\omega t\,dt$$

$$\text{even} \times \text{even} = \text{even}$$
$$\text{even} \times \text{odd} = \text{odd}$$
$$\text{odd} \times \text{odd} = \text{even}$$

(c)
$$F(\omega) = \int_{-\alpha}^{\alpha} f(t)\cos\omega t\,dt - j\int_{-\alpha}^{\alpha} f(t)\sin\omega t$$

$$\qquad \text{odd}\quad \text{even} \rightarrow \text{odd}\qquad\qquad \text{odd}\quad \text{odd}\rightarrow \text{even}$$

$$= \text{even}\ \Big|_{-\alpha}^{\alpha} - j\,\text{odd}\ \Big|_{-\alpha}^{\alpha}$$

$$= 0 - j\,\text{odd}$$

$$= -2j\int_{0}^{\alpha} f(t)\sin\omega t\,dt$$

EXERCISE 3

Find the fourier transform of x(t) using properties of fourier transform

1. $f(t) = e^{-j2t}\,x(t-3)$ given $x(\omega) = \dfrac{6}{4 + j\omega}$

   Given $\mathcal{F}\{x(t)\} = x(\omega) = \dfrac{6}{4 + j\omega}$

Using time shifting property

$$x(t-t_0) \leftrightarrow e^{-j\omega t_0} x(\omega)$$

$$x(t-3) \leftrightarrow e^{-j3\omega} x(\omega)$$

Using frequency shifting property

$$e^{j\omega_0 t} x(t) \leftrightarrow e - x(\omega - \omega_0)$$

$$e^{-j2t} x(t-3) \leftrightarrow e^{-j3(\omega+2)} \frac{6}{4+j(\omega+2)}$$

$$\therefore \quad \mathscr{F}\left\{e^{-j2t} x(t-3)\right\} = e^{-j3(\omega+2)} \frac{6}{4+j(\omega+2)}$$

2. t. $f(2t)$

Using Scaling property

$$f(at) \leftrightarrow \frac{1}{|a|} F\left(\frac{\omega}{a}\right)$$

$$f(2t) \leftrightarrow \frac{1}{2} F\left(\frac{\omega}{2}\right)$$

Using frequency differentiation property

$$-jtf(t) \leftrightarrow \frac{d}{d\omega} F(\omega)$$

$$t f(2t) = \frac{1}{2} j \frac{d}{d\omega} F\left(\frac{\omega}{2}\right)$$

3. t. $\dfrac{df}{dt}$

Using time differentiation property

$$\frac{df}{dt} \leftrightarrow j\omega F(\omega)$$

Using frequency differentiation property

$$tf(t) \leftrightarrow \frac{d}{d\omega}\{F(\omega)\}$$

$$t\frac{df}{dt} \leftrightarrow j\frac{d}{d\omega} j\omega\, F(\omega)$$

$$= -\frac{d}{d\omega}\{\omega F(\omega)\}$$

4. $f(1-t)$

Using time shifting property & scaling

$$f(t-t_0) \leftrightarrow F(\omega) e^{-j\omega t_0}$$

$$f(-t) \leftrightarrow - F(\omega)$$

$$f(-t+1) \leftrightarrow - e^{j\omega} F(\omega)$$

5. $f(t) = e^{j\omega_0 t} u(t)$

fourier transform of unit step function

$$\mathcal{F}\{u(t)\} = \frac{1}{j\omega} + \pi\delta(\omega)$$

Using frequency shifting property

$$\mathcal{F}\{e^{j\omega_0 t} u(t)\} = \frac{1}{j(\omega - \omega_0)} + \pi\delta(\omega - \omega_0)$$

6. $f(t) = \cos\omega_0 t\, u(t)$

$$F(\omega) = \mathcal{F}\left\{\frac{e^{j\omega_0 t} + e^{-j\omega_0 t}}{2} u(t)\right\}$$

$$= \frac{1}{2}\left\{f\left[e^{j\omega_0 t} u(t)\right] + f\left[e^{-j\omega_0 t} u(t)\right]\right\}$$

$$\mathcal{F}\{e^{j\omega_0 t} u(t)\} = \frac{1}{j(\omega - \omega_0)} + \pi\delta(\omega - \omega_0)$$

$$\mathcal{F}\{e^{-j\omega_0 t} u(t)\} = \frac{1}{j(\omega + \omega_0)} + \pi\delta(\omega + \omega_0)$$

$$F(\omega) = \frac{j\omega}{\omega_0^2 \omega^2} + \frac{\pi}{2}\delta(\omega - \omega_0) + \frac{\pi}{2}\delta(\omega + \omega_0)$$

7. $f(t) = \sin\omega_0 t\, u(t)$

$$F(\omega) = \mathcal{F}\left\{\frac{e^{j\omega_0 t} - e^{-j\omega_0 t}}{2j} u(t)\right\}$$

$$\mathcal{F}\{e^{j\omega_0 t} u(t)\} = \frac{1}{j(\omega - \omega_0)} + \pi\delta(\omega - \omega_0)$$

$$\mathcal{F}\{e^{-j\omega_0 t} u(t)\} = \frac{1}{j(\omega + \omega_0)} + \pi\delta(\omega + \omega_0)$$

$$F(\omega) = \frac{1}{2j}\left\{\frac{1}{j(\omega - \omega_0)} + \pi\delta(\omega - \omega_0) - \frac{1}{j(\omega + \omega_0)} - \pi\delta(\omega + \omega_0)\right\}$$

$$= \frac{1}{2j}\left\{\frac{2j\omega_0}{\omega_0^2 - \omega^2} + \pi\delta(\omega - \omega_0) - \pi\delta(\omega + \omega_0)\right\}$$

$$= \left\{\frac{\omega_0}{\omega_0^2 - \omega^2} - j\frac{\pi}{2}\delta(\omega - \omega_0) + j\frac{\pi}{2}\delta(\omega + \omega_0)\right\}$$

8. Find the inverse fourier transform of

$$F(\omega) = \frac{j\omega}{(3 + j\omega)^2}$$

we have

$$\mathcal{F}\left\{t\,e^{-at}\,u(t)\right\} = \frac{1}{(a+j\omega)^2}$$

$$\therefore \ \mathcal{F}\left\{t\,e^{-3t}\,u(t)\right\} = \frac{1}{(3+j\omega)^2}$$

Using time differentiation property

$$\mathcal{F}\left\{\frac{d}{dt}(t)\right\} = j\omega F(\omega)$$

$$\mathcal{F}\left\{\frac{d}{dt}t\cdot e^{-3t}\,u(t)\right\} = \frac{j\omega}{(3+j\omega)^2}$$

$$\therefore \qquad f(t) \ = \ \frac{d}{dt}t\cdot e^{-3t}u(t)$$

9. Find the fourier transform of

$$f(t) \ = \ 1-e^{-|t|}\cos\omega_0 t$$

$$F(\omega) \ = \ \mathcal{F}\{1\} - \mathcal{F}\left\{e^{-|t|}\cos\omega_0 t\right\}$$

$$\mathcal{F}\{1\} = 2\pi\delta(\omega)$$

$$\mathcal{F}\left\{e^{-|t|}\cos\omega_0 t\right\} = \frac{1}{2}\left\{\int_{-\alpha}^{\alpha} e^{-|t|}e^{j\omega_0 t}\,e^{-j\omega t}\,dt + \int_{-\alpha}^{\alpha} e^{|t|}e^{-j\omega_0 t}\,e^{-j\omega t}\,dt\right\}$$

$$= \frac{1}{2}\left\{\int_{-\alpha}^{0} e^{t}\,e^{j\omega_0 t}\,e^{-j\omega t}\,dt + \int_{0}^{\alpha} e^{-t}\,e^{j\omega_0 t}\,e^{-j\omega t}\,dt + \int_{-\alpha}^{0} e^{t}\,e^{-j\omega_0 t}\,e^{-j\omega t}\,dt + \int_{0}^{\alpha} e^{-t}\,e^{-j\omega_0 t}\,e^{-j\omega t}\,dt\right\}$$

$$= \frac{1}{2}\left\{\int_{-\alpha}^{0} e^{(1-j(\omega-\omega_0))t}\,dt + \int_{0}^{\alpha} e^{-(1+j(\omega-\omega_0))t}\,dt + \int_{\alpha}^{0} e^{(1-j(\omega+\omega_0))t}\,dt + \int_{0}^{\alpha} e^{-(1+j(\omega+\omega_0))t}\,dt\right\}$$

$$= \frac{1}{2}\left\{\frac{1}{1-j(\omega-\omega_0)} + \frac{1}{1+j(\omega-\omega_0)} + \frac{1}{1-j(\omega-\omega_0)} + \frac{1}{1+j(\omega+\omega_0)}\right\}$$

$$= \frac{1}{1+(\omega-\omega_0)^2} + \frac{1}{1+(\omega+\omega_0)^2}$$

$$\therefore \quad F(\omega) = 2\pi\delta(\omega) + \frac{1}{1+(\omega-\omega_0)^2} + \frac{1}{1+(\omega+\omega_0)^2}$$

10. Find the Fourier Transform of below function

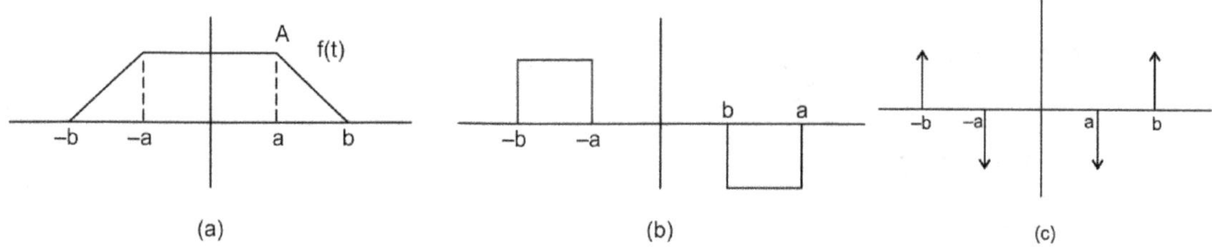

(a)   (b)   (c)

Fig. 4.26

To obtain Fourier Transform of piece wise linear signal we can apply differentiation theorem; first order derivative is of two square pulse and second order derivative is entirely unit impulses

$$\frac{d^2 f(t)}{dt^2} = k\{\delta(t+b) - \delta(t+a) - \delta(t-a) + \delta(t-b)\}$$

where $\quad k = \dfrac{A}{b-a}$

$$\mathscr{F}\left\{\frac{d^2 f(t)}{dt^2}\right\} = k\{e^{j\omega b} - e^{j\omega a} - e^{-j\omega a} + e^{-j\omega b}\}$$

$$= 2k\{\cos\omega b - \cos\omega a\}$$

$$(j\omega)^2 F(\omega) \leftrightarrow \frac{d^2 f(t)}{dt^2}$$

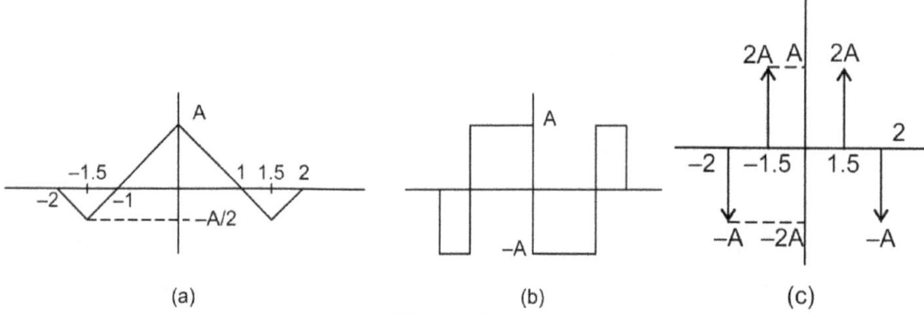

(a)   (b)   (c)

Fig. 4.27

$$\mathscr{F}\{f(t)\} = 2k\left\{\frac{\cos\omega b - \cos\omega a}{(j\omega)^2}\right\}$$

$$= k\left\{\frac{b^2 \sin^2 \frac{\omega b}{2}}{\left(\frac{\omega b}{2}\right)^2} - \frac{a^2 \sin^2 \frac{\omega a}{2}}{\left(\frac{\omega a}{2}\right)^2}\right\} = kb^2 \sin c^2\left(\frac{\omega b}{2\pi}\right) - ka^2 \sin c^2\left(\frac{\omega a}{2\pi}\right)$$

11. $k = \dfrac{A}{b-a}$

$$\dfrac{\partial^2 f(t)}{\partial t^2} = -Ae^{-j2\omega}+2Ae^{-j1.5\omega}+2Ae^{+j1.5\omega}-Ae^{+j2\omega}-2A$$

$$j(\omega)^2 f(\omega) \leftrightarrow \mathcal{F}\big(f(t)\big)$$

$$F(\omega) = -\dfrac{Ae^{-j2\omega}+2Ae^{-j1.5\omega}+2Ae^{+j1.5\omega}-Ae^{+j2\omega}-2A}{(j\omega)^2}$$

$$= \dfrac{2(-A\cos 2\omega+2A\cos 1.5\omega)-2A}{(j\omega)^2}$$

12.

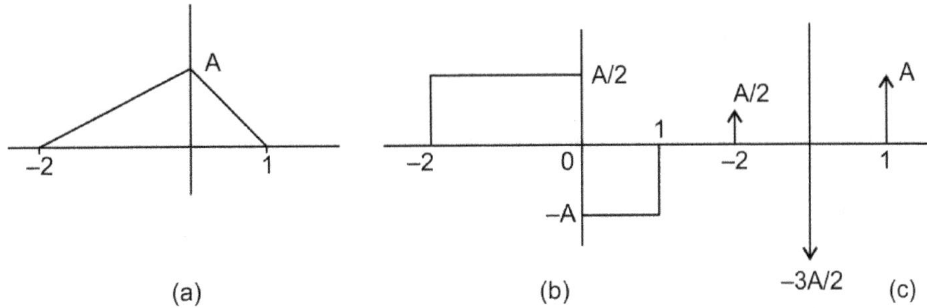

(a) (b) (c)

Fig. 4.28

$$\dfrac{d^2 f(t)}{dt^2} = \dfrac{A}{2}\delta(t+2)-\dfrac{3A}{2}\delta(t)+A\delta(t-2)$$

$$F(\omega) = \dfrac{Ae^{j2\omega}-3A+2Ae^{-j\omega}}{2(j\omega)^2}$$

13.

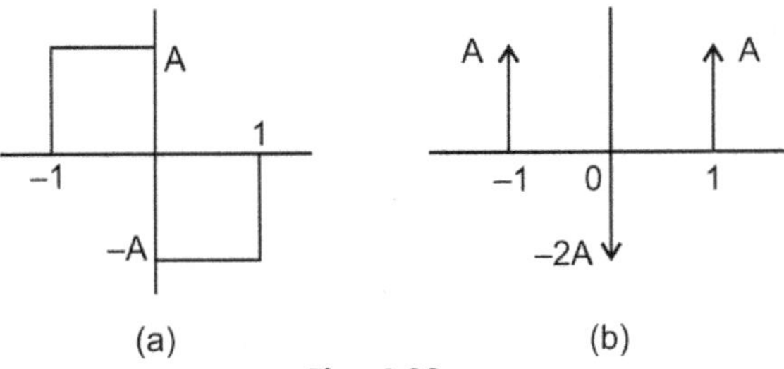

(a) (b)

Fig. 4.29

$$(+j\omega)\dfrac{df(t)}{dt} = Ae^{-j\omega}+Ae^{j\omega}-2A$$

$$F(\omega) = \dfrac{A}{+j\omega}\big\{e^{j\omega}+e^{-j\omega}-2\big\}$$

14. Find the inverse fourier transform of $\dfrac{j\omega}{(3+j\omega)^2}$

$$F(\omega) = \dfrac{j\omega}{(3+j\omega)^2}$$

we have $\mathcal{F}\left\{t\cdot e^{-at}\,u(t)\right\} = \dfrac{1}{(a+j\omega)^2}$

$\therefore\ \mathcal{F}\left\{t\cdot e^{-3t}\,u(t)\right\} = \dfrac{1}{(3+j\omega)^2}$

and   also

$$\mathcal{F}\left\{\dfrac{d}{dt}f(t)\right\} \leftrightarrow j\omega\,F(\omega) \leftrightarrow \dfrac{j\omega}{(3+j\omega)^2}$$

$$f(t) = \dfrac{d}{dt}\left[t\cdot e^{-3t}\,u(t)\right]$$

15. Find the inverse fourier transform of $e^{-|\omega|}$

$$F(\omega) = e^{-|\omega|}$$

we have $\mathcal{F}\left\{e^{-|t|}\right\} = \dfrac{2}{1+\omega^2}$

using dual property

$$\mathcal{F}\left\{\dfrac{2}{1+t^2}\right\} = 2\pi\left\{e^{-|\omega|}\right\}$$

$$\mathcal{F}^{-1}\left\{e^{-|\omega|}\right\} = \dfrac{1}{\pi(1+t^2)}$$

16. Find the inverse fourier transform of

$$F(\omega)<e^{-2\omega}\,u(\omega)$$

we have $\mathcal{F}\left\{e^{-|t|}\right\} = \dfrac{1}{1+\omega} + \dfrac{1}{1-j\omega}$

$\therefore\ \mathcal{F}\left\{e^{-t}\right\} = \dfrac{1}{1-j\omega}$

using dual property

$$\mathcal{F}\left\{\dfrac{1}{1-jt}\right\} = 2\pi\left\{e^{-\omega}\right\}$$

$$\therefore\ \mathcal{F}\left\{e-2\omega\right\} = \dfrac{1}{2\pi(2-jt)}$$

17. Find the inverse fourier transform of

$$F(\omega) = \frac{15e^{-3j\omega} - 10e^{-2j\omega} - 5e^{-j5\omega}}{\omega^2}$$

$$\omega^2 F(\omega) = 15e^{-3j\omega} - 10e^{-2j\omega} - 5e^{-j5\omega}$$

$$-\omega^2 F(\omega) = 10e^{-2j\omega} + 5e^{-j5\omega} - 15e^{-3j\omega}$$

$$(j\omega)^2 F(\omega) = 10e^{-2j\omega} + 5e^{-j5\omega} - 15e^{-3j\omega}$$

taking inverse fourier transform

$$\frac{d^2 f(t)}{dt^2} = 10\delta(t-2) + 5\delta(t-5) - 15\delta(t-3)$$

$$\frac{df(t)}{dt} = 10u(t-2) + 5u(t-5) - 15u(t-3)$$

$$f(t) = 10(t-2) + 5(t-5) - 15(t-3)$$

PROBLEMS FROM PREVIOUS QUESTION PAPERS   Nov 2008

1. $f(t) = te^{-at} u(t), \quad a > 0$

$$\mathscr{F}\{f(t)\} = \underset{T \to \alpha}{Lt} \int_0^T t \cdot e^{-at} e^{-j\omega t} \, dt$$

$$= \underset{T \to \alpha}{Lt} \int_0^T t \, e^{-(a+j\omega)t} \, dt$$

$$= \underset{T \to \alpha}{Lt} \left\{ \frac{t \cdot e^{-(a+j\omega)t}}{-(a+j\omega)} - \int \frac{e^{-(a+j\omega)t}}{-(a+j\omega)} \, dt \right\}$$

$$= \underset{T \to \alpha}{Lt} \left\{ \frac{t \cdot e^{-(a+j\omega)t}}{-(a+j\omega)} - \int \frac{e^{-(a+j\omega)t}}{-(a+j\omega)^2} \, dt \right\}$$

$$= \underset{T \to \alpha}{Lt} \frac{1}{(a+j\omega)} \left\{ -t \cdot e^{-(a+j\omega)t} - \frac{e^{-(a+j\omega)t}}{(a+j\omega)} \right\} \Bigg|_0^T$$

$$= \frac{1}{a+j\omega} + \left\{ \frac{1}{(a+j\omega)} \right\} = \frac{1}{(a+j\omega)^2}$$

2. Find out the fourier transform of the periodic pulse train.

$$f(t) = A \quad \text{for} \quad |t| \leq \frac{T}{2}$$

$$F_n = \frac{1}{T} \int_{-\frac{T_0}{2}}^{\frac{T_0}{2}} A \cdot e^{-jn\omega t}\, dt$$

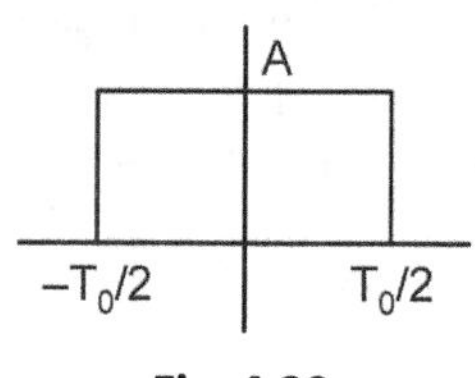

Fig. 4.30

$$= \frac{A}{T} \frac{e^{-jn\omega t}}{-jn\omega} \Bigg|_{-\frac{T_0}{2}}^{\frac{T_0}{2}} = \frac{A}{T}\left\{ \frac{e^{-jn\omega \frac{T_0}{2}} - e^{+jn\omega \frac{T_0}{2}}}{-jn\omega} \right\}$$

$$= \frac{2A\,T_0}{T}\, \frac{\sin n\omega \frac{T_0}{2}}{n\omega \frac{T_0}{2}}$$

$$= \frac{AT_0}{T} \operatorname{sinc}\left(\frac{n\omega T_0}{2}\right)$$

3. Find the fourier transform of the following

$$f(t) = e^{-at} \sin \omega_0 t\, u(t)$$

$$\mathcal{F}\{f(t)\} = \lim_{T \to \alpha} \int_0^T e^{-at}(-e^{-jn\omega_0 t} + e^{j\omega_0 t})\, e^{-j\omega t}\, dt$$

$$= \lim_{T \to \alpha} \frac{1}{2j} \int_0^T \frac{e^{-(a-j\omega_0)t}}{+j\omega} - \frac{e^{-(a+j\omega_0)t}}{+j\omega}\, dt$$

$$= \lim_{T \to \alpha} \frac{1}{2j} \left\{ \frac{e^{-(a-j\omega_0)t}}{-(a^{-j\omega_0}_{+j\omega})} + \frac{e^{-(a+j\omega_0)t}}{(a^{+j\omega_0}_{+j\omega})} \right\}_0^T$$

$$= \frac{1}{2j}\left\{ \frac{1}{(a^{-j\omega_0}_{+j\omega})} - \frac{1}{(a^{+j\omega_0}_{+j\omega})} \right\}$$

$$= \frac{1}{2j}\left\{ \frac{j(\omega_0 + \omega) + \omega_0 - j\omega}{(a + j\omega)^2 + \omega_0^2} \right\}$$

$$= \frac{\omega_0}{(a + j\omega)^2 + \omega_0^2}$$

4. Find the fourier transform

$$f(t) = \frac{2}{1+t^2}$$

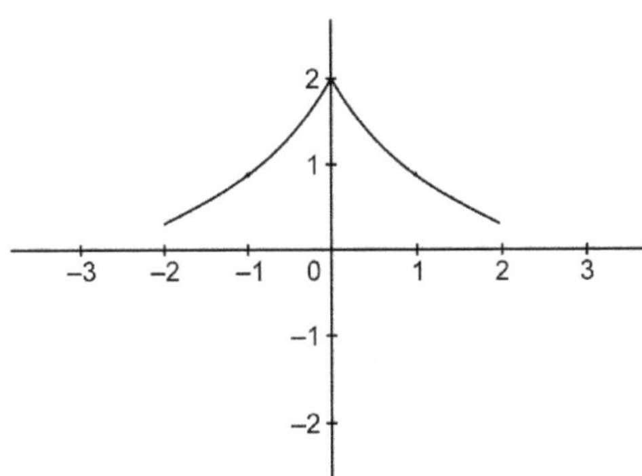

Fig. 4.31

$$\mathscr{F}\{f(t)\} = 2\pi\{e^{-|\omega|}\}$$

5. Obtain the fourier transform of   Feb 2008

$$x(t) = A\sin(2\pi f_c t)\,u(t)$$

Let $\omega_c = 2\pi f_c$

$$\mathscr{F}\{x(t)\} = \underset{T\to\alpha}{Lt} \int_{-T/2}^{T/2} A\sin\omega_c t\, e^{-j\omega t} u(t)\, dt$$

$$= \underset{T\to\alpha}{Lt} \int_{0}^{T/2} A\sin\omega_c t\, e^{-j\omega t}\, dt$$

$$= \underset{T\to\alpha}{Lt}\, A\int_{0}^{T/2} \frac{e^{j\omega_c t} - e^{-j\omega_c t}}{2j} e^{-j\omega t}\, dt = \frac{A}{2j}\underset{T\to\alpha}{Lt}\left\{\frac{e^{-j(\omega-\omega_c)t}}{-j(\omega-\omega_c)} - \frac{e^{-j(\omega+\omega_c)t}}{j(\omega+\omega_c)}\right\}_{0}^{T/2}$$

$$= \frac{A}{2j}\underset{T\to\alpha}{Lt}\left\{\frac{e^{-j(\omega-\omega_c)T/2}}{-j(\omega-\omega_c)} + \frac{1}{j(\omega-\omega_c)} - \frac{e^{-j(\omega+\omega_c)T/2}}{-j(\omega+\omega_c)} - \frac{1}{j(\omega+\omega_c)}\right\}$$

$$= \frac{A}{2j}\left\{\frac{2\omega_c}{j(\omega^2-\omega_c^2)}\right\}$$

$$= -A\frac{\omega_c}{\omega^2-\omega_c^2} = A\frac{\omega_c}{\omega_c^2-\omega^2}$$

6. Find fourier transform of a two sided exponential pulse $x(t) = e^{-|t|}$ May 2005

$$\mathcal{F}\{x(t)\} = \underset{T \to \alpha}{\text{Lt}} \int_{-T}^{0} e^{t}\, e^{-j\omega t}\, dt + \int_{0}^{T} e^{-t}\, e^{-j\omega t}\, dt$$

$$= \underset{T \to \alpha}{\text{Lt}} \int_{-T}^{0} e^{-(j\omega-1)t}\, dt + \int_{0}^{T} e^{-(j\omega+1)t}\, dt$$

$$= \underset{T \to \alpha}{\text{Lt}} \left\{ \frac{e^{(-j\omega+1)t}}{(-j\omega+1)} \Big|_{-T}^{0} + \frac{e^{-(j\omega+1)t}}{-(j\omega+1)} \Big|_{0}^{T} \right\}$$

$$= \frac{e^{t(1-j\omega)}}{(1-j\omega)} \Big|_{-\alpha}^{0} + \frac{e^{-t(1+j\omega)}}{-(1+j\omega)} \Big|_{0}^{\alpha}$$

$$= \frac{1}{1-j\omega} + \frac{1}{1+j\omega} = \frac{2}{1+\omega^2}$$

7. Find the fourier transform of signal shown

$$\mathcal{F}\{f(t)\} = \int_{-2}^{-1} e^{-j\omega t}\, dt + \int_{-1}^{1} 2e^{-j\omega t}\, dt + \int_{1}^{2} e^{-j\omega t}\, dt$$

$$= \frac{e^{-j\omega t}}{-j\omega} \Big|_{-2}^{1} + \frac{2e^{-j\omega t}}{-j\omega} \Big|_{-1}^{1} + \frac{e^{-j\omega t}}{-j\omega} \Big|_{1}^{2}$$

$$= \frac{1}{-j\omega}\left\{ e^{-j\omega} - e^{2j\omega} + 2e^{-j\omega} - 2e^{j\omega} + e^{-2j\omega} - e^{-j\omega} \right\}$$

$$= \frac{1}{-j\omega}\left\{ 2e^{-j\omega} - 2e^{j\omega} + e^{-2j\omega} - e^{2j\omega} \right\}$$

$$= \frac{1}{+j\omega}\left\{ 2(e^{j\omega} - e^{-j\omega}) + e^{2j\omega} - e^{-2j\omega} \right\}$$

$$= \frac{1}{\omega}\left\{ 4\sin\omega + 2\sin 2\omega \right\}$$

$$= 4\frac{\sin\omega}{\omega} + 2\frac{\sin 2\omega}{\omega} = 4\sin c\omega + 4\sin c2\omega$$

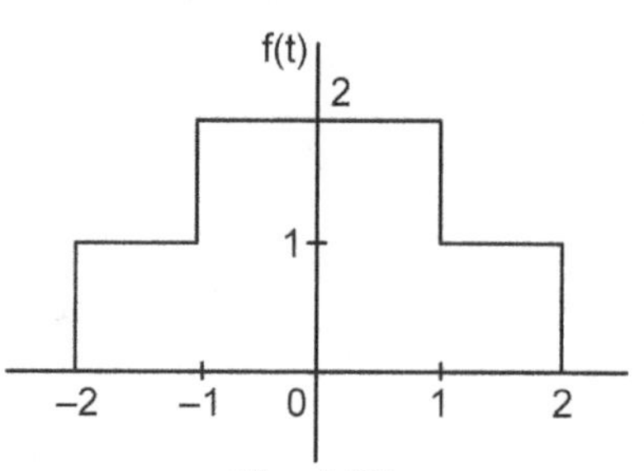

Fig. 4.32

8. Find the fourier transform of trapezoidal pulse shown in figure   May 2005

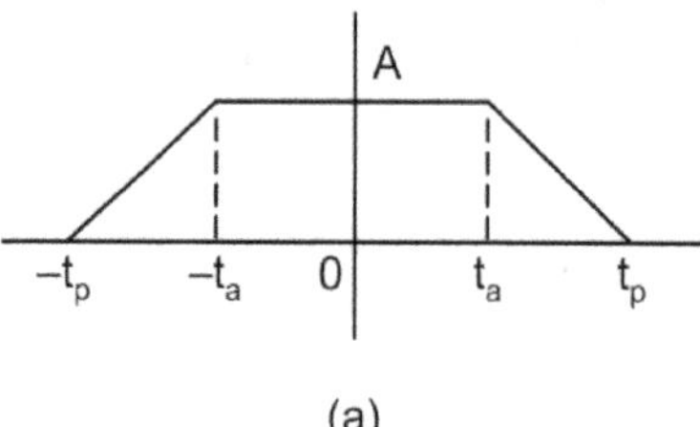

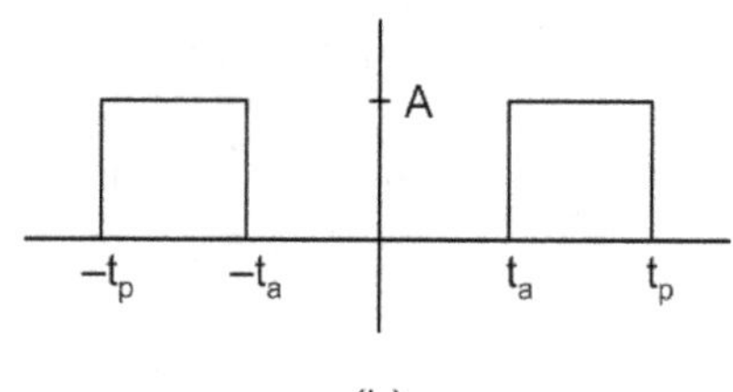

(a)   (b)

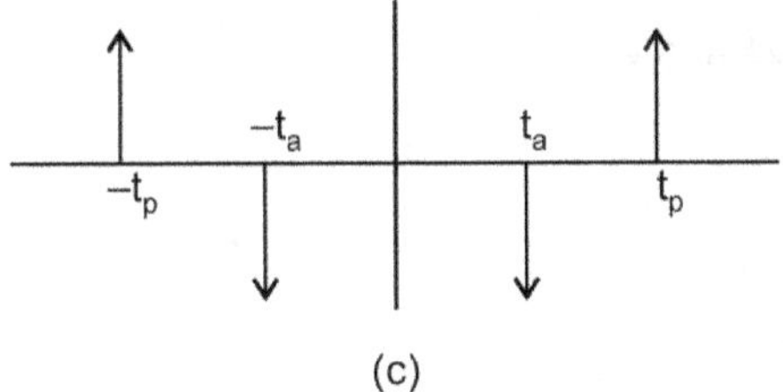

(c)

Fig. 4.33

Applying first order derivate we will have two square pulse of opposite magnitude.

The second order derivative has impulses as shown in figure.

$$\therefore \quad \frac{d^2}{dt^2} f(t) = k\left\{\delta(t+t_p) - \delta(t+t_a) - \delta(t-t_a) + \delta(t-t_p)\right\}$$

where $\quad k = \dfrac{A}{t_p - t_a}$

$$\therefore \quad \mathcal{F}\left\{\frac{d^2 f(t)}{dT^2}\right\} = k\left\{e^{j\omega t_p} - e^{j\omega t_a} - e^{-j\omega t_a} + e^{-j\omega t_p}\right\}$$

$$F(\omega) = \frac{1}{(j\omega)^2} 2k\left\{\cos\omega t_p - \cos\omega t_a\right\}$$

$$= k\left\{\frac{t_p^2 \sin^2 \frac{\omega t_p}{2}}{\left(\frac{\omega t_p}{2}\right)^2} - \frac{t_a^2 \sin^2 \frac{\omega t_a}{2}}{\left(\frac{\omega t_a}{2}\right)^2}\right\}$$

$$= k t_p^2 \sin c^2\left(\frac{\omega t_p}{2\pi}\right) - k a^2 \sin c^2\left(\frac{\omega t_a}{2\pi}\right)$$

9.   Sketch the inverse fourier transform of the waveform shown.   Sep 2007

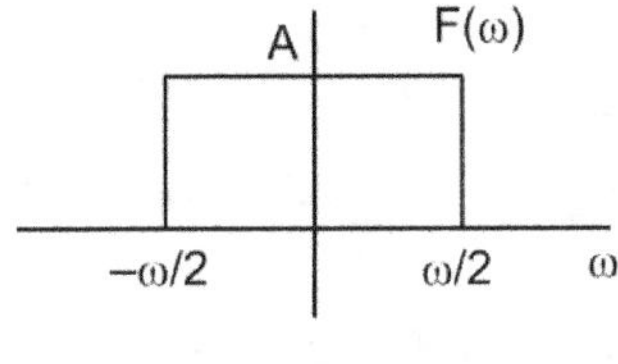

(a)

Fourier transform of gate function

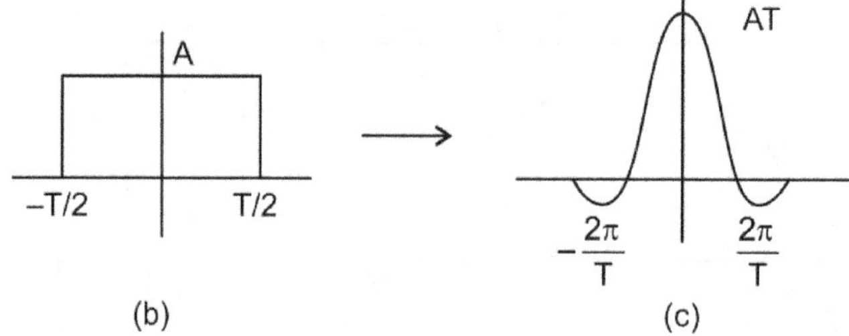

(b)   (c)

Fourier transform of sin c function must be gate

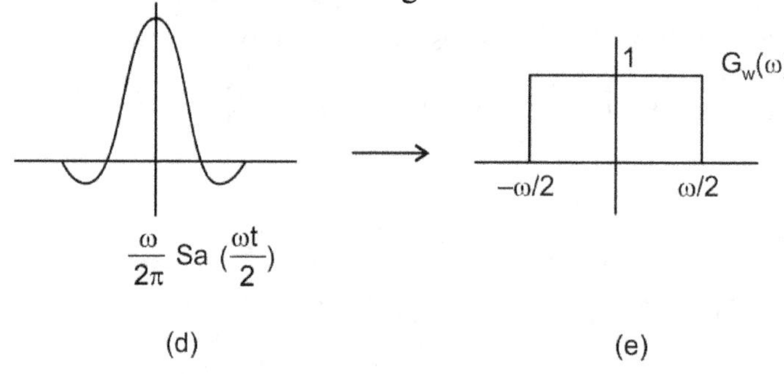

$$\frac{\omega}{2\pi} \, Sa\left(\frac{\omega t}{2}\right)$$

(d)   (e)

Fig. 4.34

$\therefore$ Inverse fourier transform of gate function $G_w(\omega)$ is sin c ()

EXERCISE PROBLEMS-1

1.   Identify the appropriate fourier representation for each of the following signals.

(a) $x[n] = \left(\frac{1}{2}\right)^n uTn$ ***Ans:*** DTFT

(b) $x(t) = 1 - \cos 2\pi t + \sin 3\pi t$ ***Ans:*** FS

(c) $x(t) = e^{-t} \cos(2\pi t) u(t)$ ***Ans:*** FT

(d) $x[n] = \sum_{n=-\alpha}^{\alpha} \delta(n-20m) - 2\delta(n-2-20m)$ ***Ans:*** DTFS

2.   Find the DTFT   of the following time domain signals.

(a) $x[n] = \begin{cases} 2^n & 0 \le n \le 9 \\ 0 & \text{otherwise} \end{cases}$ (b) $x[n] = a^{|n|} \quad |a| < |$

Ans: $x(e^{jw}) = \dfrac{1 - 2^{10} e^{-j10\omega}}{1 - 2e^{-j\omega}}$ ***Ans:*** $x(e^{j\omega}) = \dfrac{1 - a^2}{1 - 2a\cos\omega + a^2}$

3. Find the inverse DTFT of the following frequency domain signals.

(a) $x(e^{j\omega}) = 2\cos(2\omega)$

(b) $x(e^{j\omega}) = \begin{matrix} e^{-j4\omega} & \frac{\pi}{2} < |\omega| \le \pi \\ 0 & \text{otherwise} \end{matrix}$

Ans: $x[n] = \begin{matrix} 1 & n \ne \pm 2 \\ 0 & \text{otherwise} \end{matrix}$

Ans: $x[n] = \delta[n-4] - \dfrac{\sin(\pi(n-4)/2)}{\pi(n-4)}$

4. Find the FT of the following signals

(a) $x(t) = e^{2t}u(-t)$

(b) $x(t) = e^{-|t|}$

Ans: $x(j\omega) = \dfrac{1}{j\omega^{-2}}$

Ans: $x(j\omega) = \dfrac{2}{1+\omega^2}$

(c) $x(t) = e^{-2t}u(t-1)$

Ans: $e^{-(j\omega+2)/j\omega+2}$

5. Find the inverse FT of the following spectra.

(a) $x(j\omega) = \begin{matrix} 2\cos\omega & |\omega| < \pi \\ 0 & |\omega| > \pi \end{matrix}$

(b) $x(j\omega) = 3\delta(\omega-4)$

Ans: $x(t) = \dfrac{\sin(\pi(t+1))}{\pi(t+1)} + \dfrac{\sin(\pi(t-1))}{\pi(t-1)}$

Ans: $x(t) = \left(\dfrac{3}{2\pi}\right)e^{j+t}$

(c) $x(j\omega) = \pi e^{-|\omega|}$

Ans: $x(t) = \dfrac{1}{1+t^2}$

6. Find the fourier representation of following time domain signals.

(a) $x(t) = e^{-2t}u(t-3)$

(b) $y[n] = \dfrac{\sin(\pi(n+2)/3)}{\pi(n+2)}$

Ans: $x(j\omega) = e^{-6}e^{-3j\omega/j\omega+2}$

Ans: $y(e^{j\omega}) = e^{2j\omega} \quad |\omega| \dfrac{\pi}{3} 0 \dfrac{\pi}{3} < |\omega| \le \pi$

7. Find the time domain signals of following fourier representations.

(a) $x(j\omega) = \dfrac{e^{4j\omega}}{(2+j\omega)^2}$

(b) $y(k) = \dfrac{e^{-jk4\pi/5}}{10}$

Ans: $x(t) = (t+4)e^{-2(t+4)}u^{(t+4)}$

Ans: $y(n) = \displaystyle\sum_{P=-\alpha}^{\alpha} \delta(n-4-10P)$

8. Find the fourier representations of following time domain signals

(a) $x[n] = ne^{j\pi/8n}\ a^{n-3}\ u[n-3]$

(b) $x(t) = (t-2)\dfrac{d}{dt}\left[e^{-j5t}\cdot e^{-2|t-3|}\right]$

Ans: $x(e^{j\omega}) = jd/d\omega\left\{\dfrac{e^{-13(\omega-\pi/8)}}{1-a\,e^{-1(\omega-\pi/8)}}\right\}$

Ans: $x(j\omega) = \dfrac{-8j\omega e^{-j3(\omega+5)}}{4+(\omega+5)^2} + j\dfrac{d}{d\omega}\left\{\dfrac{4j\omega e^{-j3(\omega+5)}}{4+(\omega+5)^2}\right\}$

EXERCISE PROBLEMS-2

1. Find fourier transform of the below signals.

(a) $x(t) = e^{2t}\,u(-t)$

Ans: $x(j\omega) = \dfrac{-1}{(j\omega-2)}$

(b) $x(t) = e^{-|t|}$

Ans: $x(j\omega) = \dfrac{2}{(1+\omega^2)}$

(c) $x(t) = e^{-2t}\,u(t-1)$

Ans: $x(j\omega) = \dfrac{e^{-(j\omega+2)}}{(j\omega+2)}$

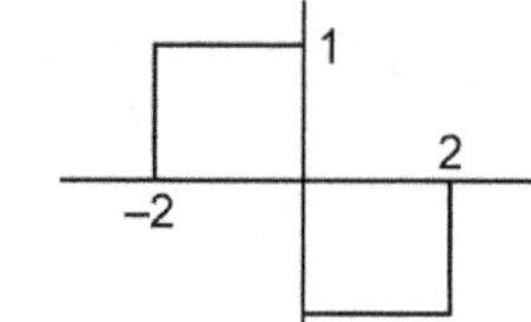

(d) $x(t) =$

Ans: $x(j\omega) = \dfrac{2j(1-\cos 2\omega)}{\omega}$

Hint for the above problems

$$x(j\omega) = \int_{-\alpha}^{\alpha} x(t)e^{-j\omega t}\,dt$$

2. Find the fourier transform of the below signals.

(a) $e^{-2(t-1)}\,u(t-1)$

Ans: $\dfrac{e^{-j\omega}}{(2+j\omega)}$

(b) $\delta(t+1) + \delta(t-1)$

Ans: $2\cos\omega$

(c) $\dfrac{d}{dt}\{u(-2-t)+u(t-2)\}$

Ans: $2j\sin 2\omega$

(d) $\sin\left(2\pi t + \pi/4\right)$

Ans: $\dfrac{\pi}{j}\left(e^{j\pi/4}\delta(\omega-2\pi) - e^{-j\pi/4}\delta(\omega+2\pi)\right)$

3. Compute the fourier transform of each of the following signals using FT properties.

   (a) $x_1(t) = x(1-t) + x(-1-t)$ **Ans:** $x_1(j\omega) = 2 \times (-j\omega)\cos\omega$

   (b) $x_2(t) = x(3t-6)$ **Ans:** $x_2(j\omega) = \dfrac{1}{3}e^{-j2\omega} \times \left(j\dfrac{\omega}{3}\right)$

   (c) $x_3(t) = \dfrac{d^2}{dt^2}x(t-1)$ **Ans:** $x_3(j\omega) = \omega^2\, e^{-j\omega} \times (j\omega)$

Hint: Use FT properties to solve the above problems.

4. Find the FT of below signals using FT properties.

   (a) $x(t) = \dfrac{1}{2}x_1(t-2.5) + x_2(t-2.5)$ **Ans:** $x(j\omega) = e^{-j5\omega/2}\left\{\dfrac{\sin\omega/2 + 2\sin 3\omega/2}{\omega}\right\}$

Hint: Use linearity and time shifting property

5. Find the FT of $x(t) = u(t)$

                                    Ans: $x(j\omega) = \dfrac{1}{j\omega} + \pi\delta(\omega)$

Hint: Use differentiation and integration property.

6. Find the FT of signal $x(t)$ shown

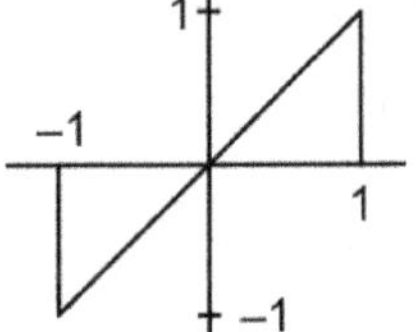

                          Ans: $x(j\omega) = \dfrac{2\sin\omega}{j\omega^2} - \dfrac{2\cos\omega}{j\omega}$

Hint: Use differentiation and integration property.

7. Find the inverse FT of below spectrums

   (a) $x(j\omega) = \begin{cases} 1 & -\omega_0 < \omega < \omega_0 \\ 0 & |\omega| > \omega_0 \end{cases}$ **Ans:** $x(t) = \dfrac{\omega_0}{\pi}\sin c\left(\dfrac{\omega_0 t}{\pi}\right)$

   (b) $x(j\omega) = 2\pi\delta(\omega)$ **Ans:** $x(t) = 1$

   (c) $x(j\omega) = 3\delta(\omega - 4)$ **Ans:** $\dfrac{3}{2\pi}e^{-j4t}$

   (d) $x(j\omega) = \pi e^{-|\omega|}$ **Ans:** $x(t) = \dfrac{1}{1+t^2}$

   (e) $x(j\omega) = 2\cos\omega$ $|\omega| < \pi$ **Ans:** $x(t) = \dfrac{\sin\pi(t+1)}{\pi(t+1)} + \dfrac{\sin\pi(t-1)}{\pi(t-1)}$

8. Use linearity property to find FT of below signals

    (a) $x(t) = 2e^{-t}u(t) - 3e^{-2t}u(t)$ **Ans:** $x(j\omega) = \dfrac{2}{j\omega+1} - \dfrac{3}{j\omega+2}$

    (b) $z(t) = \dfrac{3}{2}x(t) + \dfrac{1}{2}y(t)$ **Ans:** $z(x) = \dfrac{3}{2k\pi}\sin\dfrac{k\pi}{4} + \dfrac{1}{2k\pi}\dfrac{\sin k\pi}{2}$

9. Using convolution property find FT of below signals.

    (a) $x(t) = 3e^{-t}u(t)$ and $h(t) = 2e^{-2t}u(t)$ **Ans:** $y(j\omega) = \dfrac{2}{(j\omega+2)} \cdot \dfrac{3}{(j\omega+1)}$

    (b) $x(t) = \dfrac{3}{\pi t}\sin 2\pi t$ and $h(t) = \dfrac{1}{\pi t}\sin \pi t$ **Ans:** $y(t) = \dfrac{3}{\pi t}\sin \pi t$

10. Use differentiation property to find FT of the below signals

    (a) $x(t)\;\dfrac{d}{dt}e^{-2|t|}$ **Ans:** $x(j\omega) = \dfrac{4j\omega}{4+\omega^2}$

    (b) $x(t) = \dfrac{d}{dt}2te^{-2t}u(t)$ **Ans:** $x(j\omega) = \dfrac{2j\omega}{(2+j\omega)^2}$

11. Use frequency differentiation property to find FT of below signals.

    (a) $x(t) = te^{at}u(t)$ **Ans:** $x(j\omega) = \dfrac{1}{(a+j\omega)^2}$

    (b) $x(n) = (n+1)\,a^n\,u(n)$ **Ans:** $x(j\omega) = \dfrac{1}{\left(1-ae^{-j\omega}\right)^2}$

EXERCISE PROBLEMS-3

1. Obtain the fourier transforms of the following signals
    (a) $Ae^{-\alpha t}u(t)$ (b) $Ae^{-\alpha t}u(t) - Ae^{\alpha t}u(-t)$
    (c) $t^2 u(t)u(1-t)$ (d) $e^{-\alpha t}u(t)u(1-t)$

2. Obtain the fourier transforms of the following signals.

    (a) $\dfrac{1}{2}\left\{\delta(t+1) + \delta(t+\tfrac{1}{2}) + \delta(t-\tfrac{1}{2}) + \delta(t-1)\right\}$

    (b) $\sin c(t)u(t)$

    (c) $\sin c(t)\,\mathrm{sgn}(t)$

3. Use superposition and time delay theorems with the pair $\pi(t/T) \leftrightarrow T \sin cfT$ to obtain fourier transform for the signals shown.

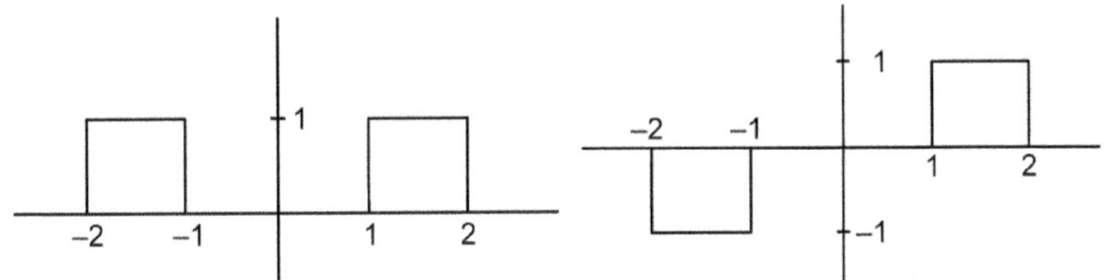

4. Use duality theorem to find the signal whose transform is

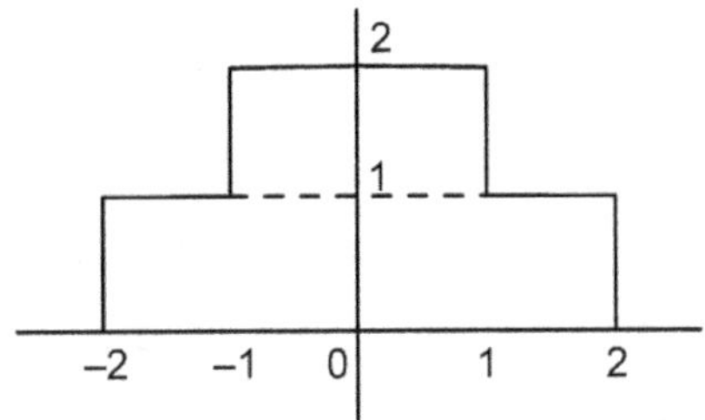

5. Obtain the inverse fourier transform of

$$x(f) = 10\frac{\sin c2f}{3 + j2\pi f}$$

6. Using the fourier transform integral

$$x(t) = te^{-\alpha t}u(t)$$

7. Obtain the fourier transforms of the following signals

$$x(t) = \frac{1}{2}\delta(t+1) + \delta(t+\tfrac{1}{2}) + \delta(t-\tfrac{1}{2}) + \delta(t-1)$$

8. Use the differentiation theorem and the transform pair $\delta(t - t_0) \ll \exp(-j2\pi ft_0)$ to find the fourier transforms of the signals shown.

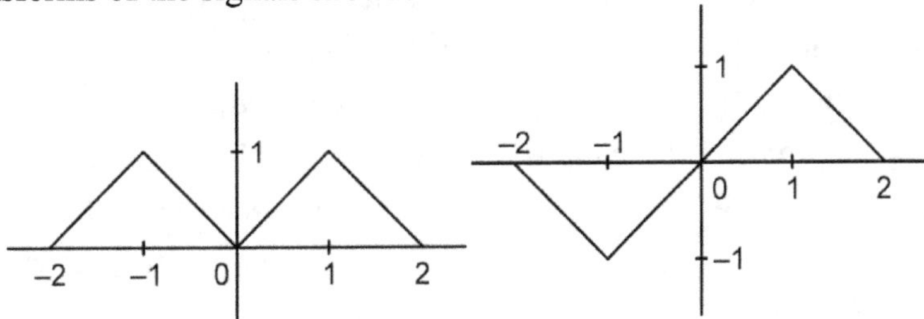

9. Find the fourier transform of

(a) Gaussian signal $x(t) = e^{-\alpha t^2}$

(b) Gaussian modulated cosine signal $x(t) = e^{-\alpha t^2}\cos \omega_0 t$

10. By using partial fraction expansion find the inverse fourier transform of

(a) $x(j\omega) = \dfrac{5j\omega + 12}{(j\omega)^2 + 5(j\omega) + 6}$

(b) $x(j\omega) = \dfrac{1 + 2j\omega}{(j\omega + 2)^2}$

OBJECTIVE QUESTIONS

1. Fourier transform of unit impulse function is ----------------------------

2. $\mathcal{F}\{sgn(t)\}$ is ----------------------------------

3. $\mathcal{F}\{\pi(t)\}$ is --------------------

4. Fourier Transform of $\delta(t)$ is ----------------

5. Fourier Transform of triangular pulse is ----------------------

6. Fourier Transform of $x(t - t_0)$ is ----------------------------

7. Fourier Transform of $\dfrac{d}{dt}x(t)$ is ----------------

8. Inverse Fourier Transform of $\delta(\omega)$ is ----------------

9. If the periodic signal f(t) is having even symmetry then fourier transform ------------

GATE QUESTIONS ON FT

1. Fourier transform of real value d time signal has   GATE 1996
    (a) Odd symmetry   (b)   Even symmetry
    (c) Conjugate symmetry   (d)   None   *Ans:* (c)

2. A signal x(t) has FT of X(w). If x(t) is real and odd, then X(w) is   GATE 1996
    (a) Real and even   (b)   Imaginary and odd
    (c) Imaginary and even   (d)   Real and odd   *Ans:* (b)

3. FT of conjugate symmetric function is always   GATE 2004
    (a) Imaginary   (b)   Conjugate symmetric
    (c) Real   (d)   Conjugate anti symmetric *Ans:* (c)

4. FT of $x(t) = e^{-3t^2}$ is   GATE 2000

    (a) $A\,e^{-B|f|}$ (b) $A + B\,|f|^2$

    (c) $A\,e^{-Bf}$ (d) $A\,e^{-Bf^2}$ *Ans:* (d)

5. A function f(t) has fourier transform g(w). Then fourier transform of g (t) is

   GATE 1997

   (a) $\dfrac{1}{2\pi}f(w)$

   (b) $\dfrac{1}{2\pi}f(-w)$

   (c) $2\pi f(-w)$

   (d) $2\pi f(w)$

   Ans: (c)

6. Fourier transform of $\dfrac{d\,x(t)}{dt}$ is

   GATE 1998

   (a) $\dfrac{dx(f)}{dt}$

   (b) $j2\pi f \times (f)$

   (c) $jf \times (f)$

   (d) $\dfrac{X(f)}{jf}$

   Ans: (b)

7. X(f) is fourier transform of x(t), what is X(f) units

   GATE 1998

   (a) Volts

   (b) Volt - sec

   (c) Volt/sec

   (d) Volt^2

   Ans: (b)

8. If f(t) has energy E, then f(2t) has an energy

   GATE 2001

   (a) E

   (b) 2E

   (c) E/2

   (d) 4E

   Ans: (c)

9. Fourier transform of $e^{-t}u(t)$ is $\dfrac{1}{1+j2\pi f}$, therefore fourier transform of $\dfrac{1}{1+j2\pi t}$

   GATE 2002

   (a) $e^{f}u(f)$

   (b) $e^{-f}u(f)$

   (c) $e^{f}u(-f)$

   (d) $e^{f}u(-)$

   Ans: (c)

10. What is the fourier transform of $x(5t-3)$ in terms of X (jw)

    GATE 2006

    (a) $\dfrac{1}{5}e^{-\frac{j3\omega}{5}}\times(\dfrac{j\omega}{5})$

    (b) $\dfrac{1}{5}e^{-\frac{j3\omega}{5}}\times\left(\dfrac{j\omega}{5}\right)$

    (c) $\dfrac{1}{5}e^{-j3\omega}\times\left(\dfrac{j\omega}{5}\right)$

    (d) $\dfrac{1}{5}e^{j3\omega}\times\left(\dfrac{j\omega}{5}\right)$

    Ans: (a)

11. Fourier transform of x(t–2) and x(t/2) are   GATE 1997

   (a) $x(f)e^{-j(4\pi f)}, 2x(2f)$

   (b) $x(f)e^{-j2\pi f}, x(2f)$

   (c) $x(f)e^{-j\pi f}, 2x(f)$

   (d) None   *Ans:* (a)

12. If x(t) and y(t) has fourier transforms X(f) and Y(f) then Y(f) in terms of X(f) is

                                                                  GATE 2004

   (a) $-\dfrac{1}{2}\times(f/2)e^{-j2\pi f}$

   (b) $-\dfrac{1}{2}\times(f/2)e^{j2\pi f}$

   (c) $-X(f/2)e^{j2\pi f}$

   (d) $-x(f/2)e^{-j2\pi f}$ *Ans:* (b)

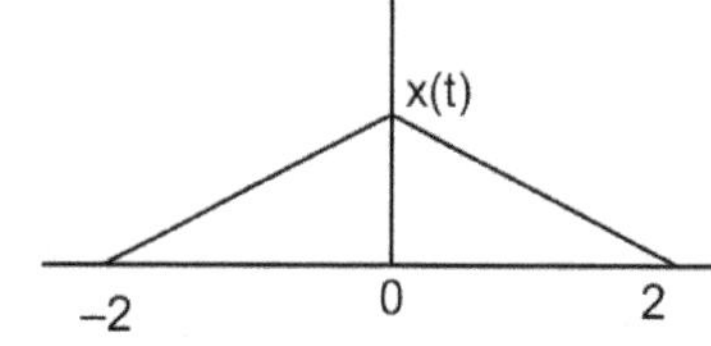

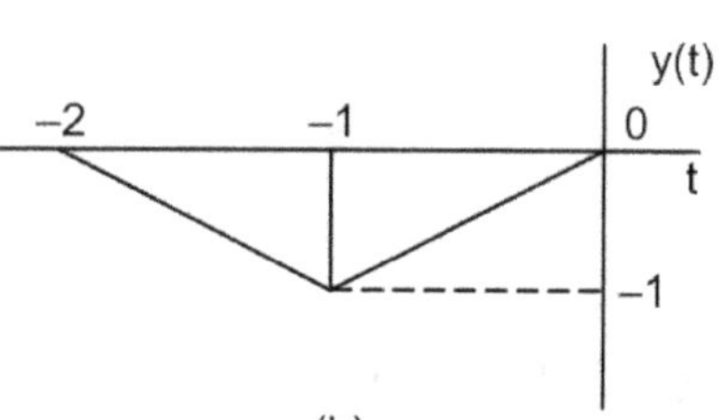

(a)   (b)

13. Inverse fourier transform of $\times(3f+2)$ is   GATE 2005

   (a) $\dfrac{1}{2}x\left(\dfrac{1}{2}\right)e^{j3\pi t}$

   (b) $\dfrac{1}{3}x\left(\dfrac{t}{3}\right)e^{-j4\pi/3}$

   (c) $3x(3t)e^{-j4\pi t}$

   (d) $x(3t+2)$ *Ans:* (b)

14. Two angular frequencies where fourier transform becomes zero are   GATE 2008

   (a) $\pi, 2\pi$

   (b) $0, \pi$

   (c) $\dfrac{\pi}{2}, \dfrac{3\pi}{2}$

   (d) $2\pi, 3\pi$ *Ans:* (a)

15. If fourier transform of h(t) is H(jω), where H(jω) = (2 cosω) (sin 2ω)/ω, then h(0) is   GATE 2012

   (a) 1/4

   (b) 1/2

   (c) 1

   (d) 2   *Ans:* (c)

16. For a function g(t) if is given that $\int_{-\alpha}^{\alpha} g(t)e^{-j\omega t}dt = \omega e^{-2\omega t}$ for any real value ∞. If y(t) = $\int_{-\alpha}^{t} g(\tau)d\tau$, then $\int_{-\alpha}^{\alpha} y(t)dt$ is

   (a) 0

   (b) –j

   (c) $-j\big/2$

   (d) $j\big/2$

17. A fourier transform pair is $\left(\dfrac{2}{3}\right)^n u[n+3] \overset{FT}{\Longleftrightarrow} \dfrac{Ae^{-j\omega f}}{1-\dfrac{2}{3}e^{-j2\pi f}}$ the values of A is _____

GATE 2014

(a) 3.36 to 3.39
(b) 3.0 to 4.0
(c) 2.0 to 3.0
(d) None

Ans: (a)

Fourier Transform Pairs

$$g(t) \overset{f}{\Longleftrightarrow} G(f) = \int_{-\alpha}^{\alpha} g(t)e^{-i2\pi ft}\,dt \qquad G(\omega) = \int_{-\alpha}^{\alpha} g(t)e^{-i\omega f}\,dt$$

1.	Square Pulse		
	$\text{rect}_T(t)$	$T\text{sinc}(f\,T)$	$T\sin C\left(\dfrac{\omega T}{2\pi}\right)$
2.	Triangle function		
	$\wedge(t)$	$\text{Sin } C^2\,(f)$	$\text{sinC}^2\left(\dfrac{\omega}{2\pi}\right)$
3.	Gaussian		
	$e^{-\pi t^2}$	$e^{-\pi f^2}$	$e^{-\left(\omega^2/4\pi\right)}$
4.	Constant		
	C	$C\delta(f)$	$2\Pi C\delta(\omega)$
5.	Dirac Delta		
	$\delta(t-a)$	$e^{-i2\pi fa}$	$e^{-i\omega a}$
6.	$\cos(2\pi At)$	$\dfrac{1}{2}\{\delta(f-A)+\delta(f+A)\}$	$\pi\{\delta(\omega-2\pi A)+\delta(\omega+2\pi A\}$
7.	$\sin(2\pi At)$	$\dfrac{1}{2}\{\delta(f-A)-\delta(f+A)\}$	$\dfrac{\pi}{i}\{\delta(\omega-2\omega A)-\delta(\omega+2\omega\pi A\}$
8.	Step function		
	$u(t)$	$\dfrac{1}{2\pi i f}+\dfrac{\delta(f)}{2}$	$\dfrac{1}{i\omega}+\pi\delta(\omega)$
9.	Signum function		
	$\text{Sgn}(t)$	$\dfrac{1}{\pi i f}$	$\dfrac{2}{i\omega}$
10.	$e^{-|a|t}u(t)$	$\dfrac{1}{|a|+2\pi i f}$	$\dfrac{1}{|a|+i\omega}$

11.			
	$e^{|a|t}u(t)$	$\dfrac{1}{|a| - 2\pi i f}$	$\dfrac{1}{|a| - i\omega}$
12.	$e^{-|at|}$	$\dfrac{2|a|}{a^2 + (2\pi f)^2}$	$\dfrac{2|a|}{a^2 + \omega^2}$

SUMMARY OF CHAPTER

1. Fourier transform of x(t) is defined by $X(\omega) = \int_{-\alpha}^{\alpha} x(t)e^{-j\omega f}\,dt$

2. Inverse Fourier transform of X (w) is defined by $x(t) = \dfrac{1}{2\pi}\int_{-\alpha}^{\alpha} X(\omega)e^{j\omega f}\,d\omega$

3. Parsvels theorem stats that $E_x = \int_{-\alpha}^{\alpha} |x(t)|^2\,dt = \dfrac{1}{2\pi}\int_{-\alpha}^{\alpha} |X(\omega)^2|\,d\omega$

4. The more a signal is localized in one domain, the less it is localized in other domain
5. Convolution and multiplications of a function are dual in time and frequency domains
6. Fourier transform of a periodic signal consists only of impulses
7. Signal energy is conserved in the Fourier transformation process
8. Fourier transform of a periodic signal consists only of impulses

CHAPTER **5**

Linear Time Invariant Systems

Objectives of the chapter

In this chapter different types of systems, LTI system response, signal transmission through linear response, Ideal filters, impulse response of filter, poly wiener criteria are discussed.

5.1 INTRODUCTION

System: A system is needed to extract additional information from the signals. A system may be a physical system or an algorithm that computes desired output. A system is an integrated physical environment that generates a response to a given input signal, governed by laws of inter connection.

Ex: Amplifier

It is a collection of individual elements which are orderly arranged for a specific purpose.

Every system accepts inputs such as voltage, current, force, pressure etc., and produces an output in response to input. Based upon input and output they are broadly characterized.

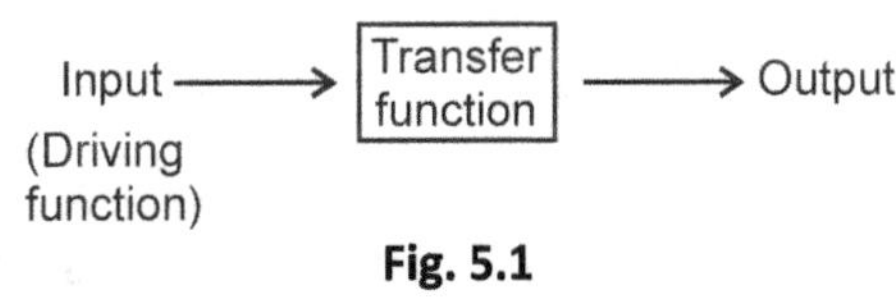

Fig. 5.1

Study of a system involves mathematical modeling, analysis and design.

5.2 TYPES OF SYSTEMS

1. Continuous time and discrete time systems
2. Lumped parameter and distributed systems
3. Static and dynamic systems
4. Causal and Non Causal systems
5. Linear and Non linear systems
6. Time invariant and time variant systems
7. Stable and unstable systems
8. Invertible and Non invertible systems
9. Analog and digital systems

1. **Continuous time system:** If the input/output are continuous time signals, then it is said to be continuous time system.

$$y\,(t) = \; T\{x(t)\}$$

2. **Discrete time system:** Systems, whose inputs and outputs are discrete signals, are discrete time systems.

$$T\{x(n)\} = y(n)$$

3. **Lumped parameter systems:** If the energy stored or distributed is independent of time and space and lumped at one point in space, then it is said to be lumped parameter system.

4. **Distributed parameter system:** If the energy stored or dissipated depends on time and space and distributed over space, then it is said to be distributed parameter systems.

   Ex: transmission lines,

5. **Static and dynamic Systems:** If output depends on present input only then it is static system.

$$y\,(n) = \; T\{x(n)\}$$

$$y\,(t) = \; T\{x(t)\}$$

These are also known as memory less systems.

If output depends on present input, past or future inputs, then it is called as dynamic system.

$$y\,(t) = \; \frac{d}{dt}x(t)$$

These are also known as systems with memory.

6. **Causal and Non Causal Systems:** If output depends on present and past, values of input $x(t)$ then it is said to be causal. It doesn't depend on future values.

$$T\{x(t) + x(t-1)\} = y(t)$$

$$y\,(n) = \; T[x(n) + x(n-1)]$$

If output depends on both past, present and future then it is called as non causal.

$$y\,(t) = \; T\{x(t) + x(t+1)\}$$

7. **Linear and Non Linear Systems:** If a system satisfies super position principle it is said to be linear.

$$T\{ax_1(t) + bx_2(t)\} = aT\{x_1(t)\} + bT\{x_2(t)\}$$

If system doesn't satisfies super position then it is said to be non linear.

8. **Time variant and time invariant System:** If input/output characteristics doesn't change with time, then it is said to be time invariant.

$$T\{x(t)\} = y(t)$$

$$T\{x(t-T)\} = y(t-T) \ \text{ if it is discrete } \ T\{x(n-n_0)\} = y(n-n_0)$$

Otherwise it is called time invariant.

9. **Stable/unstable system:** If input/output are bounded within a given limits then it is stable system, otherwise it is unstable

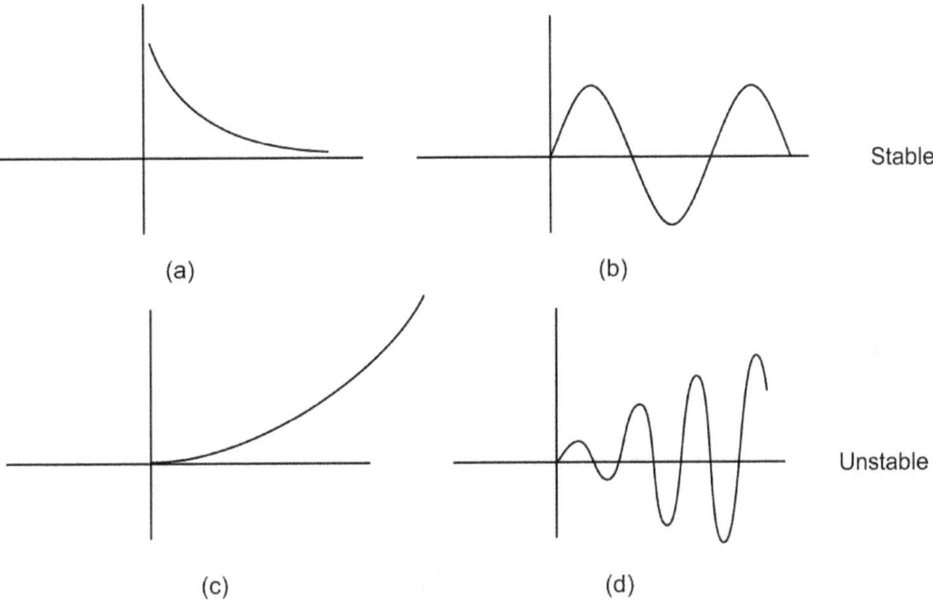

Fig. 5.2

10. **Invertible and Non invertible system:** If there is one to one mapping between input and output, then the systems are invertible.

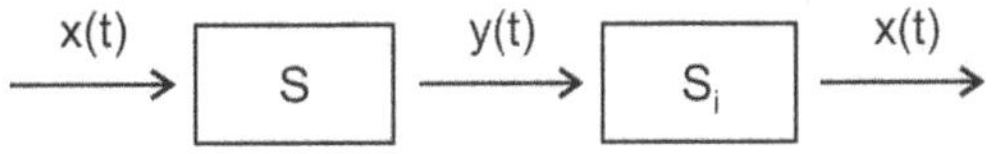

Fig. 5.3

If there is a system S to produce output y(t) for input x(t), then there exists another system S_i which produces x(t) from y(t).

5.3 LINEAR TIME INVARIANT SYSTEM

The LTI system investigates the response of a linear and time invariant system to an arbitrary input signal.

Linearity means that the relationship between the input and output of the system is a linear map. If input $x_1(t)$ produces response $y_1(t)$ and input $x_2(t)$ produces response $y_2(t)$ then the scaled and summed input.

$a_1 x_1(t) + a_2 x_2(t)$ produce the scaled and summed response $a_1 y_1(t) + a_2 y_2(t)$ where a_1 and a_2 are real scalars.

Time invariance means that whether we apply an input to the system now or T seconds from now, the output will be identical except for a time delay of the T seconds.

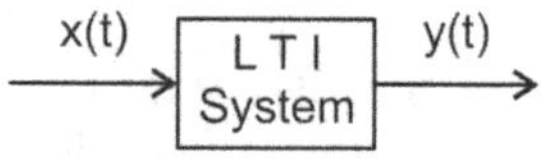

If y(t) is output of x(t)

Then y (t–T) is output of x(t–T)

$$ax_1(t-t_0) + bx_2(t-t_0) \rightarrow ay_1(t-t_0) + by_2(t-t_0)$$

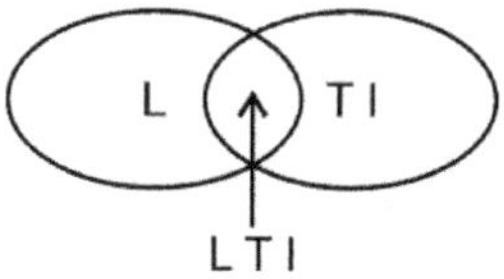

Fig. 5.4

Example:

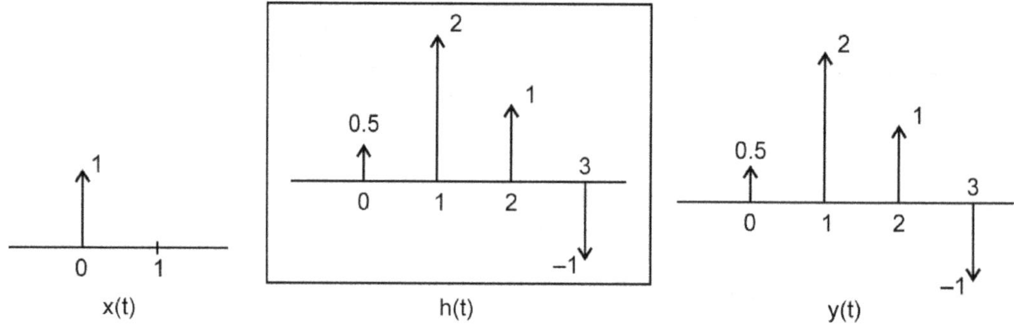

Fig. 5.5

we can write at any moment

$$y\,(n) = x(0)h(n) + x(1)h(n-1) + x(2)h(n-2) + x(3)h(n-3)$$

$$y\,(n) = \sum_{k=-\alpha}^{\alpha} x[k]h[n-k]$$

5.3.1 LINEAR SYSTEM

A linear system exhibits the additive property. It must satisfy homogeneity or scaling property. These two can be combined into the property of super position. A linear system may have multiple inputs and multiple outputs (MIMO).

A linear system output $(t \geq 0)$ depends on zero input response and zero state response.

$$\text{Total response} = \text{Zero input response} + \text{zero state response}$$

5.3.2 CAUSAL SYSTEM

Any real time system is causal. Any physically realizable or nonanticipatory system whose output at any time to or n_0 depends only on the values of the input at the present time and in the past. All memory less system are causal.

If $\qquad x_1(n) = x_2(n) \; n \leq n_0$

$\qquad\qquad y_1(n) = y_2(n) \; n \leq n_0$

Non Causal Systems are

1. Realizable when the independent variable is something other than time i.e., space.

2. Study upper bound on the performance of a causal system.

SOLVED PROBLEMS - 1

EXERCISE

1. Check the following systems for linearity

(a) $\qquad y(n) = e^{x(n)}$

   If input $x(n) = 0 \qquad$ output $y(n) = e^0 = 1$

   $\therefore$ system is non linear, because it produces non zero output when input is zero.

(b) $\qquad y(n) = 3x(n+3)$

   If $x(n+3) = 0$ then $y(n) = 0$

   Linear as output is 0 input is 0.

   Let

   $\qquad y_1(n) = 3x_1(n+3)$

   $\qquad y_2(n) = 3x_2(n+3)$

   $\qquad y^1(n) = y_1(n) + y_2(n) = 3\big[x_1(n+3) + x_2(n+3)\big]$

   $\qquad\qquad = H\big\{x_1(n) + x_2(n)\big\}$

   $\qquad\qquad = 3x_1(n+3) + 3x_2(n+3)$

   Hence the system is linear

(c) $\qquad y(n) = \cos(x(n))$

   If $x(n) = 0; \; y(n) = \cos 0 = 1$

   $\qquad y^1(n) = y_1(n) + y_2(n) = \cos(x_1(n)) + \cos(x_2(n))$

$$y^{11}(n) = \cos\left[x_1(n) + x_2(n)\right] = \cos\left(x_1(n)\right)\cos\left(x_2(n)\right) + \sin\left(x_1(n)\right)\sin\left(x_2(n)\right)$$

$$y^1(n) \neq y^{11}(n)$$

The system is non linear

(d) $\qquad y(n) = 5n\left[x(n)\right]^2$

$$y^1(n) = y_1(n) + y_2(n) = 5n\,x_1(n)^2 + 5n\,x_2(n)^2$$

$$y^{11}(n) = 5n\left\{x_1(n) + x_2(n)\right\}^2 = 5n\left\{x_1^2(n) + x_2^2(n) + 2x_1(n)x_2(n)\right\}$$

$$y^1(n) \neq y^{11}(n)$$

The system is non linear

2. Test time invariance of below systems.

(a) $\qquad y(n) = x(n+1)e^{-n}$

$$y(n-1) = x\left[n+1-1\right]e^{-(n-1)} = x(n)e^{-(n-1)}$$

Thus it is time invariant

(b) $\qquad y(n) = K\left\{x(n+1) - x(n)\right\}$

$$y(n-1) = K\left\{x(n) - x(n-1)\right\}$$

Thus it is time invariant

(c) $\qquad y(t) = x(t)\cos 50\pi t$

$$y(t-t_0) = x(t-t_0)\cos 50\pi(t-t_0)$$

$$x(t-t_0)\cos 50\pi t \neq y(t-t_0)$$

$\therefore$ it is time variant

(d) $\qquad y(t) = x(2t)$

$$x(2t-t_0) = y^1(t-t_0)$$

$$y^{11}(t-t_0) = x\left(2(t-t_0)\right)$$

$\therefore \qquad y^{11}(t-t_0) \neq y^1(t-t_0)$

$\therefore$ it is time variant

(e) $\qquad y(n) = x^2(n-1)$

$$y^1(n-K) = x^2(n-1-K)$$

$$y^{11}(n-K) = x^2(n-K-1)$$

$$\therefore \quad y^1(n-K) = y^{11}(n-K)$$

$\therefore$ it is time invariant

3. Check the following systems are causal or non-causal.

(a) $y(t) = \text{odd}\{x(t)\}$

$$= \frac{1}{2}\{x(t) - x(-t)\}$$

$$t = -1$$

$$y(-1) = \frac{1}{2}\{x(-1) - x(1)\}$$

Since output depends on future values for t is negative, it is non causal

(b) $y(n) = \displaystyle\sum_{K=-\alpha}^{n+1} x(k)$

for n = 0

$$y(0) = \sum_{K=-\alpha}^{1} x(u)$$

The output depends on future, hence it is non causal.

4. Check whether the following systems are static or dynamic.

(a) $y(n) = x(n+1) - x(n-1)$

it is dynamic since present output depends on past and future and also non causal.

(b) $y(n) = a^n x(n)$

the system is static as present output depends on present input.

5.4 CONVOLUTION INTEGRAL

To obtain the output of an LTI system or system impulse response this integral is useful.

$$y(t) = \int_{-\alpha}^{\alpha} x(T)h(t-T)dT = x(t) * h(t)$$

Example:

$$y(t) = x(t) * h(t) = u(t) * u(t)$$

$$y(t) = \int_{-\alpha}^{\alpha} u(T)u(t-T)dT$$

$$y\,(t) = \int_0^t dT \qquad t \geq 0$$

$$= 0$$

Convolution can be described as a weighted average of function x(T) at the moment t where the weighting is given by h(–T) simply shifted by amount t. As t changes weighting function emphasizes different parts of input function.

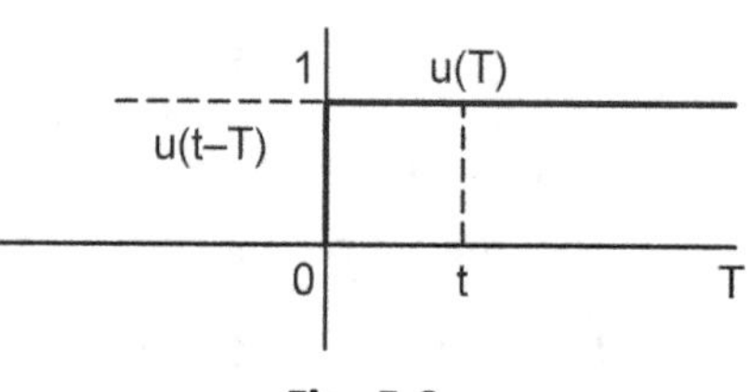

Fig. 5.6

We know that any arbitrary signal x(t) can be represented as

$$x\,(t) = \int_{-\alpha}^{\alpha} x(T)\,\delta(t-T)dT$$

$$\left[\text{even property of impulse } \delta(t-T) = \delta(T-t)\right]$$

but $\qquad y(t) = T\{x(t)\}$

$$\therefore \qquad y\,(t) = T\left\{\int_{-\alpha}^{\alpha} x(T)\delta(t-T)dT\right\}$$

for a linear system

$$y\,(t) = K \cdot \int_{-\alpha}^{\alpha} x(T)\delta(t-T)dT$$

$$\text{if} \quad K = 1$$

$$y(t) = \int_{-\alpha}^{\alpha} x(T)\,\delta(t-T)dT$$

if response of the system to impulse is h(t), then response of system due to delayed impulse is h(t, T)

$$h\,(t,\,T) = T\{\delta(t-T)\}$$

$$y\,(t) = \int_{-\alpha}^{\alpha} (T)\,h(t,T)dT$$

for time invariant systems if input is delayed by T secs, output is also delayed by T secs.

$$y\,(t) = \int_{-\alpha}^{\alpha} x(T)\,b(t-T)dT$$

5.5 REPRESENTATION OF CONTINUOUS TIME SIGNAL IN TERMS OF IMPULSE

Any signal x(t) can be approximated by the sum of a series of narrow rectangular pulses $x_n^1(t)$ each of height x(t) and width ΔT.

Consider a pulse $\delta_\Delta(t)$ and width ΔT.

As limit $\Delta T \to 0$, the pulse approaches an impulse to the area under that pulse.

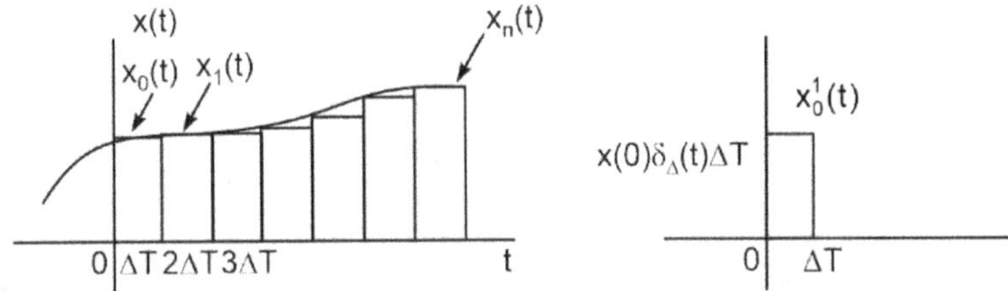

Fig. 5.7

Each rectangular pulse $x_n^1(t)$ can be represented as a product of the magnitude of x(t) and $\delta_\Delta(t)\Delta T$ where

$$\delta_\Delta(t) \;=\; \begin{cases} \dfrac{1}{\Delta T} & 0 \le t \le \Delta T \\[2mm] 0 & \text{otherwise} \end{cases}$$

thus $\delta_\Delta(t)\Delta T$ has unit amplitude

Now the pulse $x_0^1(t)$ can be expressed as

$$x_0^1(t) \;=\; x(0)\delta_\Delta(t)\Delta T$$

for any n^{th} pulse

$$x_n^1(t) \;=\; x(n\Delta T)\delta_\Delta(t-n\Delta T)\Delta T$$

$$x^1(t) \;=\; \sum_{n=-\alpha}^{\alpha} x_n^1(t) = \sum x(n\Delta T)\delta_\Delta(t-n\Delta T)\Delta T$$

As the signal is continuous

$$x(t) = \int_{-\alpha}^{\alpha} x(T)\delta(t-T)dT$$

Impulse response of LTI System

since response of $\delta(t)$ is 1.

5.5.1 TIME DOMAIN ANALYSIS

$$H(s) = \int_{-\alpha}^{\alpha} h(t)e^{-st}dt \quad \text{where } S = \sigma + j\,\omega$$

$\therefore$ y (s) for impulse is y (s) = h(s)

where $T\{\delta(t)\} = h(t)$

we can also write

$$\int x(n) \;=\; \int x(k)\delta(n-k)dk$$

$$\therefore \; y\,(n) \;=\; {}^{T}\!\left\{\sum_{k=-\alpha}^{\alpha} x(k)\,\delta(n-k)\right\}$$

$$=\; \sum x(k)\,T\{\delta(n-k)\}$$

$$=\; \sum_{k=-\alpha}^{\alpha} x(k)\,h(n-k)$$

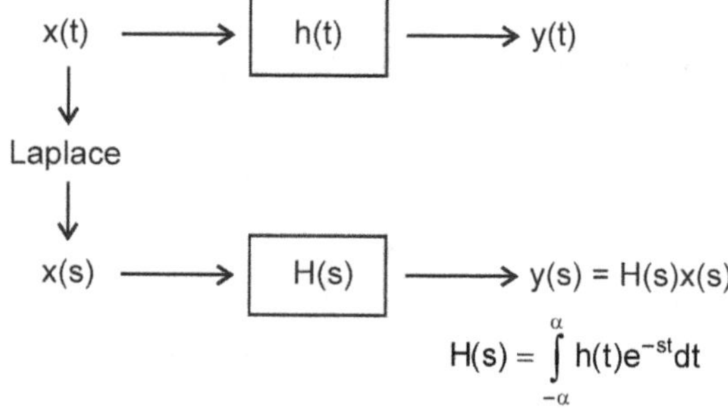

Fig. 5.8

This is known as convolution sum and represented as y(n) = x(n) * h(n)

for continuous signals

$$y\,(t) = T\left\{\int_{-\alpha}^{\alpha} x(T)\,\delta(t-T)dT\right\}$$

$$= \int_{-\alpha}^{\alpha} x(T)\,h(t-T)dT$$

$$y\,(t) = \int_{-\alpha}^{\alpha} x(t-T)\,h(t)dT$$

where

$$H(s) = \int_{-\alpha}^{\alpha} h(t)e^{-st}dt$$

$$= \mathcal{L}\{h(t)\}$$

$$H\{j\omega\} \;=\; F\{h(t)\}$$

$$y\,(t) \;=\; \mathcal{L}^{-1}\{H(s)x(s)\}$$

for any lumped linear system

$$H(s) = \frac{N(s)}{D(s)}$$

$$= \frac{a_1}{S-P_1} + \frac{a_2}{S-P_2} \cdots \frac{a_n}{S-P_n}$$

$$h(t) = \mathcal{L}^{-1}(H(s)) = a_1 e^{P_1 t} + a_2 e^{P_2 t} + \cdots a_n e^{P_n t}$$

5.5.2 RELATION BETWEEN LTI SYSTEM AND IMPULSE RESPONSE

Impulse response totally changes the input-output behavior of an LTI system. Hence the properties of the system related to impulse response should be studied.

(a) LTI systems with and without memory: Output of a memory less LTI system depends only on the present input.

Using commutative property of convolution

$$y(n) = h(n) \times x(n)$$

$$= \sum_{K=-\alpha}^{\alpha} h(k) \times (n-k)$$

$$= \ldots + h(-2) \times (n+2) + h(-1) \times (n+1) + h(0) \times (n) + \ldots h(1) \times (n-1) + \ldots$$

For memory less systems $y(n)$ depends only on $x(n)$

$\therefore$ $x(n-k)$ for $k \neq 0$ can be neglected

$\therefore$ $y(n) = h(0) \times (n)$

This condition implies that $n(x) = 0$ for $k \neq 0$

and thus system is memory less if and only if $n(n) = k\delta(n)$ where $k = h(0)$

$\therefore$ $y(n) = kx(n)$

(b) Causality for LTI system: Causality system depends only on the present and past values of the input

$$y(n) = \sum_{K=-\alpha}^{\alpha} h(k) \times (n-k)$$

$$= \ldots + h(-2) \times (n+2) + h(-1) \times (n+1) + h(0) \times (n) + \ldots h(1) \times (n-1) + \ldots$$

Past and present values of the input $x(n)$, $x(n-1)$, $x(n-2)$ are associated with $k \geq 0$ in the impulse response $h(k)$, while the future values of the input $x(n+1)$, $x(n+2)$... are associated with $k < 0$.

If y(n) depends only on past and present values of input we require $h(k) = 0$ for $k < 0$

$\therefore \quad h(n) = 0$ for $n < 0$

$\therefore$ we can rewrite $y(n) = \displaystyle\sum_{K=0}^{\alpha} h(k) \times (n - k)$

or $\qquad y(n) = \displaystyle\sum_{K=-\alpha}^{n} x(k) h(n - k)$

and $\qquad y(t) = \displaystyle\int_{-\alpha}^{t} x(T) h(t - T) dT$

(c) Stability for LTI systems: A system is stable if every bounded input produces a bounded output

$$|y(n)| = |h(n) \times x(n)|$$

$$= \left| \sum_{K=-\alpha}^{\alpha} h(k) \times (n - k) \right|$$

$$\leq \sum_{K=-\alpha}^{\alpha} |h(k)| |x(n - k)|$$

If we assume that input is bounded

$$|x(n)| \leq B_x < \alpha \quad \text{then} \quad |x(n - k)| \leq B_x$$

$$|y(n)| \leq B_x \sum_{K=-\alpha}^{\alpha} |h(k)|$$

From the above equation we can conclude

$$\sum_{K=-\alpha}^{\alpha} |h(k)| \, k$$

Then $y(x)$ is bounded and hence system is stable.

(d) Investibility for LTI System: If an LTI system is invertible only if an inverse system exists that, when connected in series with the original system produces an output equal to input to the first system.

$x(t) h(t) h_1(t) x(t)$

$\therefore \quad x(t) \times [h(t) \times h_1 (t)] = x(t)$

This requirement implies that

$$h(t) \times h_1(t) = \delta(t)$$

EXAMPLES

1. Find the response of LTI System

$$x(n) = 2^n u(-n)$$

$$h(n) = u(n)$$

$$y(n) = \sum x(k)\, h(n-k)$$

$$\quad\; = \sum 2^k u(-k)\, u(n-k)$$

for $\quad k=0$

$$y(0) = \sum u(0)\, u(n) = 1$$

$$k = -1$$

$$y(-1) = \sum 2^{-1} u(1)\, u(n+1)$$

$$\quad\quad = \frac{1}{2}$$

$$y(1) = 0$$

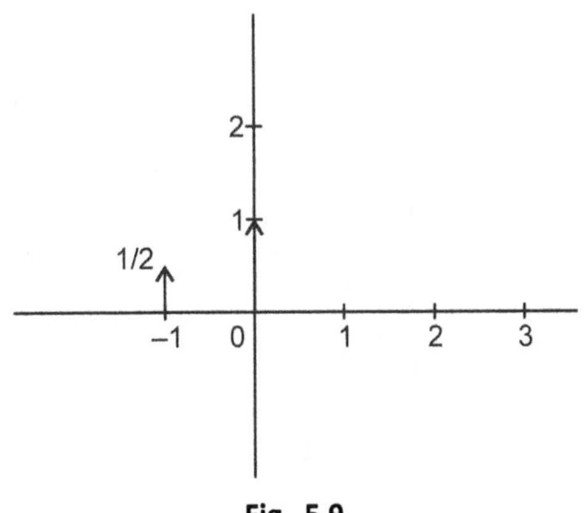

Fig. 5.9

$$y(2) = 0$$

2. Find the impulse response of RL circuit

$$x(t) = -Ri(t) - \frac{L\,di(t)}{dt} = 0$$

$$x(s) = RI(s) + LsI(s)$$

$$I(s) = \frac{x(s)}{R+LS}$$

$$y(t) = + \frac{L\,di(t)}{dt}$$

$$y(s) = \frac{Ls}{R+Ls} x(s)$$

$$H(s) = \frac{Ls}{R+LS}$$

$$\quad\quad = 1 - \frac{R}{R+LS}$$

$$\quad\quad = 1 - \frac{R}{L} \frac{1}{\frac{R}{L}+S}$$

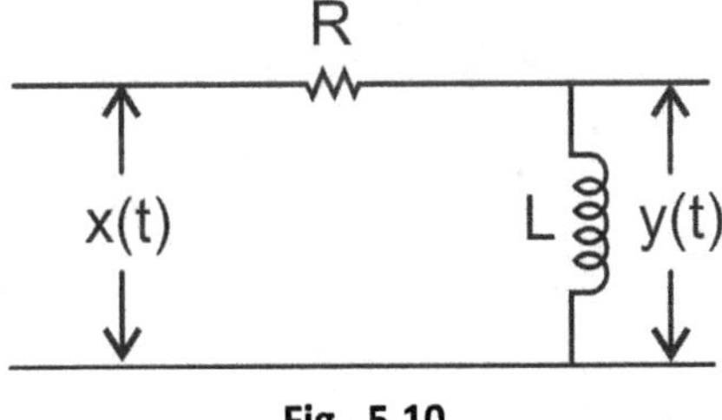

Fig. 5.10

$$= 1 - \frac{R}{L}\left\{\frac{1}{S + R/L}\right\}$$

$$h(t) = \delta(t) - \frac{R}{L}e^{-R/L\,t}$$

3. Find the impulse response of LC circuit.

$$x(t) = \frac{L\,di}{dt} - \frac{1}{C}\int i\,dt = 0$$

$$y(t) = \frac{1}{C}\int i\,dt$$

$$x(s) = \left(LS + \frac{1}{CS}\right)i(s)$$

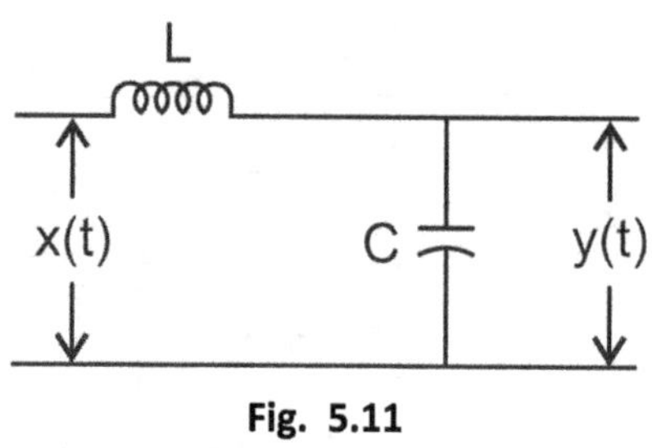

Fig. 5.11

$$y(s) = \frac{1}{CS}i(s)$$

$$H(s) = \frac{1/CS}{LS + \frac{1}{CS}} = \frac{1}{LCS^2 + 1}$$

$$= \frac{1}{S^2 + \frac{1}{LC}} \cdot \frac{1}{LC}$$

Let $\omega_0 = \dfrac{1}{\sqrt{LC}}$

$$H(s) = \omega_0 \frac{\omega_0}{S^2 + \omega_0^2}$$

$$= \omega_0 \sin(\omega_0 t)u(t)$$

4. Find the impulse response of the system whose $H(s) = \dfrac{e^s}{S+1}$ using time shifting property

$$x(t - t_0) \leftrightarrow e^{-st_0} \times (s)$$

$$e^{-t}u(t) \leftrightarrow \frac{1}{S+1}$$

$$e^{-(t+1)}u(t+1) \leftrightarrow \frac{e^s}{S+1}$$

5. Find impulse response of $\dfrac{dy}{dt} + 3y = x$

$$\frac{dy}{dt} + 3y = x$$

$$Sy(s) + 3y(s) = x(s)$$

$$H(s) \ = \ \frac{1}{S+3}$$

$$h(t) \ = \ e^{-3t}u(t)$$

6. Find the impulse response of below function

$$\frac{d^2y}{dt^2} + \frac{4dy}{dt} + 3y = \frac{dx}{dt} + 2x$$

$$S^2y(s) + 4Sy(s) + 3y(s) \ = \ Sx(s) + 2x(s)$$

$$H(s) \ = \ \frac{S+2}{S^2 + 4S + 3}$$

$$\frac{S+2}{S^2 + 4S + 3} = \frac{S+2}{(S+1)(S+3)} = \frac{A}{S+1} + \frac{B}{S+3}$$

$$AS + 3A + BS + B = S + 2$$

$$S(A+B) = S$$

$$3A + B = 2$$

$$A + B = 1$$

$$2A + 1 = 2$$

$$2A = 1$$

$$A = \frac{1}{2} \qquad B = \frac{1}{2}$$

$$\therefore \ h(t) = \ \frac{1}{2}e^{-t} + \frac{1}{2}e^{-3t}$$

5.6 SIGNAL TRANSMISSION THROUGH LINEAR SYSTEMS

For any given input signal x(t), the output y(t) is the convolution of input signal and impulse response h(t) of the system

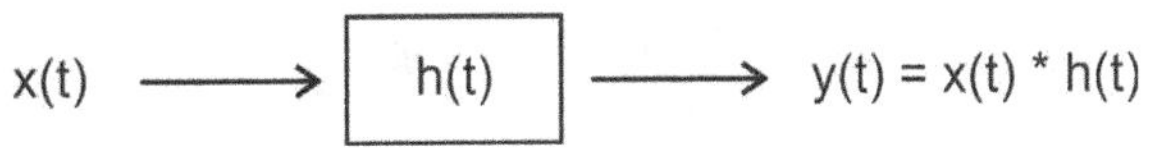

$$x(t) \longrightarrow \boxed{h(t)} \longrightarrow y(t) = x(t) * h(t)$$

Fig. 5.12

In frequency domain

$$Y(S) \ = \ X(S)H(S)$$

i.e., $|Y(j\omega)| = |X(j\omega)||H(j\omega)|$

and $\lfloor y(j\omega) = \lfloor x(j\omega) + \lfloor H(j\omega)$

The input amplitude $|X(j\omega)|$ is changed to $|X(j\omega)||H(j\omega)|$ and phase is changed to $\underline{X(j\omega)} + \underline{H(j\omega)}$.

As the input signals are normally sinusoids, the system h(t) is modifying these frequency components. It may boost certain frequency components and attenuate others, it also may change relative phases of these components. Therefore output may not be same as input. Then the signal is distorted.

5.6.1   DISTORTION LESS TRANSMISSION

If the system attenuates or boosts all frequency components equally and allows same phase shift to all frequency components then we can call as distortion less transmission. Any how a time delay may be allowed.

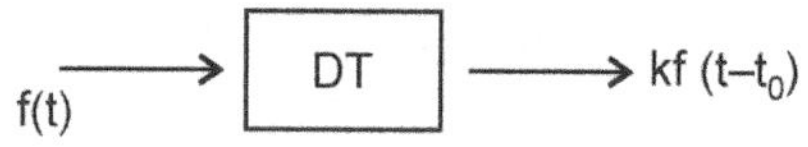

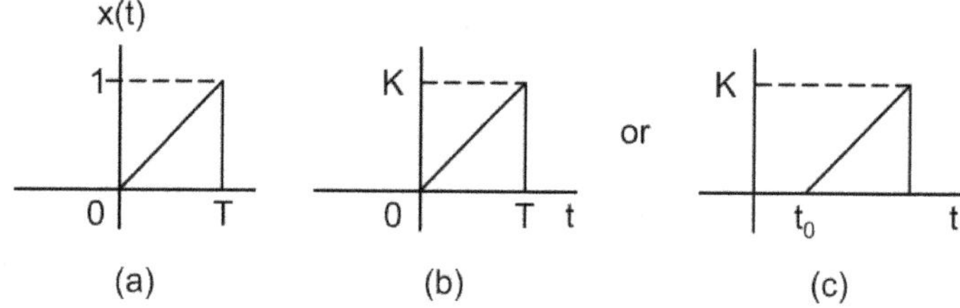

Fig. 5.13

A transmission is said to be distortion less if output is same as input, except that the magnitude is scaled by a constant k and a time delay t_0.

Fig. 5.14

$$\mathcal{L}\{y(t)\} = \mathcal{L}\{kf(t-t_0)\} = k \cdot F(s)e^{-st_0}$$

$$F(s)H(s) = F(s)ke^{-st_0}$$

$$H(s) = ke^{-st_0}$$

$$H(j\omega) = ke^{-j\omega t_0}$$

where $$|H(j\omega)| = k$$

$$\underline{H(j\omega)} = n\pi - \omega t_0$$

5.6.2   LINEAR PHASE SYSTEM

If the system doesn't introduce any phase distortion, then the transfer function of linear phase system is

$$H(j\omega) = |H(j\omega)|e^{-j\omega t_0}$$

for pass band $|H(j\omega)| = 1$

$$\mathcal{L}^{-1}\{H(j\omega)\} = \frac{1}{2\pi}\int_{-\omega_0}^{\omega_0} e^{-j\omega t_0} e^{j\omega t}\, d\omega$$

$$= \frac{1}{2\pi}\int_{-\omega_0}^{\omega_0} e^{j\omega(t-t_0)}\, d\omega$$

$$= \frac{e^{j\omega(t-t_0)} - e^{-j\omega(t-t_0)}}{2\pi j(t-t_0)} = \frac{\sin\omega_0(t-t_0)}{\pi(t-t_0)}$$

$$= \frac{\omega_0}{\pi}\frac{\sin\omega_0(t-t_0)}{\omega_0\pi(t-t_0)} = \frac{\omega_0}{\pi}\sin c\,\omega_0(t-t_0)$$

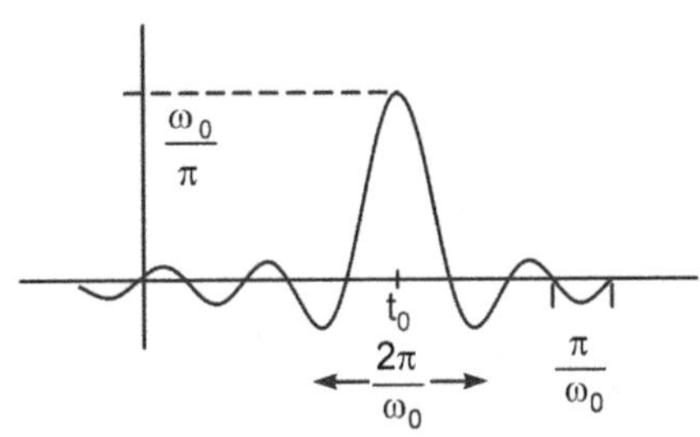

Fig. 5.15

5.7 IDEAL FILTERS

An ideal filter is a system that allows distortion less transmission for all frequencies in a band.

LPF: Low pass filter allows to pass all frequency components less than cutoff frequency ω_0.

$$H(j\omega) = \begin{cases} e^{-j\omega t_0} & \omega \le |\omega_0| \\ = 0 & |\omega > \omega_0| \end{cases}$$

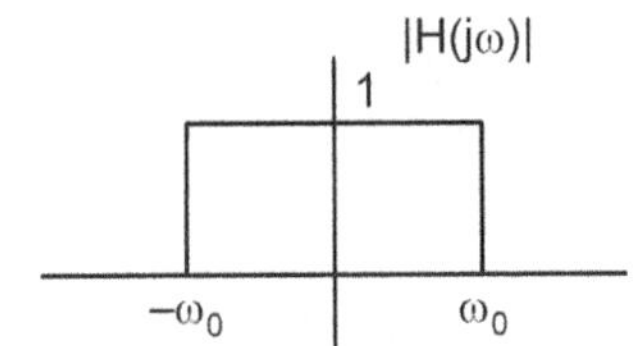

Fig. 5.16

HPF: It allows to pass all frequency components that are greater than cutoff frequency ω_0.

$$H(j\omega) = \begin{cases} e^{-j\omega t_0} & \omega \ge |\omega_0| \\ 0 & |\omega < \omega_0| \end{cases}$$

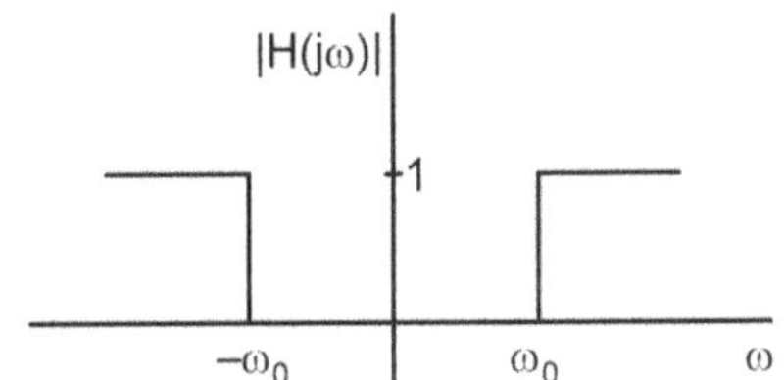

Fig. 5.17

BPF: It allows to pass all frequency components that are between two specified cutoff frequencies.

$$H(j\omega) = \begin{cases} e^{-j\omega t_0} & |\omega_1| \le \omega \le |\omega_2| \\ 0 & \text{otherwise} \end{cases}$$

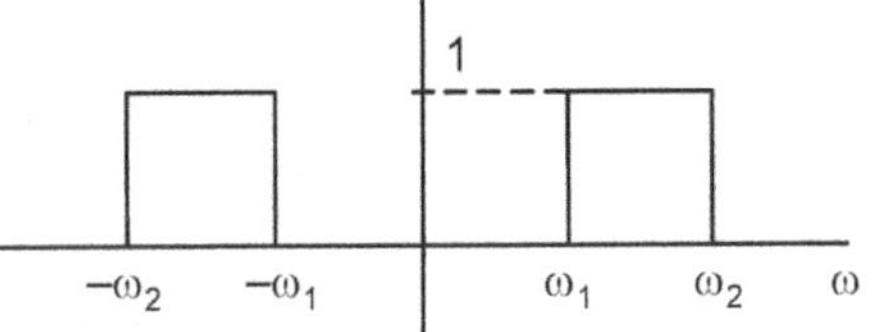

Fig 5.18

BRF: It rejects to pass all frequencies that are between two specified cutoff frequencies

$$H(j\omega) = \begin{cases} e^{-j\omega t_0} & \text{otherwise} \\ 0 & |\omega_1| \le \omega \le |\omega_2| \end{cases}$$

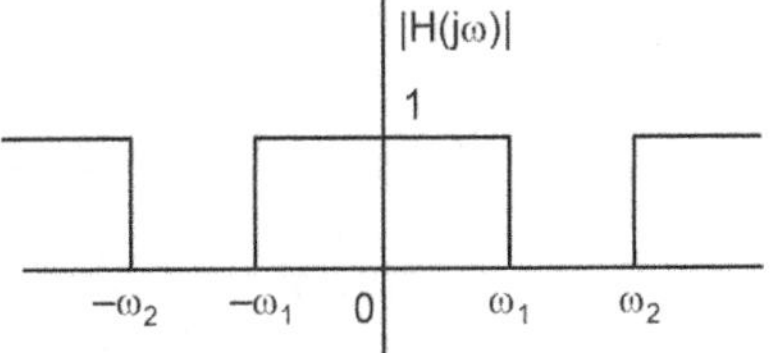

Fig 5.19

5.7.1 IMPULSE RESPONSE OF FILTERS

1. LPF: Transfer function of LPF is similar to gate function of width $2\omega_0$.

$$H(j\omega) \;=\; e^{-j\omega t_0}$$

or

$$H(j\omega) \;=\; G_{2\omega_0}(\omega)e^{-j\omega t_0}$$

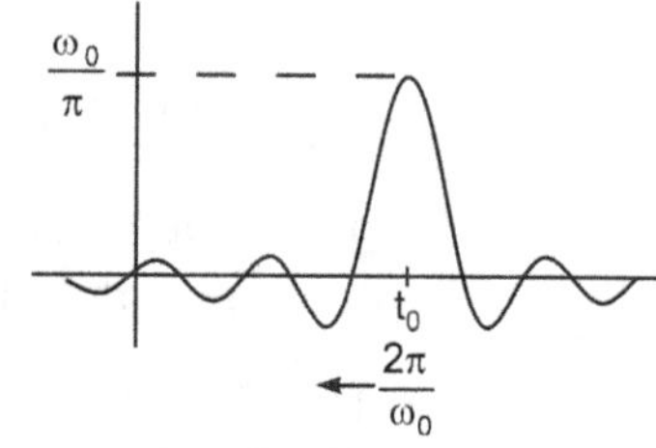

Fig. 5.20

$$\therefore \quad h(t) \;=\; \mathcal{L}^{-1}\{H(j\omega)\} = \mathcal{L}^{-1}\left\{G_{2\omega_0}(\omega)e^{-j\omega t_0}\right\}$$

$$=\; \frac{\omega_0}{\pi}\,\mathrm{Sa}\{\omega_0(t-t_0)\}$$

$$\text{since } \mathcal{F}^{-1}\left[\left\{G_{\omega_0}(\omega)\right\}=\frac{\omega_0}{2\pi}\mathrm{Sa}\frac{\omega_0 t}{2}\right]$$

Proof:

$$H(j\omega) \;=\; |H(j\omega)|e^{-j\omega t_0}$$

for pass band $|H(j\omega)| = 1$

$$H(j\omega) = 1$$

$$\mathcal{L}^{-1}\{H(j\omega)\} \;=\; \frac{1}{2\pi}\int_{-\omega_0}^{\omega_0} e^{-j\omega t_0}\, e^{j\omega t}\, d\omega$$

$$=\; \frac{1}{2\pi}\int_{-\omega_0}^{\omega_0} e^{j\omega(t-t_0)}\, d\omega$$

$$=\; \frac{e^{j\omega_0(t-t_0)} - e^{-j\omega_0(t-t_0)}}{2\pi j(t-t_0)}$$

$$=\; \frac{\sin\omega_0(t-t_0)}{\pi(-t-t_0)} = \frac{\omega_0}{\pi}\sin c\big(\omega_0(t-t_0)\big)$$

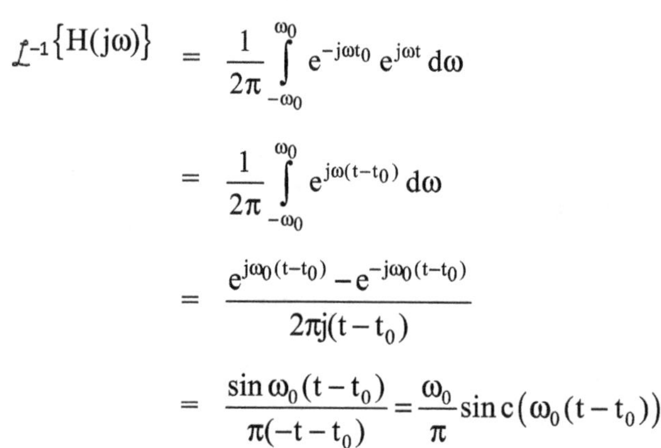

Fig. 5.21

2. HPF:

$$H(j\omega)_{HPF} \;=\; 1 - H(j\omega)_{LPF}$$

$$h(t) \;=\; \mathcal{L}^{-1}\{1 - H(j\omega)_{LPF}\}$$

$$h(t) = \; \delta(t) - \frac{\omega_0}{\pi}\sin c\big(\omega_0(t-t_0)\big)$$

Ideal filters are not physically realizable since they exist before the driving function is applied.

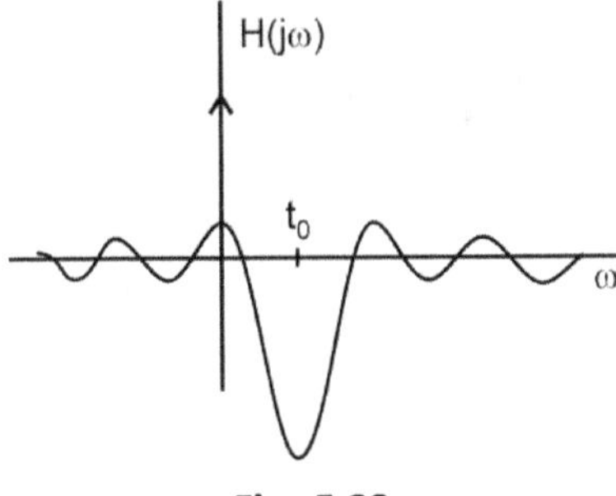

Fig. 5.22

5.7.2 BANDWIDTH

Bandwidth of a system can be defined as range of frequencies where there occurs distortion less transmission. Ideally it is not possible to provide such a transmission for all ranges of frequencies. It is only possible for certain frequencies.

Bandwidth of a system having a transfer function $H(j\omega)$ is equal to the range of frequencies for which $|H(j\omega)|$ remain within $\dfrac{1}{\sqrt{2}}$

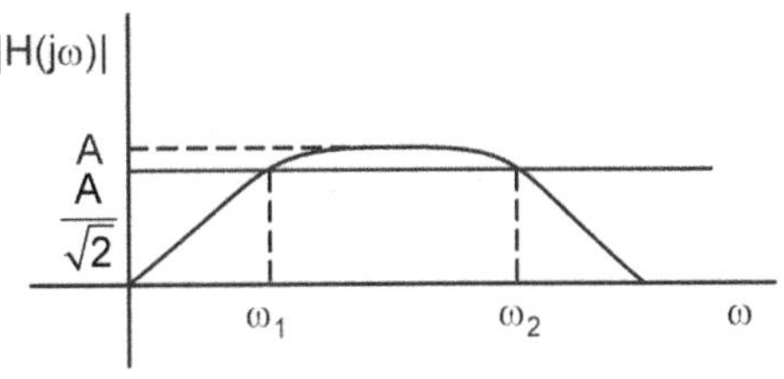

Fig. 5.23

of its maximum value.

For a signal bandwidth is defined as the range of frequencies (spectral components) that have a constant and maximum energy.

5.7.3 RISE TIME

For a given system, it is the time taken to reach the response from 10% to 90% of final value.

Rise time is inversely proportional to the bandwidth.

Bandwidth × Rise time = constant.

Relationship between Bandwidth and Rise time

Let us consider step response of LPF

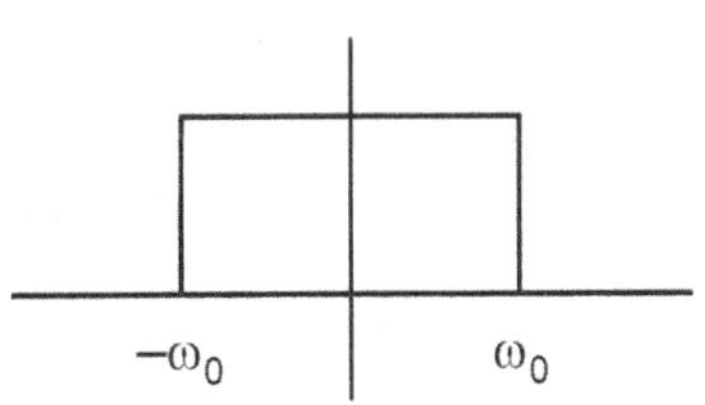

Fig. 5.24

$$h(t) = \frac{\omega_0}{\pi} \operatorname{sinc}\left(\omega_0(t - t_0)\right)$$

$$= \frac{\omega_0}{\pi} \frac{\sin\left(\omega_0(t - t_0)\right)}{\omega_0(t - t_0)}$$

Response of this filter

$$y(t) = h(t) * u(t)$$

$$\therefore \quad \left[y(t) = \int_{-\alpha}^{\alpha} u(t)\, h(t-T)\, dT \right] = \int_{-\alpha}^{t} h(T)\, dT$$

$$= \int_{-\alpha}^{t} \frac{\omega_0}{\pi} \frac{\sin\big(\omega_0(T-t_0)\big)}{\omega_0(T-t_0)}\, dT$$

Let $\quad x = \omega_0(T-t_0)$

$$dx = \omega_0\, dT$$

$$dT = \frac{dx}{\omega_0}$$

$$y(t) = \frac{1}{\pi} \int_{-\alpha}^{\omega_0(t-t_0)} \frac{\sin x}{x}\, dx$$

$$= \frac{1}{\pi} Si(x) \Big/_{-\alpha}^{\omega_0(t-t_0)}$$

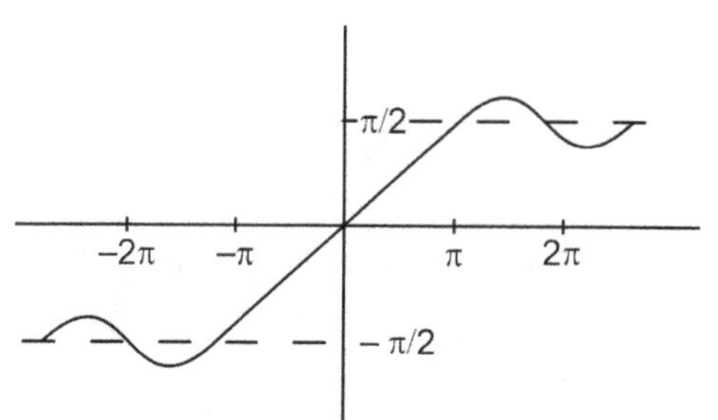

Fig. 5.25

where Si is sine integral function

(a) Si (x) is an odd function Si $(-x) = -Si(x)$ (b)   Si $(0) = 0$

(c) Si $(\alpha) = \dfrac{\pi}{2}$

$$\therefore \quad y(t) = \frac{1}{\pi}\Big\{ Si\big[\omega_0(t-t_0)\big] - Si(-\alpha) \Big\}$$

$$= \frac{1}{\pi}\Big\{ Si\big[\omega_0(t-t_0)\big] + \dfrac{\pi}{2} \Big\}$$

$$= \frac{1}{2} + \frac{1}{\pi} Si\big[(\omega_0(t-t_0)\big]$$

if $\quad \omega_0 = \alpha \qquad y(t) = \dfrac{1}{2} + \dfrac{1}{\pi} Si(\alpha) = \dfrac{1}{2} + \dfrac{1}{2} = 1$

$$\omega_0 = -\alpha \qquad y(t) = \dfrac{1}{2} - \dfrac{1}{2} = 0$$

Rise time, t_r: It is defined as the time required for the response to rise from 10% to 90% of final value.

$$\frac{d}{dt}y(t) = \frac{1}{t_r} = \frac{\omega_0}{\pi}$$

$$t_r = \frac{\pi}{\omega_0}$$

$$\omega_0 \times t_r = \pi$$

$$\therefore \quad BW \times RT = constant$$

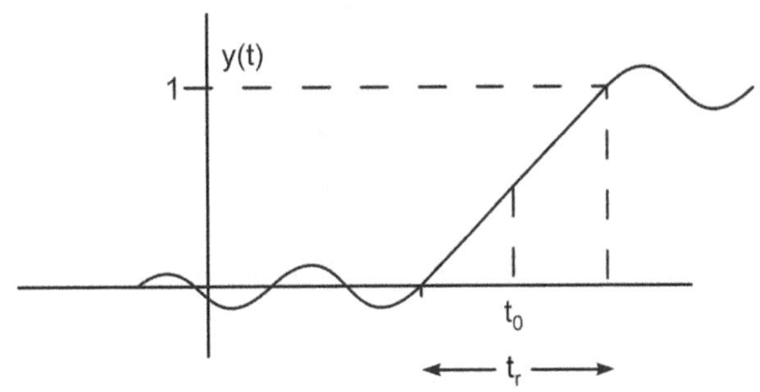

Fig. 5.26

5.8 POLY WIENER CRITERIA

A physically realizable system cannot have a response before the driving function is applied. This is known as causality condition. Ideal filters cannot be physically realizable.

Impulse response h(t) of a physically realizable system must be causal, then in time domain.

$$h(t) = 0 \quad \text{for all } t < 0$$

in frequency domain

$$\int_{-\alpha}^{\alpha} \left| \frac{\left\| L_n \right| H(j\omega) \right\|}{1 + \omega^2} \right| d\omega < \alpha$$

or magnitude function

$$\int_{-\alpha}^{\alpha} \left| H(j\omega) \right|^2 d\omega < \alpha$$

A system which violates the above criteria is non causal. The response of the system exists in past and present, but not in future. Magnitude function may be zero at some discrete frequencies, but not over a range of frequencies. The amplitude function cannot full of zero faster than exponential order.

$$\left| H(j\omega) \right| = k e^{-\alpha |\omega|}$$

SOLVED PROBLEMS - 1

1. Transfer function of a **LPF** is

$$H(\omega) = \begin{array}{ll} 1 & |\omega| < \omega_0 \\ 0 & |\omega| > \omega_0 \end{array}$$

Show that its impulse response is non causal? What do you do to make it causal? What is its physical significance?

Ans: Transfer function of **LPF**   Nov – 2008

$$H(\omega) = \begin{cases} 1 & \omega \le |\omega_0| \\ 0 & \omega > |\omega_0| \end{cases}$$

Is same as that of gating function $G_{2\omega_0}(\omega)$ shown in figure

The impulse response of this filter is a sine function extending to negative axis which indicates it is also present in future. Thus it is non causal and physically not realizable.

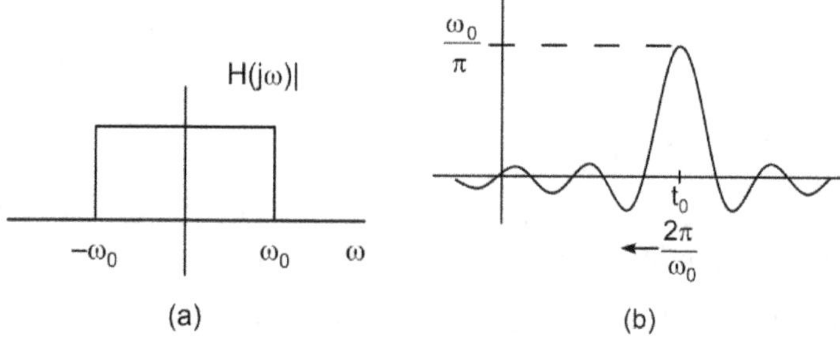

Fig. 5.27

To make it physically realizable it must satisfy Paley wiener criteria given by

$$\int_{-\alpha}^{\alpha} \frac{\left| L_n \left| H(j\omega) \right| \right|}{1+\omega^2} d\omega < \alpha$$

If $\left| H(j\omega) \right| = 1$ for all $\omega < |\omega_c|$

$$\int_{-\alpha}^{\alpha} \frac{L_n(1)}{L+\omega^2} d\omega = \alpha \qquad \text{and system is non causal. To make it causal}$$

$$\left| H(j\omega) \right| > 1$$

2. Find the response e(t) of the electric circuit shown in figure if $i(t) = e^{-t}u(t)$

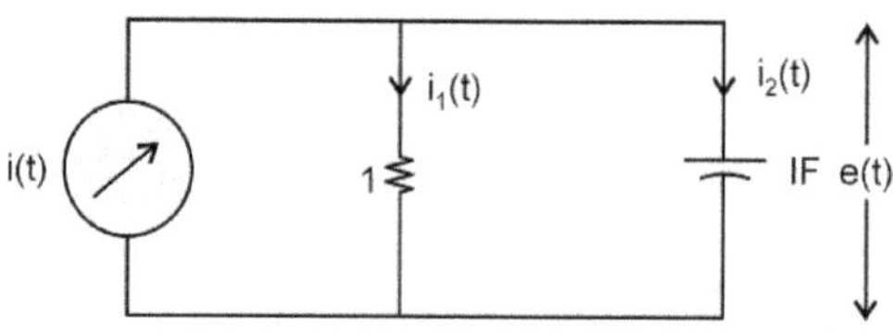

Fig. 5.28

Ans: $i(t) = e^{-t}u(t)$

$$I(s) = \frac{1}{S+1}$$

$$I_2(s) = \frac{1}{1+\frac{1}{s}}I(s)$$

$$= \frac{S}{S+1} \times \frac{1}{S+1} = \frac{S}{S^2+1}$$

$$\mathcal{L}^{-1}\{e(s)\} = \therefore e(s) = \frac{1}{s} \cdot I_2(s)$$

$$= \propto^{-1}\left\{\frac{1}{S^2+1}\right\} = t \cdot e^{-t}u(t)$$

3. Find the impulse response of the system shown in figure. Find the transfer function. What would be its frequency response?   Feb 2008

$$H(s) = \frac{V_0(s)}{V_i(s)} = \frac{R}{2+\frac{1}{cs}} = \frac{RCS}{1+RCS}$$

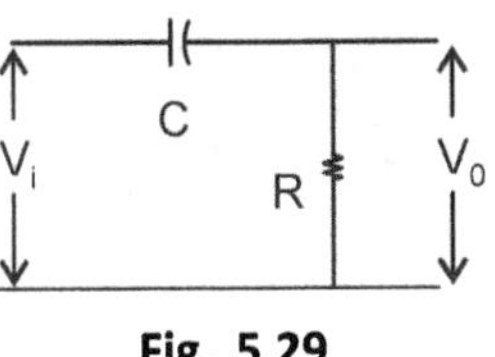

Fig. 5.29

$$h(t) = \mathcal{L}^{-1}\{H(s)\} = \mathcal{L}^{-1}\left\{1-\frac{1}{1+RCS}\right\}$$

$$= \delta(t) - \frac{1}{RC}e^{-t/RC}$$

$$= \delta(t) - \frac{1}{RC}e^{-t/RC}$$

4. Consider a stable LTL system characterized by the differential equation   Feb 2008

$$\frac{d}{dt}y(t) + 2y(t) = x(t) \text{ find its impulse response}$$

Ans: $Sy(s) + 2y(s) = x(s)$

$$H(s) = \frac{y(s)}{x(s)} = \frac{1}{S+2}$$

$$\mathcal{L}^{-1}\{H(s)\} = h(t) = e^{-2t}u(t)$$

5. Find the impulse response for RL filter shown in figure   August 2007

   Ans: Applying Laplace transform

$$x(s) - i(s)R - SLi(s) = 0$$

$$y(s) = i(s)SLi(s)$$

$$\therefore \quad h(s) = \frac{y(s)}{x(s)} = \frac{SL}{R+SL} = \frac{S}{R/L + S}$$

$$= -\frac{R/L}{R/1 + S}$$

$$\mathcal{L}^{-1}\{h(s)\} = \quad h(t) = \delta(t) - \frac{R}{L}\,e^{-R/Lt}$$

$$= \quad \delta(t) - \frac{R}{L}\,e^{-R/Lt}$$

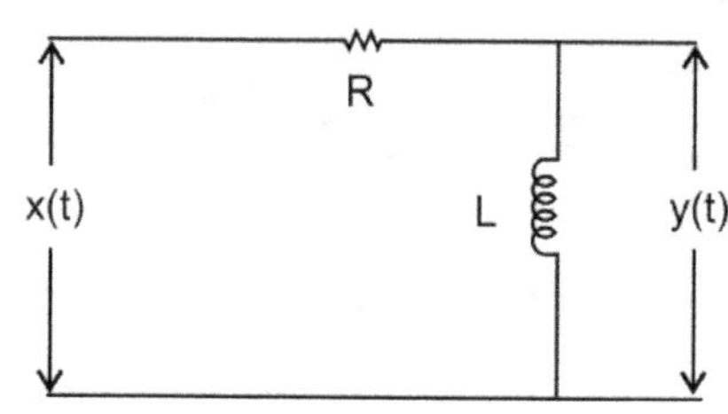

Fig. 5.30

6. The transfer function of an ideal band pass filter is given by

$$H(j\omega) = K\{G_\omega(\omega-\omega_0)+G(\omega+\omega_0)\}e^{-j\omega t_0}.$$ Sketch the magnitude and phase function of this transfer function. Evaluate the impulse response of this filter. Sketch this response and state whether the filter is physically realizable. May 2005

Ans: Transfer function of band pass filter is

$$H(j\omega) = K\{G_\omega(\omega-\omega_0)+G_\omega(\omega+\omega_0)\}e^{-j\omega t_0}$$

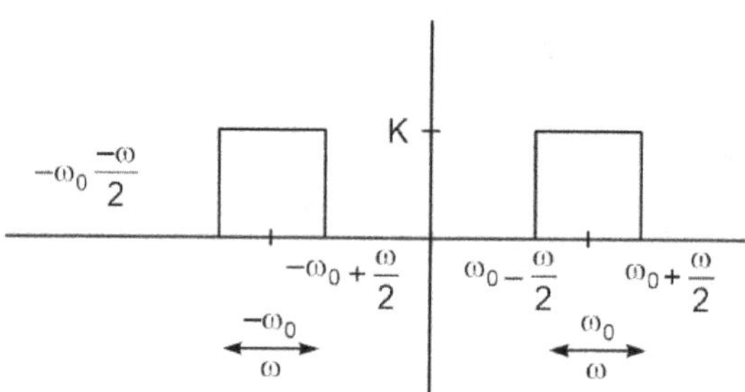

Fig. 5.31

$$\mathcal{L}^{-1}\{H(j\omega)\} = \frac{1}{2\pi}\int_{-\omega_0-\omega/2}^{-\omega_0+\omega/2} K\cdot e^{-j\omega t_0}\cdot e^{j\omega t}\,d\omega + \int_{\omega_0-\frac{\omega}{2}}^{\omega_0+\frac{\omega}{2}} K\cdot e^{-j\omega t_0}\cdot e^{j\omega t}\,d\omega$$

$$= \frac{K}{2\pi}\left\{\int_{-\omega_0-\omega/2}^{-\omega_0+\omega/2} e^{j\omega(t-t_0)}\,d\omega + \int_{\omega_0-\frac{\omega}{2}}^{\omega_0+\frac{\omega}{2}} e^{j\omega(t-t_0)}\,d\omega\right\}$$

$$= \frac{K}{2\pi}\left\{\frac{e^{j\left(-\omega_0+\omega/2\right)(t-t_0)} - e^{j\left(-\omega_0-\omega/2\right)(t-t_0)}}{j(t-t_0)} + \frac{e^{j\left(\omega_0+\frac{\omega}{2}\right)(t-t_0)} - e^{j\left(\omega_0-\omega/2\right)(t-t_0)}}{j(t-t_0)}\right\}$$

$$= \frac{K}{2\pi} \left\{ \frac{e^{-j\omega_0(t-t_0)}\left\{ e^{\frac{\omega}{2}(t-t_0)} - e^{-j\frac{\omega}{2}(t-t_0)}\right\}}{j(t-t_0)} + \frac{e^{j\omega_0(t-t_0)}\left\{ e^{j\frac{\omega}{2}(t-t_0)} - e^{-j\frac{\omega}{2}(t-t_0)}\right\}}{j(t-t_0)} \right\}$$

$$= \frac{K}{j2\pi(t-t_0)}\left\{ e^{-j\omega_0(t-t_0)} + e^{j\omega_0(t-t_0)}\right\}\left\{ e^{j\frac{\omega}{2}(t-t_0)} - e^{-j\frac{\omega}{2}(t-t_0)}\right\}$$

$$= \frac{2K}{\pi(t-t_0)}\left\{ \cos\left[\omega_0(t-t_0)\right]\sin\left[\frac{\omega}{2}(t-t_0)\right]\right\}$$

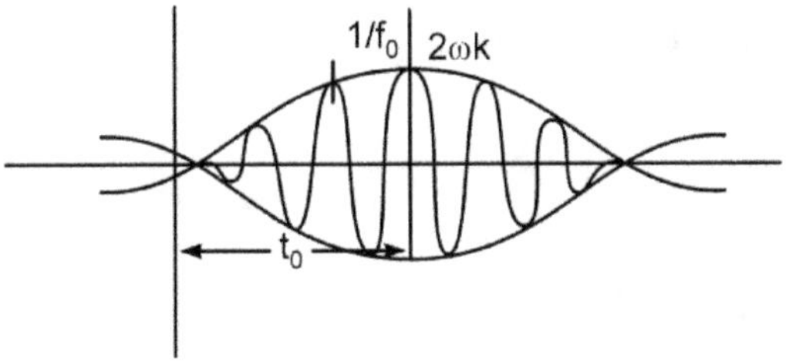

Fig. 5.32

7. Let the system function of a LTI systems be $\dfrac{1}{2+j\omega}$. What is the output of the system for an input $(0.8)^t\, u(t)$.

Nov 2009

Ans:

$$h(j\omega) = \frac{1}{2+j\omega}$$

$$h(s) = \frac{1}{S+2}$$

$$x(t) = = 0.8^t$$

$$x(s) = \frac{1}{S-\log_e 0.8}$$

$$\therefore \quad y(s) = x(s) \quad h(s) = \frac{1}{S+2} \times \frac{1}{S-\log 0.8}$$

We have

$$a^t = e^{t\log a}$$

$$\mathcal{L}^{-1}\left\{ e^{t\log a} = \frac{1}{S-\log_e^a}\right\}$$

8. Find the impulse response LC circuit shown in figure

> *Ans:* For impulse input the response of the circuit is given by

$$h(t) = \frac{y(t)}{x(t)}; \qquad h(s) = \frac{1/cs}{SL + 1/cs} = \frac{1}{Lcs^2 + 1}$$

$$\therefore \qquad \mathcal{L}^{-1}\{h(s)\} = \mathcal{L}^{-1}\left\{\frac{1/Lc}{S^2 + 1/Lc}\right\}$$

$$= \frac{1}{\sqrt{LC}}\sin\left(\frac{1}{\sqrt{LC}}\right)t$$

$$h(t) = \omega_0 \sin\omega_0 t\, u(t)$$

$$\text{where} \quad \omega_0 = 1/\sqrt{LC}$$

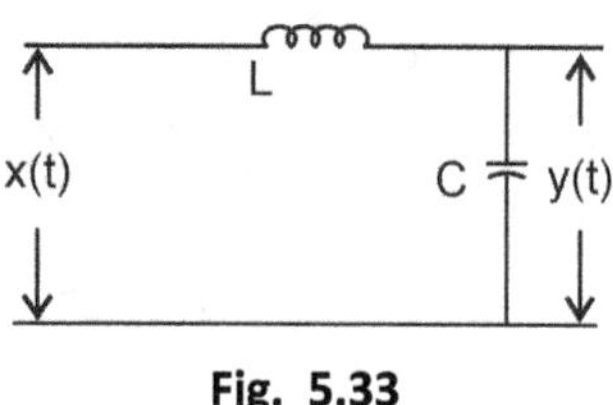

Fig. 5.33

SOLVED PROBLEMS - 2

1. For each of the following find whether system is linear or not

$$\text{(a) } y(t) = t\,x(t) \quad \text{(b) } y(t) = x^2(t) \quad \text{(c) } \frac{dy(t)}{dt} + 3y(t) = x(t)$$

Ans:

(a) Let $y_1(t) = t\,x_1(t)$

$y_2(t) = t\,x_2(t)$

$y_3(t) = t\,x_3(t)$

Let $x_3(t) = ax_1(t) + bx_2(t)$

where a & b are arbitrary variables.

If system is linear $\therefore$ $y_3(t) = ay_1(t) + b\,y_2(t)$

but $y_3(t) = t\,x_3(t) = t\{ax_1(t) + bx_2(t)\}$

$$= a\,tx_1(t) + b\,tx_2(t)$$

$$= ay_1(t) + by_2(t)$$

Hence system is linear

(b) Let $y_1(t) = x_1^2(t)$

$y_2(t) = x_2^2(t)$

$y_3(t) = x_3^2(t)$

Let $\qquad x_3(t) = ax_2(t) + bx_1(t)$

$\therefore \qquad y_3(t) = ay_2(t) + by_1(t)$ if system is linear

But we have $y_3(t) = x^2{}_3(t)$

$$= \{ax_2(t) + bx_1(t)\}^2 = a^2x_2{}^2(t) + b^2x^2{}_1(t) + 2abx_2(t)x_1(t)$$

$$= ay_2(t) + by_1(t) + 2y_2(t)y_1(t)$$

Hence system is non linear.

(c) $\qquad\qquad \dfrac{dy(t)}{dt} + 3y(t) = x(t)$

Let $\qquad \dfrac{dy_1(t)}{dt} + 3y_1(t) = x_1(t)$

$$\dfrac{dy_2(t)}{dt} + 3y_2(t) = x_2(t)$$

$$\dfrac{dy_3(t)}{dt} + 3y_3(t) = x_3(t)$$

Let $\qquad x_3(t) = a\, x_2(t) + b\, x_1(t)$

If system is linear

$$\frac{d}{dt}\{ay_2(t) + by_1(t)\} + 3\{ay_2(t) + by_1(t)\} = ax_2(t) + bx_1(t)$$

$$a\frac{d}{dt}y_2(t) + b\frac{d}{dt}y_1(t) + a3y_2(t) + b3y_1(t)$$

$$= a\left\{\frac{d}{dt}y_2(t) + 3y_2(t)\right\} + b\left\{\frac{d}{dt}y_1(t) + 3y_1(t)\right\}$$

$$\frac{d}{dt}\left(y_3(t)\right) + 3y_3(t) = \frac{d}{dt}\{ay_2(t) + by_1(t)\} + 3\{ay_2(t) + by_1(t)\}$$

$$= ax_2(t) + bx_1(t) \ = x_3(t)$$

Hence system is linear.

2. Find the following relations are time variant or not

    (a) $y(x) = n(n)$ $\qquad\qquad$ (b) $y(t) = \sin\{x(t)\}$ $\qquad$ (c) $y(t) = x(2t)$

Ans:

(a) Let $\qquad y_1(n) = x\, x_1(n)$

$\qquad\qquad\qquad y_2(n) = n\, x_2(n)$

Let $\qquad x_2(n)$ is time shifted version of $x_1(n)$

$\qquad\qquad\qquad x_2(n) = x_1(n - n_0)$

if system is time invariant

$$y_2(n) = y_1(n - n_0)$$

$$y_2(n) = nx_2(n) = n\,x_1(n - n_0)$$

we have　　$y_1(n) = n\,x_1(n)$

$$y_2(n) = y_1(n - n_0) = (n - n_0)\,x_1(n - n_0)$$

$\therefore$ system is not time invariant

(b)　Let　　$y_1(t) = \sin(x_1(t))$

$$y_2(t) = \sin(x_2(t))$$

　　Let　　$x_2(t)$ is time shifted version of $x_1(t)$

$$x_2(t) = x_1(t - t_0)$$

if system is time invariant

$$y_2(t) = y_1(t - t_0)$$

$$= \sin(x_2(t)) = \sin(x_1(t - t_0))$$

$$y_2(t) = y_1(t - t_0) = \sin x(t - t_0)$$

Hence system is time invariant

(c)　Let　　$y_1(t) = x_1(2t)$

$$y_2(t) = x_2(2t)$$

Let $x_2(t)$ is time shifted version of $x_1(t)$

$$x_2(t) = x_1(t - t_0)$$

if system is time invariant

$$y_2(t) = y_1(t - t_0)$$

$$= x_1(2(t - t_0))$$

but　　$y_2(t) = x_2(2t)$

$$= x_1(2t - t_0)$$

Hence system is not time invariant.

3.　Find the below relations are causal or not

(a) $y(n) = nx(n)$　　　(b) $y(t) = \sin(x(t))$　　(c) $y(t) = x(2t)$

Ans:

(a)　Let $y(n)$ is at a time n_0

$$y(n)\big|_{n=n_0} = y(n_0) = n_0 x(n_0)$$

Then $y(n)$ at a time $- n_0$

$$y(n)\big|_{n=-n_0} = y(-n_0) = -n_0 x(-n_0)$$

in both cases output depends on present input hence it is causal.

(b) $y(t) = \sin x(t)$ also current output depends on current input hence it is causal

(c) $y(t)$ for $t_0 \Rightarrow y(t)\big|_{t=-t_0} = y(t_0) = x(2t_0)$

$y(t)$ for $-t_0 \Rightarrow y(-t)\big|_{t=t_0} = y(-t_0) = x(-2t_0)$

In first case output depends on future in fact and in second case past input. Hence system is non causal.

4.   Check whether below systems are stable or not.

(a) $y(t) = t\,x(t)$ (b) $y(t) = e^{x(t)}$

Ans:

(a)   Let $|x(t)| \le B_x < \alpha$ for all t

We have $y(t) = t\,x\,(t)$

$|y(t)| = |t\,x(t)| = |t|\,|x(t)| = |t|\,B_x$

As $t \to \alpha$, $|y\,(t)| \to \alpha$ which is unbounded hence system is unstable.

(b)   Let $|x(t)| \le B_x < \alpha$ for all t

Or $-B_x \le x\,(t) \le B_x$ for all t

we have $y(t) = e^{x(t)}$

∴ $e^{-Bx} \le y(t) \le e^{Bx}$

as system is bounded by a constant, the system is stable.

5.   Find the convolution of two signals

$$f(t) = e^{-t^2} \text{ and } y(t) = 3t^2 \text{ for all t}$$

Ans:

$$y(t) = f(t) \times y(t) = \int_{-\alpha}^{\alpha} f(\tau)y(t-\tau)d\tau \quad = \int_{-\alpha}^{\alpha} e^{-\tau^2}\left\{3(t-\tau)^2\right\}d\tau$$

$$= 3t^2 = \int_{-\alpha}^{\alpha} e^{-\tau^2}d\tau - 6t\int_{-\alpha}^{\alpha}\tau e^{-\tau^2}d\tau\tau\, 3\int_{-\alpha}^{\alpha}\tau^2 e^{-\tau^2}d\tau$$

$$= 3t^2\sqrt{\pi} - 0 + 3\sqrt{\pi}\Big/2$$

$$= 5.31\,t^2 + 2.659$$

6.   Find the convolution of

$$x(t) = e^{-L+1} \text{ and } h(t) = \begin{array}{ll} e^{-2t} & t \ge 1 \\ 0 & t < 1 \end{array}$$

Ans:

$$y(t) = x(t) \times h(t)$$

$$= \int_{-\alpha}^{\alpha} x(\tau) h(t-\tau) d\tau$$

$$x(t) = \begin{cases} e^{-t} & t \geq 0 \\ e^{t} & t < 0 \end{cases} \qquad h(t) = \begin{cases} e^{-2t} & t \geq 1 \\ 0 & t < 0 \end{cases}$$

rewriting the equations

$$x(\tau) \quad \text{and} \quad h(t-\tau)$$

By precedence rule

$$h(t) \rightarrow h(\tau) \rightarrow h(\tau + t) \rightarrow h(-\tau + t) = h(t-\tau)$$

Case 1: $(t-1) < 0 \rightarrow t < 1$

$$y(t) = \int_{-\alpha}^{t-1} e^{T} e^{-2(t-\tau)} dT + \int_{t-1}^{\alpha} x(\tau) \times 0 \, d\tau$$

$$= e^{-2t} \int_{-\alpha}^{t-1} e^{3\tau} dT + 0 = \frac{1}{3} \quad e^{(t-3)} + < 1$$

Case 2: $(t-1) \geq 0 \rightarrow t \geq 1$

$$y(t) = \int_{-\alpha}^{0} e^{T} e^{-2(t-\tau)} d\tau + \int_{0}^{t-1} e^{-\tau} e^{-2(t-\tau)} d\tau + \int_{t-1}^{\alpha} e^{-\tau} d\tau$$

$$= e^{-(t+1)} - \frac{2}{3} e^{-2t}, \quad t \geq 1$$

$$\therefore \qquad y(t) = e^{-(t+1)} - \frac{2}{3} e^{-2t} \; t \geq 1 \quad \frac{1}{3} e^{(t-3)} \; t < 1$$

OBJECTIVE QUESTIONS

1. A system is ---------------- if every bounded input produces a bounded output.

2. A system whose behavior and characteristics are fixed over a time is called ----------------

3. The system is stable if the impulse response is absolutely integrable, if ------------------

4. A transmission is said to be ----------- if the response of the system is exact replica of the input signal.

5. The response of the system, with zero initial conditions, to a unit impulse or delta function $\delta(t)$ applied to the input of the system is called ---------------

6. A ------------------ system is one whose response does not begin before the input function is applied.

7. The condition for stability of the system in continuous time case is -------------

8. The BIBO stability criteria in discrete time is --------------------

9. A system is said to be ------------------------- if O/P depends on 1/P at that time only.

10. If the system informs about past samples, such a system is said to be with -------------------

11. A linear system obeys the property of ----------------------

12. A non causal system depends on ----------------- values of 1/P signal

13. A system that violates the principle of super position is ------------- system.

14. A cascade if 3 linear time invariant systems is causal and unstable. From this we conclude that   Gate 2009

   (a) each system in the cascade is individually causal and unstable

   (b) at least one system is unstable and at least one system is causal

   (c) at least one system is causal and all systems are unstable

   (d) the majority are unstable and the majority are causal   *Ans:* (b)

15. A system having $y(t) = \int\limits_{n}^{5t} x(\tau)d\tau \ t > 0$ is   Gate 2010

   (a) Linear and causal   (b) Linear but not causal

   (c) Causal but not linear   (d) Neither causal nor linear   *Ans:* (b)

16. Given x[n] and y[n] find h[n]   Gate 2010

$$x[n] = \{1,-1\} \rightarrow \boxed{h(n)} \rightarrow y(n) = \{1,0,0,0,-1\}$$

   (a) $h[n] = \{1, 0, 0, 1\}$ (b) $h[n] = \{1, 0, 1\}$

   (c) $h[n] = \{1, 1, 1, 1\}$ (d) $h[n] = \{1, 1, 1\}$ *Ans:* (c)

17. Given $x(t) = e^{-t}$ and $y(t) = e^{-2t}$ for $t > 0$, the convolution of $x(t) \times y(t)$ is   Gate 2011

   (a) $e^{-t} - e^{-2t}$ (b) e^{-3t}

   (c) e^{+t} (d) $e^{-t} + e^{2t}$ *Ans:* (a)

18. Consider $\dfrac{d^2y(t)}{dt^2} + \dfrac{2dy(t)}{dt} + y(t) = (t)$ with $y(t)\big|_{t=0} = -2$ Gate 2012

$\dfrac{dy}{dt} = 0\Big|_{t=0}$ then $\dfrac{dy}{dt}\Big|_{t=0^+}$ is

(a) -2 (b) -1

(c) 0 (d) 1 *Ans:* (d)

19. If $y(t) = \int_{-\alpha}^{t} x(\tau)\cos(3\tau)\,dT$, the system is

(a) T1 and stable   (b) NTI and stable

(c) T1 and unstable   (d) NTI and unstable   *Ans:* (d)

20. If $h(t) = tu(t)$. The for $u(t-1)$ output is   Gate 2013

(a) $\dfrac{t^2}{2}u(t)$ (b) $\dfrac{t(t-1)}{2}u(t-1)$

(c) $\dfrac{(t-1)^2}{2}u(t-1)$ (d) $\dfrac{t^2-1}{2}u(t-1)$ *Ans:* (c)

21. Two systems $h_1(t)$ and $h_2(t)$ are in cascade then overall impulse response is Gate 2013

(a) Product of $h_1(t)$ and $h_2(t)$ (b) $h_1(t) + h_2(t)$

(c) $h_1(t) \times h_2(t)$ (d) $h_2(t) - h_1(t)$ *Ans:* (c)

22. If $h(t) = \delta(t-1) + \delta(t-3)$. Its output at $t-z$ is   Gate 2013

(a) 0 (b) 1

(c) 2 (d) 3 *Ans:* (b)

23. If $y[n]$ is $h[n] \times g[n]$, $h[n] = \left(\dfrac{1}{2}\right)^n u[x]$ and $g[n]$ is causal sequence. If $y[0] = 1$, $y[1] = \frac{1}{2}$

the $g[1]$ is   Gate 2012

(a) 0 (b) $\frac{1}{2}$

(c) 1 (d) $3/2$ *Ans:* (a)

24. If $h(t) = e^{-5t}u(t)$ is impulse response. Then what is input if output is $y(t) = e^{-3tu}(t) - e^{-5t}$
u(t)   Gate 2014

(a) $e^{-3t}u(t)$ (b) $2\,e^{-3t}u(t)$

(c) $e^{-5}u(t)$ (d) $2e^{-5t}u(t)$ *Ans:* (b)

SUMMARY OF CHAPTER

1. An LTI system is completely characterized by its impulse response.
2. Output of LTI system is convolution of input with impulse response of the system

$$y(t) = x(t) \times h(t)$$

3. Convolution operation gives only zero state response of the system.
4. LTI system is causal if $h(t) = 0$ for $t < 0$.

5. LTI system is stable if $\displaystyle\int_{-\alpha}^{\alpha} |h(T)|\, dT < \alpha$

CHAPTER 6

Convolution and Correlation

Objectives of the chapter

Convolution in time domain and frequency domain, graphical representation of convolution, convolution property of fourier transforms, cross correlation and auto correlation, properties of correlation functions, energy density spectrum, Parsevals theorem, power density spectrum, relation between convolution and correlation. Extraction of signal from noise by filtering, detection of signals in the presence noise by correlation.

6.1 INTRODUCTION

Convolution is a formal mathematical operation like addition, multiplication and integration. Addition takes two numbers and produces a third number, while convolution takes two signals and produces a third signal. Convolution is used to describe the relation between three signals of interest; the input signal, impulse response and the output signal. Correlation is also same as convolution, which produces third signal using two signals. This third signal is called cross-correlation of two input signals. It extracts the similarities between two signals.

6.2 CONVOLUTION IN TIME DOMAIN

The response of a system having a transfer function h(t) for a driving function f(t).

$$r(t) = f(t) * h(t)$$

* is known as time convolution and is given by, convolution integral

$$r(t) = \int_{-\alpha}^{\alpha} f(T) h(t-T) dT$$

If h(t) and f(t) are causal

then
$$r(t) = \int_{0}^{t} f(T) h(t-T) dT$$

If h(T) is causal then h(t − T) is 0 for $T > t$

Any arbitrary driving function can be expressed as a continuous sum of exponential functions, impulse functions.

r (t) = continuous sum of impulse response components

Let f(T) is sum of narrow pulses of width ΔT.

If ΔT is small then each pulse becomes impulse of strength f(T) ΔT.

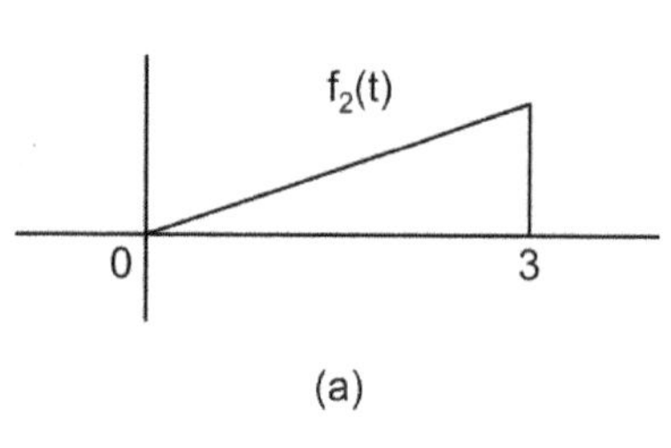

$\therefore$ r (t) = sum of responses of these impulses up to r = t.

Let h(t – T) is the response of the system for impulse at T.

$\therefore$ response of system for impulse of strength f(T) ΔT

$$r_T(t) = \lim_{\Delta T \to 0} f(T) h(t - T) \Delta T$$

$$r(t) = \int_0^t f(T) h(t - T) dT$$

Since the responses occuring at T > t, has no influence, hence limits are from 0 to t.

6.3 GRAPHICAL REPRESENTATION OF CONVOLUTION

Let us consider $f_1(t)$ and $f_2(t)$ as rectangular and triangular pulses

$$f_1(t) * f_2(t) = \int_{-\alpha}^{\alpha} f(T) f_2(t - T) dT$$

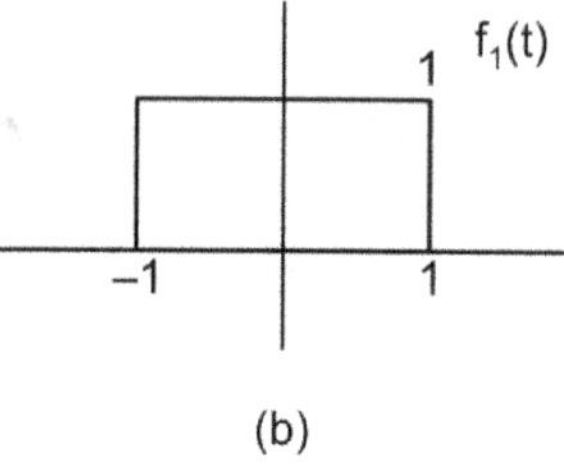
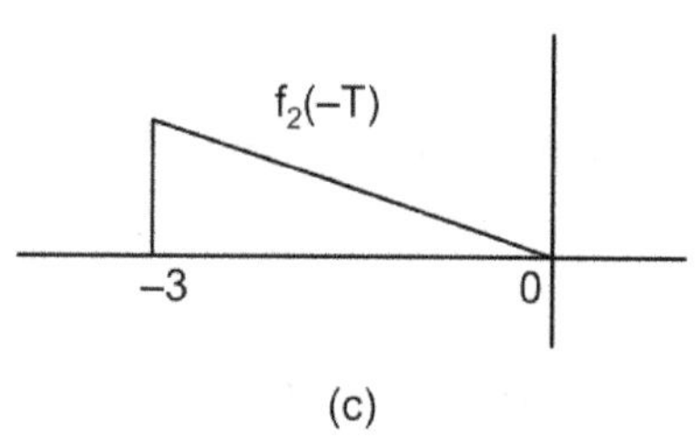

(a)   (b)   (c)

Fig. 6.2

1. $f_2(-T)$ is obtained by folding $f_2(T)$ about vertical axis as shown in fig. 6.2c

2. $f_2(t_1 - T)$ is obtained by shifting of $f_2(-T)$ along positive axis

3. $f_2(t_1 - T)$

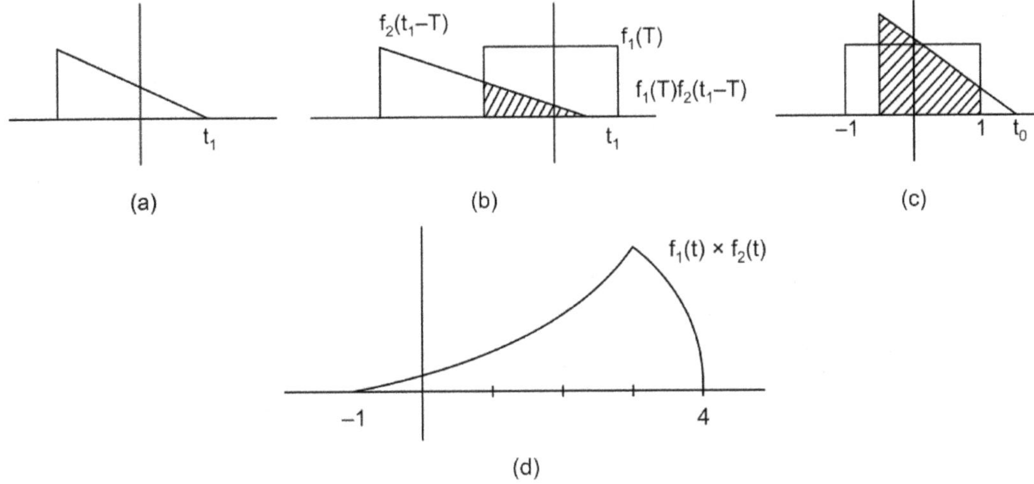

Fig. 6.3

1. Fold the function $f_2(T)$ about the vertical axis passing through the origin of T axis and obtain $f_2(-T)$.

2. Progress the folded function along T axis by an amount t_0. The rigid frame now represents $f_2(t_0 - T)$.

3. The product of the function represented by this $f_2(t)$ with $f_1(T)$ represents $f_1(T)\,f_2(t_0 - T)$ and the area under the curve is

$$\int f_1(T)\,f_2(t - T)\,dT$$

4. Repeat the procedure for different values of t by successively progressing the frame by different amounts.

Example: Find the convolution of x(t) and h(t) shown below

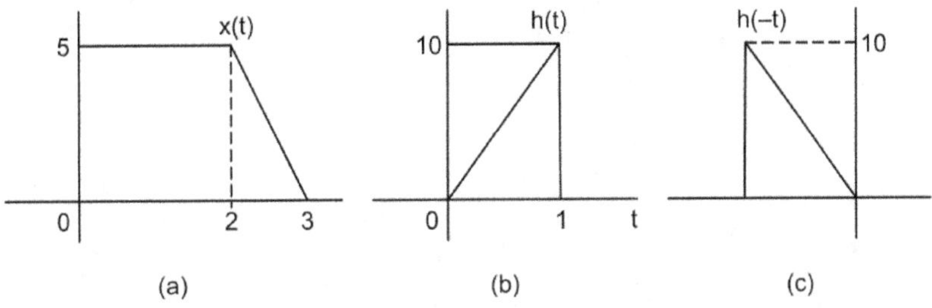

Fig 6.4

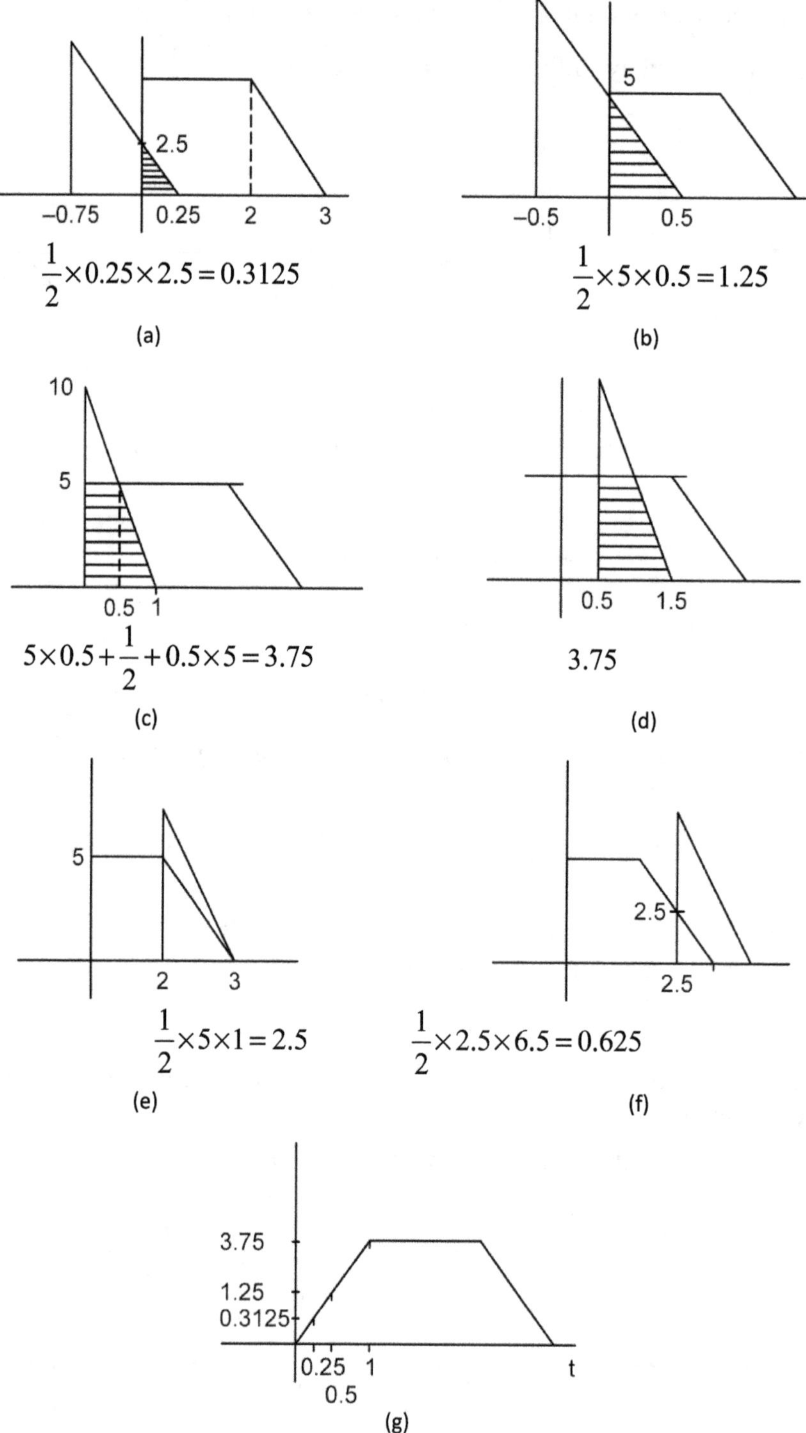

$$\frac{1}{2}\times 0.25\times 2.5 = 0.3125$$

(a)

$$\frac{1}{2}\times 5\times 0.5 = 1.25$$

(b)

$$5\times 0.5+\frac{1}{2}+0.5\times 5 = 3.75$$

(c)

$$3.75$$

(d)

$$\frac{1}{2}\times 5\times 1 = 2.5$$

(e)

$$\frac{1}{2}\times 2.5\times 6.5 = 0.625$$

(f)

(g)

Fig. 6.5

SOLVED PROBLEMS

1. Find the convolution of e^{-2t} and e^{-t}

$$e^{-2t} \cdot u(t) * e^{-t} u(t)$$

$$f_1(t) = e^{-2t} u(t)$$

$$f_2(t) = e^{-t} u(t)$$

$$f_1(t) * f_2(t) = \int_{-\alpha}^{\alpha} e^{-2T} u(T) e^{-(t-T)} u(t-T) dT$$

$$\int_0^t e^{-2T} e^{-(t-T)} dT = e^{-t} \int_0^t e^{-T} dT$$

$$= e^{-t}\left\{1 - e^{-t}\right\}$$

2. Find the convolution of t^2 and e^{-t}

$$t^2 * e^{-t} u(t)$$

$$f_1(t) * f_2(t) = \int_{-\alpha}^{\alpha} T^2 e^{-(t-T)} dT$$

$$= e^{-t} \int_0^t T^2 \cdot e^T dT$$

$$\int_0^T T^2 e^T dT = T^2 e^T \Big|_0^t - 2\int_0^t e^T \cdot T \cdot dT$$

$$= T^2 e^t - 2\int_0^t T e^T dT$$

$$= T^2 e^t - \left[2Te^t\right]_0^t + 2\int_0^t e^T dT$$

$$= T^2 e^t - 2te^t + 2\left(e^t - 1\right)$$

6.4 PROPERTIES OF CONVOLUTION

1. Commutative law:

$$f_1(t) * f_2(t) = f_2(t) * f_1(t)$$

$$f_1(t) * f_2(t) = \int_{-\alpha}^{\alpha} f_1(T) f_2(t-T) dT \qquad T = t - x = \int_{-\alpha}^{\alpha} f_1(t-x) f_2(T) dx$$

2. Distributive law:

$$f_1(t) * \{f_2(t) + f_3(t)\} = f_1(t) * f_2(t) + f_1(t) * f_3(t)$$

3. Associative law:

$$f_1(t) * \{f_2(t) * f_3(t)\} = \{f_1(t) * f_2(t)\} * f_3(t)$$

6.5 CONVOLUTION IN FREQUENCY DOMAIN

From convolution theorem, convolution of two time functions results in multiplication in frequency domain.

$$\mathcal{L}^{-1}\{F_1(w) = f_1(t)\} \text{ and }$$
$$\mathcal{L}^{-1}\{f_2(w)\} = f_2(t)$$

Then

$$f_1(t) * f_2(t) = \mathcal{F}^{-1}\{F_1(\omega)F_2(\omega)\}$$
$$= \mathcal{L}^{-1}\{F_1(s)F_2(s)\}$$

SOLVED PROBLEMS

1. Find convolution of $2.e^{-2t}$ and e^{-t}

$$h(s) = \frac{2}{S+2}$$
$$h(t) = 2 \cdot e^{-2t} u(t)$$

If

$$x(t) = e^{-t} u(t)$$
$$x(s) = \frac{1}{S+1}$$

then

$$r(s) = x(s)h(s) = \frac{1}{S+1}\frac{2}{S+2}$$
$$= \frac{2}{S+1} - \frac{2}{S+2}$$
$$r(t) = 2 \cdot e^{-t} - 2 \cdot e^{-2t}$$
$$r(t) = \int_0^t e^{-T} \cdot 2e^{-2(t-T)} dT$$
$$= 2e^{-2t}\int_0^t e^{T} dT = 2e^{-2t}(e^t - 1)$$
$$= 2e^{-t} - 2e^{-2t}$$

2. Find f(t) if $F(s) = \dfrac{1}{\left(S^2 + a^2\right)^2}$

$$F_1(s) = \dfrac{1}{S^2 + a^2}$$

$$F_2(s) = \dfrac{1}{S^2 + a^2}$$

$$\mathcal{L}^{-1}\left\{\dfrac{1}{S^2 + a^2}\right\} = \sin at$$

$$\mathcal{L}^{-1}\left\{\dfrac{1}{\left(S^2 + a^2\right)^2}\right\} = \int_0^t f_1(T)f_2(t-T)dT$$

$$= \dfrac{1}{a^2}\int_0^t \sin aT \sin(at - T)dT$$

$$= \dfrac{1}{2a^3}\left\{\sin at - at\cos at^{-1}\right\}$$

3. Solve $\displaystyle\int_{-\alpha}^{\alpha} \dfrac{\sin 4x}{x} \cdot \dfrac{\sin(t-x)}{(t-x)}dx$

$$f_1(t) = \dfrac{\sin 4t}{t} \qquad\qquad f_2(t) = \dfrac{\sin t}{t}$$

$$f_1(t) * f_2(t) = \mathcal{F}^{-1}\left\{F_1(\omega)F_2(\omega)\right\}$$

$$= \dfrac{\pi \sin t}{t}$$

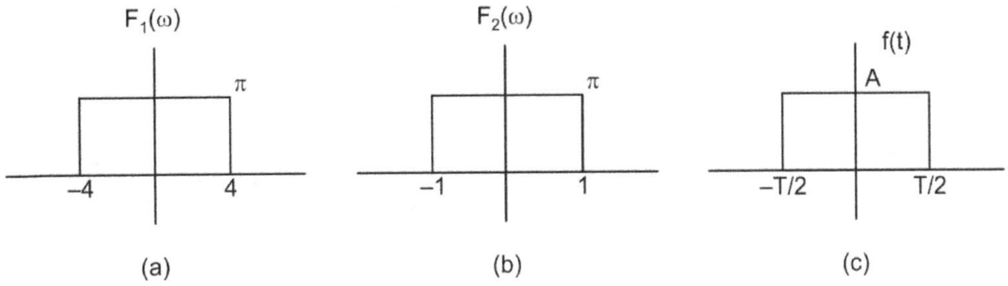

$F_1(\omega)$ — (a), $F_2(\omega)$ — (b), $f(t)$ — (c)

Fig. 6.6

6.6 CORRELATION

It is defined as measure of similarity between two signals or to compare similarities between them.

If there are no similarities

$$\int_{-\alpha}^{\alpha} f_1(t)\, f_2(t)\, dt = 0$$

Sometimes, even the two signals are similar, if one is time delayed then the above assumption will fail.

Thus we must search the similarity at different time shifts, then the integral becomes

$$\int_{-\alpha}^{\alpha} f_1(t)\, f_2(t-T)\, dt$$

which is known as cross correlation

It is similar to convolution except the function $f_2(t)$ is not folded back.

Correlation of $f_1(t)$ and $f_2(t)$ is similar to convolution of $f_1(t)$ and $f_2(-t)$

$$\text{con } v = \int_{-\alpha}^{\alpha} f_1(T)\, f_2(t-T)\, dT$$

$$\text{corr} = \int f_1(t) f_2(t-T)\, dt$$

6.7 AUTO CORRELATION

It defines similarity between $f_1(t)$ and $f_1(t-T)$ i.e., between same function but shifted by time.

$$\text{Auto correlation} = \int f_1(t) f_1(t-T)\, dt$$

SOLVED PROBLEMS

1. Find auto correlation for

$$x(t) = \sin \omega t$$

$$\phi_{11}(t) = \frac{1}{T} \int_{-T/2}^{T/2} \sin \omega t \sin \omega(t-T)\, dt$$

$$= \frac{1}{2T} \int_{-T/2}^{T/2} \cos \omega T - \cos(2\omega t - \omega T)\, dt$$

$$= \frac{1}{2T} \left\{ t \cos \omega T - \frac{\sin(2\omega t - \omega T)}{2\omega} \right\}_{-T/2}^{T/2}$$

$$= \frac{\cos \omega T}{2}$$

2. Find the auto correlation of $x(t) = A\cos(\omega_c t + \theta)$

$$\phi_{11}(T) = \frac{1}{T} \int_{-T/2}^{T/2} A\cos(\omega_c t + \theta) A\cos(\omega_c t + \omega_c T + \theta) dT$$

$$= \frac{A^2}{T} \int_{-T/2}^{T/2} \frac{\cos\omega T + \cos(2\omega_c t + \omega T + 2\theta) dT}{2}$$

$$= \frac{A^2}{2}\cos\omega T$$

6.8 COMPUTATION OF CORRELATION

To find the cross correlation of two functions $f_1(t)$ and $f_2(t)$,

1. Obtain the function $f_2(t - T)$ by shifting the function right by a time lag T.
2. Multiply the functions $f_2(t - T)$ with $f_1(t)$ and integrate to obtain $\phi_{12}(T)$.

The correlation of two functions $f_1(t)$ and $f_2(t)$ can be obtained by using convolution procedure also.

$$\phi_{12}(T) = \int f_1(t) f_2(t - T) dt$$
$$= \int f_1(t) f_2(-(T - t)) dt$$
$$= f_1(t) * f_2(-T)$$

1. Fold the function $f_2(t)$ to obtain $f_2(-t)$.
2. Then find convolution between $f_1(T)$ and $f_2(T)$, to obtain correlation.

6.9 DECONVOLUTION

Whenever we need to recover the input signal by the help of h(t) and y(t), deconvolution is used.

We have $y(t) = \int x(T) h(t - T) dT$

$$E = \frac{1}{2\pi} \int F(\omega) \int f(t) e^{j\omega t} dt \, d\omega$$

$$\int_{-\alpha}^{\alpha} f^2(t) dt = \frac{1}{2\pi} \int_{-\alpha}^{\alpha} F(\omega) \, F(-\omega) \, d\omega$$

or $\qquad \displaystyle\int_{-\alpha}^{\alpha} f^2(t) dt = \frac{1}{2\pi} \int_{-\alpha}^{\alpha} |F(\omega)|^2 \, d\omega$

Energy of a signal is given by $\dfrac{1}{2\pi}$ times area under $F(\omega)^2$ curve.

6.10 EXTRACTION OF SIGNAL FROM NOISE BY FILTERING AND CORRELATION

Periodic signals are often effected by random noise. Radar signals, brain waves etc are prone to such noise.

Such random noise is un correlated with any other signal.

Let $s(t)$ is periodic signal

$n(t)$ noise signal

$$\lim_{T \to \alpha} \frac{1}{T} \int_{-T/2}^{T/2} S(t) n(t-T) dt = 0 \quad \text{for all } T$$

$$\phi_{sn}(T) = 0$$

Let $\phi_{ff}(T)$, $\phi_{ss}(T)$, $\phi_{nn}(T)$ are AC functions of $f(t)$ received signals, $s(t)$ transmitted and $n(t)$ noise signals.

$$\phi_{ff}(T) = \lim_{T \to \alpha} \frac{1}{T} \int_{-T/2}^{T/2} f(t) f(t-T) dt$$

$$= \lim_{T \to \alpha} \frac{1}{T} \int_{-T/2}^{T/2} [S(t) + n(t)][S(t-T) + n(t-T)] dt$$

$$= \phi_{ss}(T) + \phi_{m}(T) + \phi_{sn}(T) + \phi_{ns}(T)$$

$$= \phi_{ss}(T) + \phi_{nn}(T)$$

$$\because \quad \phi_{sn} = \phi_{ns} = 0 \text{ since they are un correlation}$$

$$\therefore \quad \phi_{ff}(T) = \phi_{ss}(T) + \phi_{nn}(T)$$

AC of a non periodic function is 0 if $T \to \alpha$.

$\therefore$ If it is made large then it is easy to detect and separate $S(t)$ from $n(t)$.

Limitations

1. Auto correlation is applicable for signals with finite interval
2. Cross correlation is not zero for finite interval
3. The operation of cross correlation is equal to filtering in frequency domain which allows to pass through only frequency components of fundamental frequency of $S(t)$.
4. A part from filtering noise, relative attenuation will be there for harmonics of $S(t)$.
5. Cross correlation of $f(t)$ with uniform periodic impulse function is equal to passing $f(t)$ through a filter which allows all frequency components.

6.10.1 DETECTION OF NOISE BY CROSS CORRELATIONS

Let us consider a locally generated signal c(t)

$$\therefore \ \phi_{fc}(T) = \lim_{T \to \alpha} \int_{-T/2}^{T/2} \{S(t) + n(t)\} c(t - T) dt$$

$$= \phi_{sc}(T) + \phi_{nc} T$$

$$= \phi_{sc}(T)$$

If c(t) is a train of impulse with same frequency

$$\text{then} \quad \phi_{fc}(T) = \frac{1}{T} \int S(t) \delta(t - T) dt$$

$$= \frac{1}{T} S(T)$$

6.10.2 DETECTION BY USING FILTERING

$$f_1(t) \leftrightarrow F_1(\omega)$$

$$f_2(t) \leftrightarrow F_2(\omega)$$

$$\phi_{12}(T) \leftrightarrow F_1(\omega) F_2(-\omega)$$

$$c(t) \leftrightarrow c(\omega)$$

$$c(-t) \leftrightarrow c(-\omega)$$

since c(t) is a train of impulses

$$c(t) = \sum c_n \delta(\omega - n\omega_0)$$

6.11 POWER DENSITY SPECTRUM

Some Signals known as Power Signals have Infinite Energy

That is
$$\int_{-\alpha}^{\alpha} f(t) 2 dt = \alpha$$

$\therefore$ correlation function doesn't occur for these signals

Therefore it is better to take average correlation

$$\overline{\phi_{11}(t)} = \lim_{T \to \alpha} \frac{1}{T} \int_{-T/2}^{T/2} f_1(t) f_1(t - T) dT$$

$$\overline{\phi}_{12}(t) = \lim_{T \to \alpha} \frac{1}{T} \int_{-T/2}^{T/2} f_1(t) f_2(t - T) dT$$

Let
$$f_T(t) \leftrightarrow F_T(\omega)$$

$$\left| F_T(\omega) \right|^2 = \pi S_T(\omega)$$

$\therefore$
$$\overline{\phi}_{ff}(t) \leftrightarrow \lim_{T \to \alpha} \frac{1}{T} \left| F_T(\omega) \right|^2$$

$$= \lim_{T \to \alpha} \pi \frac{S_T(\omega)}{T}$$

As $T \to \alpha$, the energy density spectrum $S_T(\omega) \to \alpha$ and hence $\dfrac{S_T(\omega)}{T} \to \alpha$

$\therefore$ Average power density $P(\omega) = \lim\limits_{T \to \alpha} \dfrac{1}{T} S_T(\omega)$

$$\overline{\phi}_{ff}(t) \leftrightarrow \pi P(\omega)$$

or
$$\int_0^{\alpha} P(\omega) d\omega = \lim_{T \to \alpha} \frac{1}{T} \int_{-T/2}^{T/2} f^2(t) dt$$

$$= \frac{1}{2\pi} \int_{-\alpha}^{\alpha} S(\omega) d\omega$$

Power density spectrum $S_f(\omega) = \lim\limits_{T \to \alpha} \dfrac{\left| F(\omega) \right|^2}{T}$

In power density spectrum we have information about magnitude but not about phase. Signals with identical frequency spectrum but different phase functions will have identical power density spectrum.

6.12 ENERGY DENSITY SPECTRUM

We have
$$\phi_{11}(T) = f_1(t) * f_1(-t)$$

If $f_1(t)$ is even signal

$$f_1(t) = f_1(-t)$$

$\therefore$ for even signals correlation and convolution are same
$$f_1(t) * f_2(-t) = X_1(\omega) X_1(-\omega)$$

$$\mathscr{F}\left\{ \phi_{11}(T) \right\} = X_1(\omega) x_1(-\omega)$$

$$\phi_{11}(T) = \int f_1(t) f_1(t - T) e^{-j\omega T} dT$$

Let T = 0

$$\phi_{11}(O) = \int f_1(t)f_1(t)dT$$

$$= \int f_1^2(t)dT$$

At the origin AC is equal to energy of the signal.

For non periodic signals energy is finite and power is zero.

For a non periodic signal energy

$$E = \int_{-\alpha}^{\alpha} f^2(t)dt$$

also we have

$$f(t) = \frac{1}{2\pi} \int_{-\alpha}^{\alpha} F(\omega)e^{j\omega t}\, d\omega$$

$$\therefore \quad E = \int_{-\alpha}^{\alpha} f(t)\left\{\frac{1}{2\pi} \int_{-\alpha}^{\alpha} F(\omega)e^{j\omega t}\, d\omega\right\} dt$$

$$= \frac{1}{2\pi} \int_{-\alpha}^{\alpha} F(\omega)\left\{\int_{-\alpha}^{\alpha} f(t)e^{j\omega t}\, dt\right\} d\omega = \frac{1}{2\pi} \int_{-\alpha}^{\alpha} F(\omega)F(-\omega)\, d\omega$$

$$\text{If} \quad F(\omega) = |F(\omega)|e^{j\theta\omega}, \qquad F(-\omega) = |F(\omega)|e^{-j\theta\omega}$$

$$\therefore \quad F(\omega)\cdot F(-\omega) = |F(\omega)|^2$$

$$\therefore \quad E = \int_{-\alpha}^{\alpha} f(t^2)dt = \frac{1}{2\pi} \int_{-\alpha}^{\alpha} |F(\omega)|^2\, d\omega$$

The energy of a signal is $\dfrac{1}{2\pi}$ times the area under the curve $|F(\omega)^2|$.

Energy spectrum is continuous and the energy associated with a particular frequency component is zero. But there is finite energy associated with finite band of frequencies.

The energy contained in the frequency band (ω_1,ω_2) is

$$E = 2\cdot\frac{1}{2\pi} \int_{\omega_1}^{\omega_2} |F(\omega)|^2\, d\omega$$

$$= \frac{1}{\pi} \int_{\omega_1}^{\omega_2} |F(\omega)|^2\, d\omega$$

[As the same band occurs on negative side hence energy is double]

$$\therefore \text{ Energy / unit bandwidth } = S(\omega) = \frac{1}{\pi}\left|F(\omega)\right|^2$$

$$\therefore \qquad E = \int_{\omega_1}^{\omega_2} S(\omega)\,d\omega$$

Ex: For gate function energy density function is given by

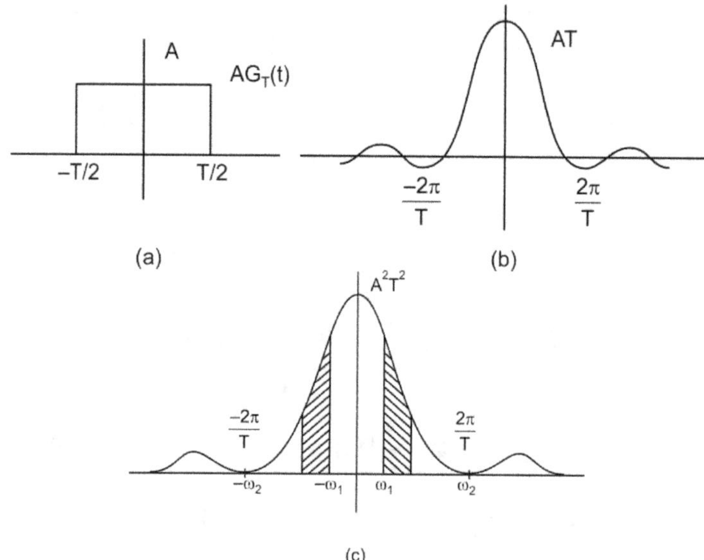

Fig. 6.7

PROBLEM

Find the auto correlation function of gate function and hence find its energy density spectrum of
the function.

$$\phi_{11}(T) = \int_{-\alpha}^{\alpha} AG_T(t)\,AG_T(t-T)\,dt$$

$$= A^2(T - |T|)$$

The energy density spectrum of gate function is $\dfrac{1}{\pi}$ time of $\mathscr{F}T$ of $\phi_{11}(t)$

$$F(\omega) = ATSa\left(\frac{\omega T}{2}\right)$$

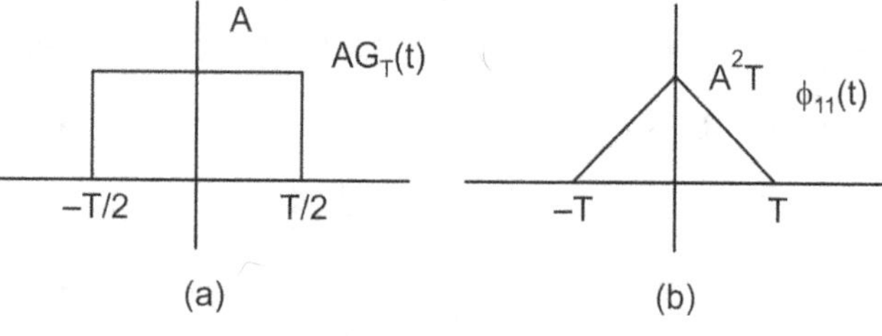

Fig. 6.8

$$\phi_{11}(\omega) \;=\; \int\limits_{-\alpha}^{\alpha} \phi_{11}(t)e^{-j\omega T}\,dT$$

$$\therefore\; \phi_{11}(\omega) \;=\; A^2 T^2 \left[Sa\left(\frac{\omega T}{2}\right)\right]^2$$

$$S(\omega) \;=\; \frac{1}{\pi}A^2 T^2 \left\{Sa\left(\frac{\omega}{2}T\right)^2\right\}$$

6.13 PARSEVALS THEOREM

Energy of a signal is given by

$$E = \int\limits_{-\alpha}^{\alpha} f^2(t)\,dt$$

We have
$$f(t) = \frac{1}{2\pi}\int F(\omega)e^{j\omega t}\,d\omega$$

$$= \int f(t)\left\{\frac{1}{2\pi}\int F(\omega)e^{j\omega t}\,d\omega\right\}dt$$

6.14 PROPERTIES OF CORRELATION

1. If $\quad f(t) \leftrightarrow F(\omega)\;$ then $\; f(-t) \leftrightarrow F(-\omega)$

$$F(\omega) = \int\limits_{-\alpha}^{\alpha} f(t)e^{-j\omega t}\,dt \;\; \text{and} \;\; F(-\omega) = \int\limits_{-\alpha}^{\alpha} f(t)e^{j\omega t}\,dt$$

Auto correlation at origin is equal to energy of signal

2. $\phi_{11}(0) = \int\limits_{-\alpha}^{\alpha} f_1^2(t)\,dt$

3. $\phi_{11}(0) > \left|\phi_{11}(T)\right| \qquad\qquad T \neq 0$

4. $\phi_{11}(T) = \phi_{11}(-T)$

5. $\phi_{11}(T) \leftrightarrow \left|F_1(\omega)\right|^2$

6. $\phi_{12}(T) \leftrightarrow F_1(\omega)F_2(-\omega)$

7. $\phi_{11}(0) = \phi_{22}(0) > 2\left|\phi_{12}(T)\right|$

8. $\phi_{12}(T) = \phi_{21}(-T)$

9. Auto correlation is maximum when $T = 0$

$$\int_{-\alpha}^{\alpha} \{f_1(t) \pm f_1(t-T)\}^2 \, dt > 0 \qquad \text{for} \quad T \neq 0$$

$$\int f_1(t)^2 + \int f_1(t-T)^2 \pm 2 \int f_1(t) f_1(t-T) \, dt > 0$$

$$2\phi_{11}(0) \pm 2\phi_{11}(t) > 0$$

$$\phi_{11}(0) > \phi_{11}(T) \qquad \text{at} \quad T \neq 0$$

10. Auto correlation becomes smaller at a larger value of T for non periodic signals.

$$\underset{T \to \alpha}{Lt} \, \phi_{11}(t) = 0$$

SOLVED PROBLEMS: EXERCISE 1

1. Find the convolution between

$$x(t) = \pi \frac{(t-1)}{2}; \qquad h(t) = u(t-10)$$

Ans: $x(T)$

$$\pi(t) \Rightarrow t \leq |t|$$

$$\pi\left(\frac{t-1}{2}\right) \Rightarrow \pi t \qquad t \leq 2$$

$h(T)$ $\qquad\qquad$ $h(t-T)$

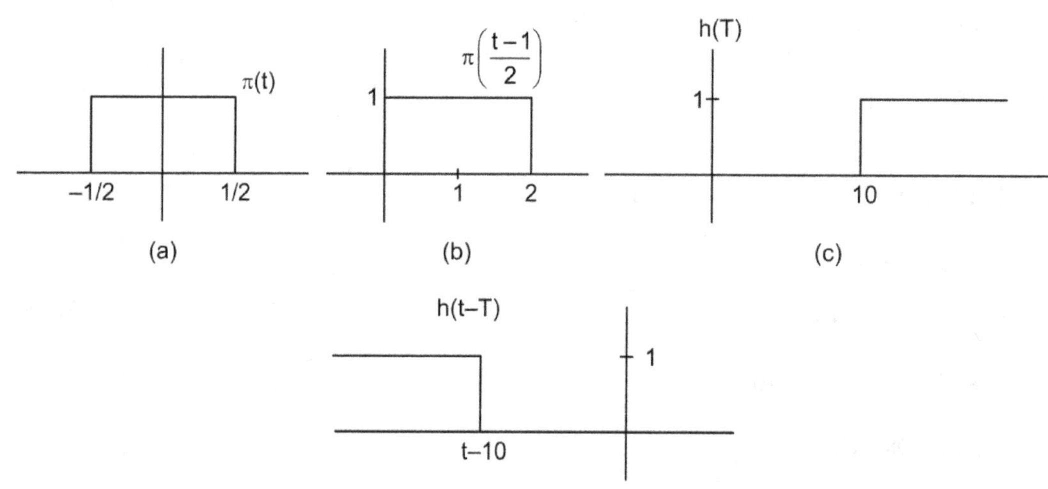

for

$t \le 10$

$y(t) = 0$

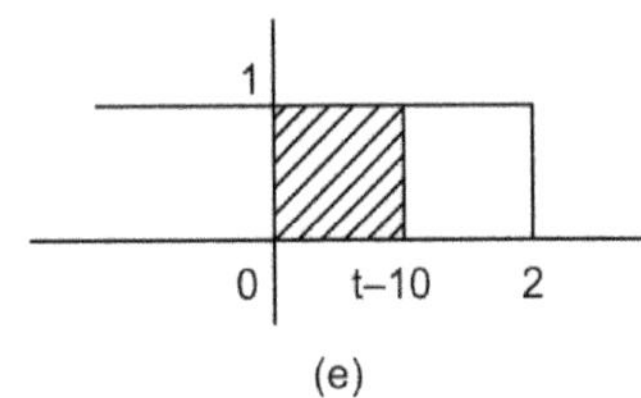

(e)

$10 \le t \le 12$

$y(t) = (t-10) \times 1$

$\quad = t - 10$

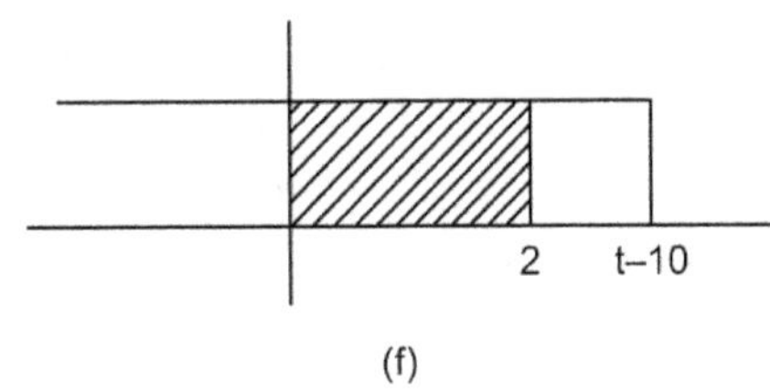

(f)

$t > 12$

$y(t) = 1 \times 2 = 2$

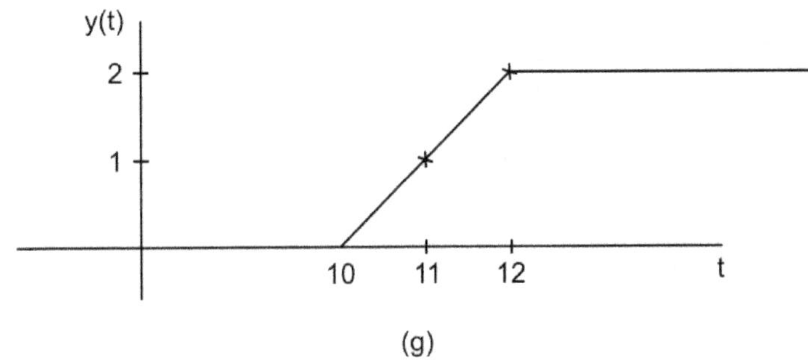

(g)

Fig. 6.9

2. Find the Convolution of the Signals x(T) and h(t)

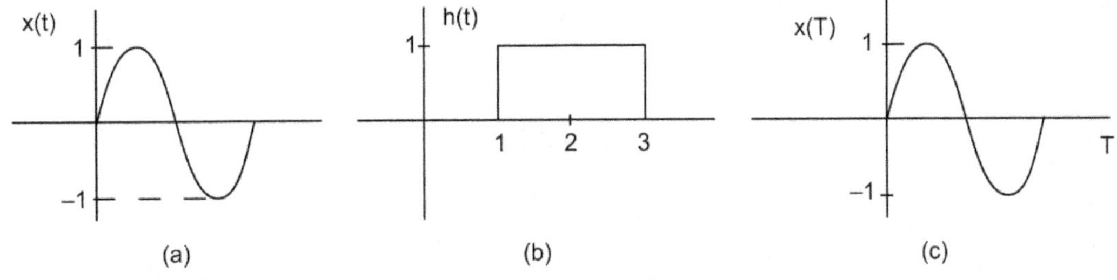

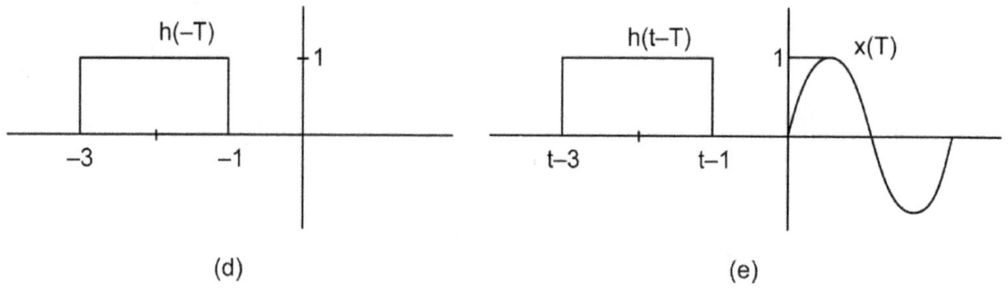

(d)   (e)

for $t \leq 1$

$$y\,(t) = 0$$

for $1 \leq t \leq 3$

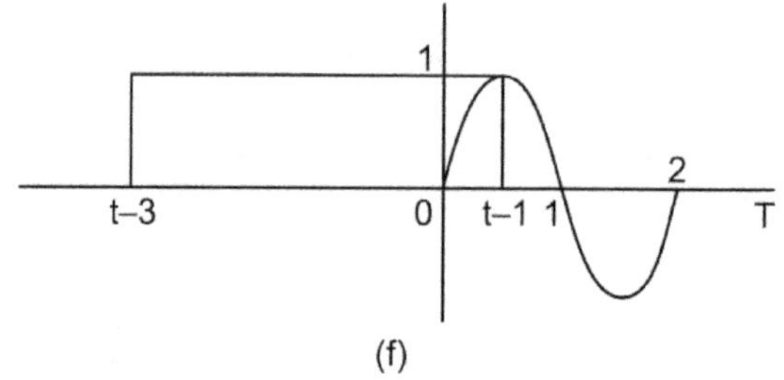

(f)

$$y\,(t) = \int_{0}^{t-1} \sin \pi T \, dT$$

$$= \left. \frac{\cos \pi T}{\pi} \right/_{0}^{t-1} = \frac{-1}{\pi}\{\cos \pi (t-1) - 1\}$$

$$= \frac{1}{\pi}\{1 - \cos \pi (t-1)\}$$

for $3 < t \leq 5$

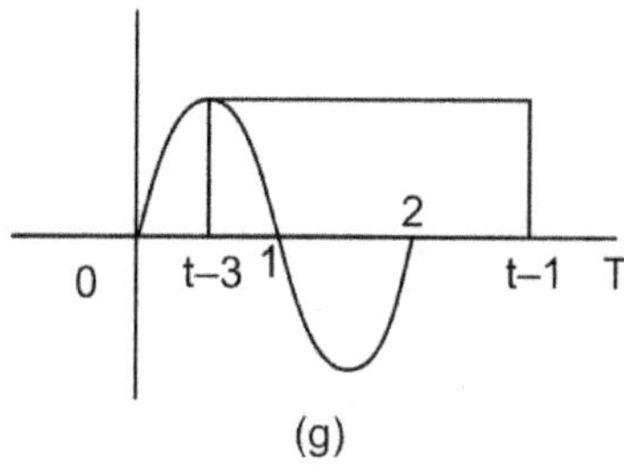

(g)

Fig. 6.10

$$y(t) = \int_{t-3}^{2} \sin \pi \, dT$$

$$= -\frac{\cos \pi T}{\pi} \Big/ \Big._{t-3}^{2} = \frac{-1}{\pi}\{\cos 2\pi - \cos \pi(t-3)\}$$

$$= \frac{-1}{\pi}\{1 - \cos \pi(t-3)\}$$

for $\quad 5 < t$

$$y(t) = 0$$

3. Find the correlation of x(t) and h(t) given by

$$x(t) = t \qquad\qquad \text{for} \quad 0 \le t \le 1$$
$$\quad\;\; = 2 - t \qquad\qquad \text{for} \quad 1 \le t \le 2$$
$$h(t) = 1 \qquad\qquad \text{for} \quad 0 \le t \le 2$$

Ans:

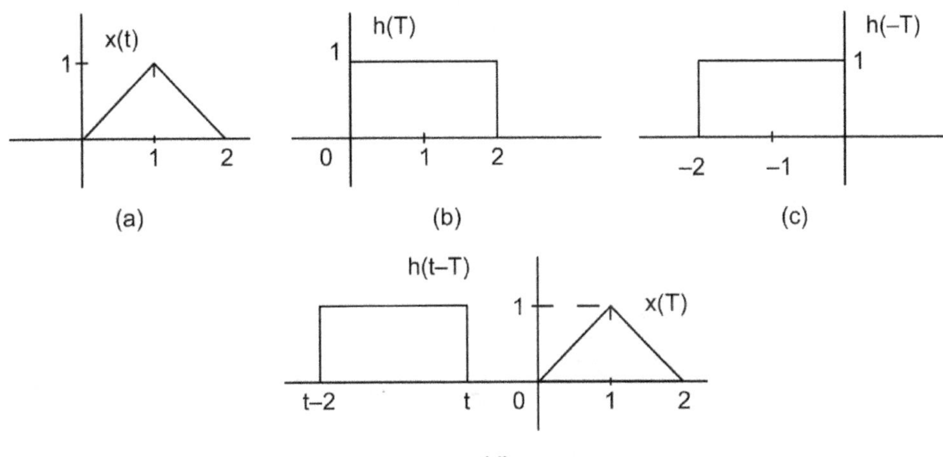

for $\quad t = 0 \quad y(t) = 0$

for $\quad t = 1$

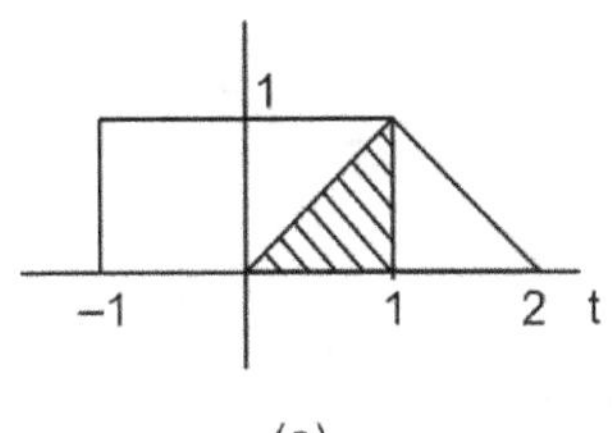

$$y(t) \;=\; \frac{1}{2} \times 1 \times 1 = 0.5$$

for $t = 2$

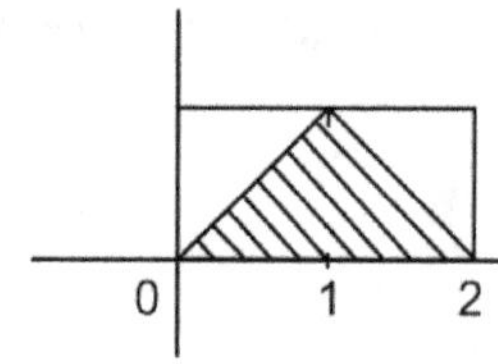

(f)

$$y(t) \;=\; 2\left(\frac{1}{2} \times 1 \times 1\right) = 1$$

for $t = 3$

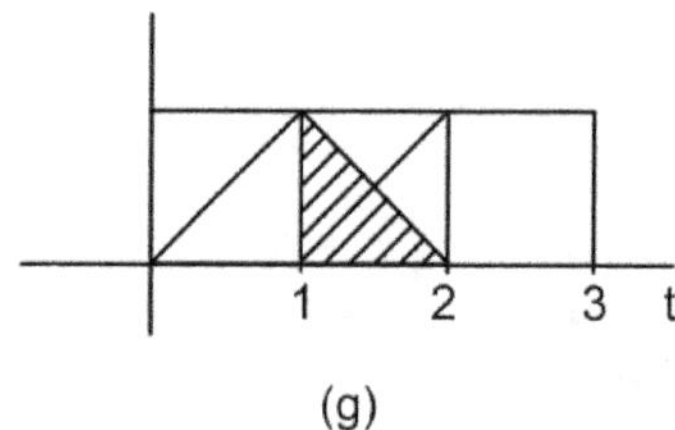

(g)

$$y(t) = 0.5$$

for $t = 4$

$$y(t) = 0$$

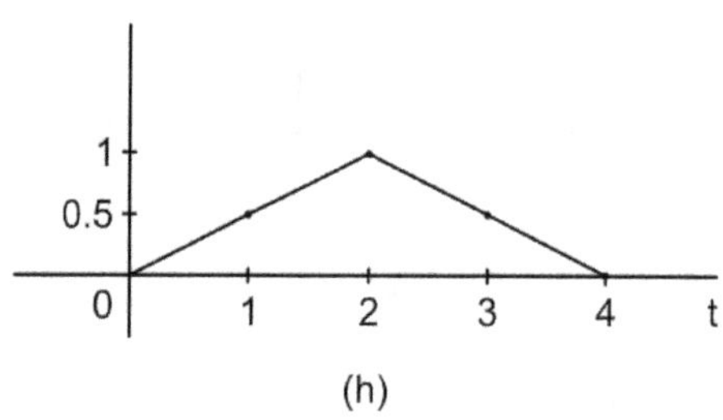

(h)

Fig. 6.11

4. Find the convolution of $x(t)$ and $h(t)$

$$x(t) \;=\; 1 \qquad 0 \leq t \leq 2$$
$$\quad\;\; =\; 0 \qquad \text{other wise}$$
$$h(t) \;=\; 1 \qquad 0 \leq t \leq 3$$
$$\quad\;\; =\; 0 \qquad \text{other wise}$$

Ans:

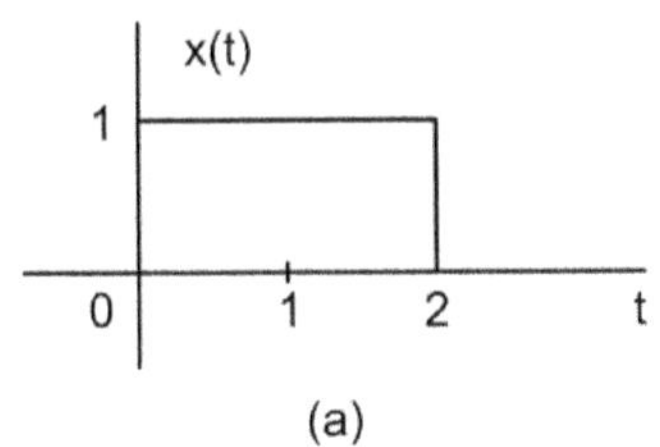

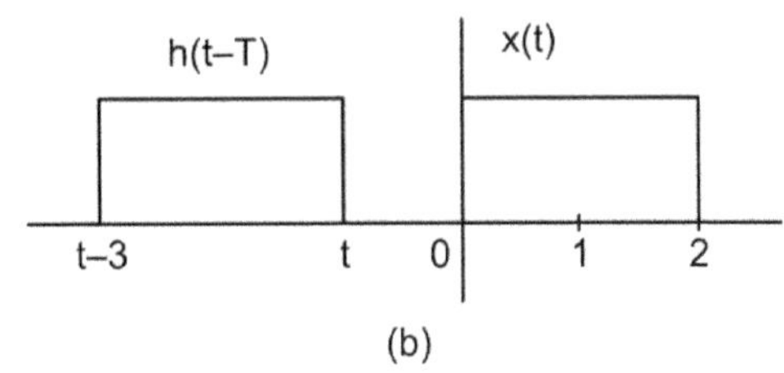

(a)   (b)

for $t = 0$ $y(t) = 0$

for $0 < t < 2$

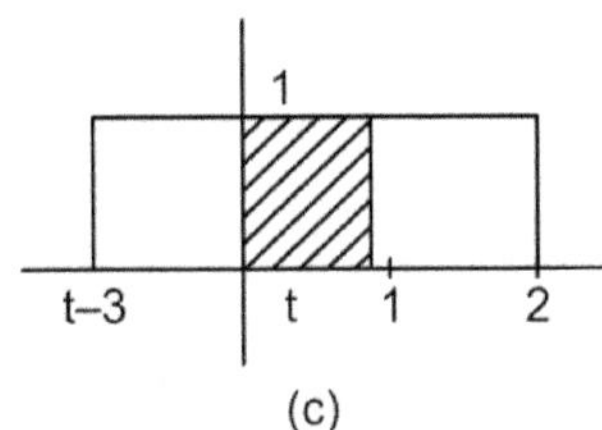

(c)

$$y(t) = 1 \times t = t$$

for $2 \le t \le 3$

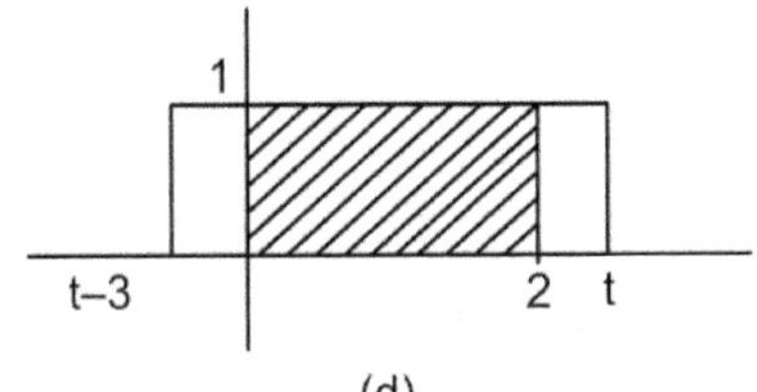

(d)

$$y(t) = 2$$

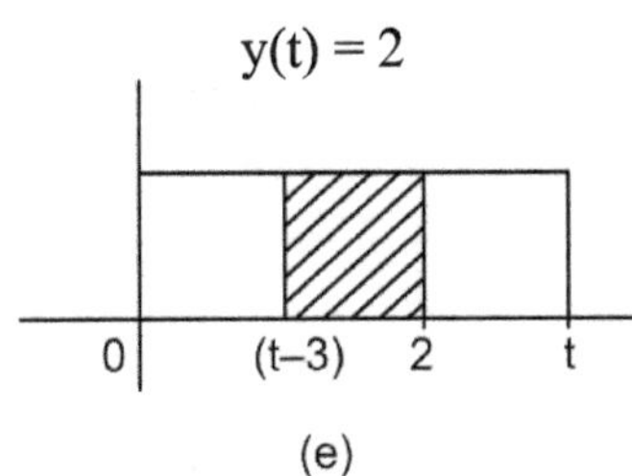

(e)

for $3 < t \le 5$

$$y(t) = 2 - (t - 3)$$
$$= 5 - t$$

for $t > 5$
$$y(t) = 0$$

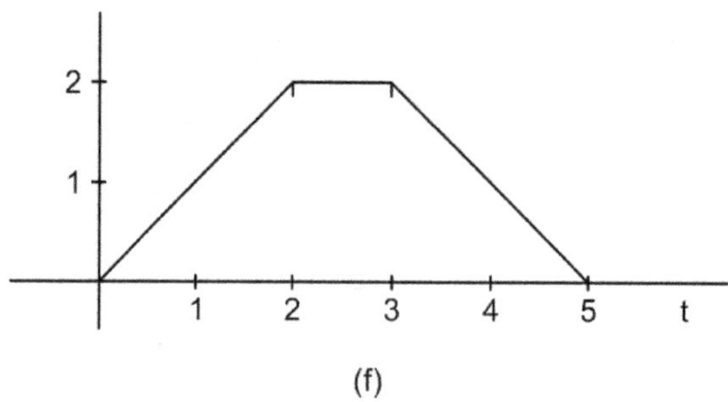

(f)

Fig. 6.12

5. Find the convolution of

$$x(t) = 5\cos 2t, \qquad h(t) = e^{-|t|} = \begin{cases} e^{t} & t < 0 \\ e^{-t} & t > 0 \end{cases}$$

Ans: $y(t) = \int\limits_{-\alpha}^{t} 3\cos 2T\, e^{(T-t)} dT + \int\limits_{t}^{\alpha} 3\cos 2T\, e^{(t-T)} dT$

$\because\quad h(t) = e^{t-T} \qquad t - T < 0$

$\qquad\qquad\qquad T > t$

$\qquad = e^{T-t} \qquad t - T \geq 0$

$$= 3e^{-t}\int\limits_{-\alpha}^{t} e^{T}\cos 2T\, dT + 3e^{t}\int\limits_{t}^{\alpha} e-T\cos 2T\, dT$$

$$= 3e^{-t}\left\{ e^{T}\frac{(\cos 2T + 2\sin 2T)}{5}\right\}\Bigg|_{-\alpha}^{t} + 3e^{t}\left\{\frac{e^{-T}(-\cos 2T + 2\sin 2T)}{5}\right\}\Bigg|_{t}^{\alpha}$$

$$= \frac{6}{5}\cos 2t$$

6. Find the convolution of

$$x(t) = t^{2}u(t) \quad \text{and} \quad h(t) = e^{-t}u(t)$$

$$y(t) = \int\limits_{-\alpha}^{\alpha} T^{2}u(T)e^{-(t-T)}u(t-T)dT$$

$$= \int\limits_{0}^{t} T^{2}\, e^{-(t-T)} dT$$

$$= e^{-t}\int\limits_{0}^{t} T^{2}e^{T} dT$$

$$= e^{-t}\left\{ T^2 \cdot e^T \Big/_0^t - 2\int_0^t c^T \cdot T \cdot dT \right\}$$

$$= e^{-t}\left\{ t^2 e^t - 2t\, e^t + 2(e^t - 1) \right\}$$

$$= t^2 - 2t + 2 - 2e^{-t} = (t-1)^2 + (1 - 2e^{-t})$$

7. Find f(t)

If $f_1(s) = \dfrac{1}{s^2}$ and $f_2(s) = \dfrac{1}{s+2}$ find f(t)

$$f(t) = L^{-1}\left\{ f_1(s) \cdot f_2(s) \right\}$$

Ans: $f(s) = \dfrac{1}{s^2(s+2)}$

$$= \frac{k_1}{s} + \frac{k_2}{s^2} + \frac{k_3}{s+2}$$

$$= -\frac{1}{4s} + \frac{1}{2s^2} + \frac{1}{4(s+2)}$$

$$f(t) = \frac{1}{4}\left\{ 1 - 2t - e^{-2t} \right\}$$

8. Find f(t) if $F(s) = \dfrac{1}{\left(s^2 + a^2\right)^2}$

Let $F(s) = \dfrac{1}{\left(s^2 + a^2\right)} \cdot \dfrac{1}{\left(s^2 + a^2\right)}$

$$\mathcal{L}^{-1}\left\{ \frac{1}{\left(s^2 + a^2\right)^2} \right\} = \int_0^t f_1(T)f_2(t-T)dT$$

$$= \frac{1}{a^2}\int_0^t \sin aT \sin(at - T)dT$$

$$= \frac{1}{2a^3}\left(\sin at - at\cos at\right)$$

SOLVED PROBLEMS: EXERCISE-2

1. Find the auto correlation of $A_0 + A_1 \sin(\omega t + \theta_1) + A_2 \sin(\omega t + \theta_2)$

$$\phi_{11}(T) \ =$$

$$= \frac{1}{T}\left\{ \int_{-T/2}^{T/2} A_0 \cdot A_0(1-T)dt + \int_{-T/2}^{T/2} A_1^2 \sin(\omega t + \theta_1)\sin(\omega t + \theta_1 - T)dt + \int_{-T/2}^{T/2} A_2^2 \sin(\omega t + \theta_2)\sin(\omega t + \theta_2 - T)dt \right\}$$

$$= \frac{A_0^2}{T}\{t - tT\} \Big/_{-T/2}^{T/2} + \frac{A_1^2}{2T}\int_{-T/2}^{T/2}\{\cos\omega T - \cos(2\omega t - \omega T + 2\phi_1)\}dt + \frac{A_2^2}{2T}\int_{-T/2}^{T/2}\{\cos\omega T - \cos(2\omega t - \omega T + 2\phi_2)dt\}$$

$$= A_0^2\left\{1 - \frac{2\pi}{\omega}\right\} + A_1^2 \cos\omega T$$

2. Find the auto correlation function for $A\sin(\omega t + \phi_1) + B\sin(2\omega t + \phi_2)$

$$\text{for}\quad \omega t = 2\pi$$

$$t = \frac{2\pi}{\omega}$$

$$\phi_{11}(T)$$

$$= \frac{1}{T}\int_{-T/2}^{T/2} A\sin(\omega t + \phi_1)\sin(\omega t - \omega T + \phi_1)dt + \frac{1}{T}\int_{-T/2}^{T/2} B^2 \sin(2\omega t + \phi_2)\sin(2\omega t - 2\omega T + \phi_2)dt$$

$$= \frac{A^2}{2}\cos\omega T + \frac{B^2}{2}\cos 2\omega T$$

3. Find the cross correlation function of two periodic signals $\{u(t) - u(t-T)\}$ and $e^{-t}u(t)$

$$f_1(t) \ = \ u(t) - u(t-T)$$

$$F_1(\omega) = \left(\pi\delta(\omega) + \frac{1}{j\omega}\right) - e^{-j\omega T}\left(\pi\delta(\omega) + \frac{1}{j\omega}\right)$$

$$f_2(t) \ = \ e^{-t}u(t)$$

$$F_2(\omega) = \frac{1}{1 + j\omega}$$

$$\therefore \ \phi_{12}(T) = F_1(\omega)F_2(-\omega) = \left\{\left(\pi\delta(\omega) + \frac{1}{j\omega}\right)\left(1 - e^{-j\omega T}\right)\right\}\left\{\frac{1}{1 - j\omega}\right\}$$

4. Show that the cross correlation of any function $f(t)$ with unit impulse function $\delta(t)$ yields the function $f(t)$ itself.

    Ans: Cross correlation between $f(t)$ and $\delta(t)$ $= \int_{-\alpha}^{\alpha}\delta(t)f(t-T)dt$

Using shifting property of the impulse function the above integral $f(t+T)\big|_{t=0}$ i.e., $f(T)$ which is same as f(t).

5. Show that correlation can be written in terms of convolution as

$$\phi(t) \;=\; \underset{T\to\alpha}{\mathrm{Lt}}\,\frac{1}{T}\{x(t)*x(-t)\}$$

average auto correlation is given by

$$\overline{\phi(t)} = \underset{T\to\alpha}{\mathrm{Lt}}\,\frac{1}{T}\int x(t)x(t-T)dt$$

$$= \underset{T\to\alpha}{\mathrm{Lt}}\,\frac{1}{T}\int x(t)x\big(-(T-t)\big)dt$$

$$= \underset{T\to\alpha}{\mathrm{Lt}}\,\frac{1}{T}\{x(t)*x(-t)\}$$

SOLVED PROBLEMS: EXERCISE-3

1. Consider a filter with $H(\omega)=\dfrac{1}{1+j\omega}$ and input $x(t)=e^{-2t}u(t)$. Find the energy spectral density of output.

Ans:

$$x(t) = e^{-2t}\,u(t)$$

$$H(\omega) = \frac{1}{1+j\omega}$$

$$\therefore\; X(\omega) = \frac{1}{2+j\omega}$$

$$\left|X(\omega)\right|^2 = \frac{1}{4+\omega^2}$$

$$\left|H(\omega)\right|^2 = \frac{1}{1+\omega^2}$$

$\therefore$ Energy spectral density

$$S(\omega) = \left|x(\omega)\right|^2 * \left|H(\omega)\right|^2 = \frac{1}{(4+\omega^2)(1+\omega^2)}$$

2. The waveform shown below is passed through an ideal LPF whose response is

$$H(f) = 1 \qquad -\omega_0 \le \omega_0$$

$$= 0 \qquad \text{other wise}$$

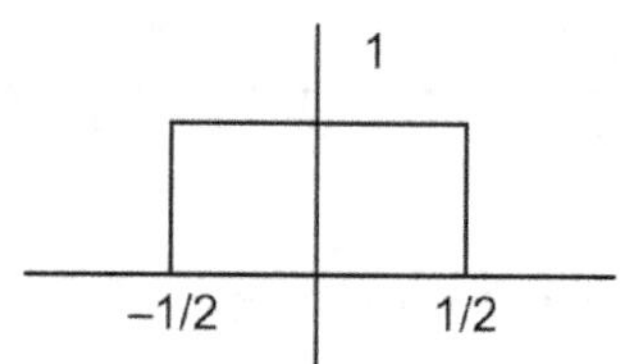

Fig. 6.13

find the energy spectral density of out put of the filter and also energy of the out put of the filter

$$x(t) = 1 \quad -\frac{1}{2} \le t \le \frac{1}{2}$$

$$x(\omega) = Sa\left(\frac{\omega}{2}\right)$$

$$\therefore \quad S(\omega) = |X(\omega)|^2 |H(\omega)|^2 = Sa\left(\frac{\omega}{2}\right)^2$$

As the transfer function is of LPF its energy must be found between $-\omega_0$ & ω_0.

$$E = \frac{1}{2\pi} \int_{-\omega_0}^{\omega_0} Sa^2\left(\frac{\omega}{2}\right) d\omega$$

3. Find the power density spectrum of signal $\dfrac{df}{dT}$.

   Ans: $\mathcal{F}T$ of $\dfrac{df}{dt}$ is $j\omega F(\omega)$ for any input signal f(t).

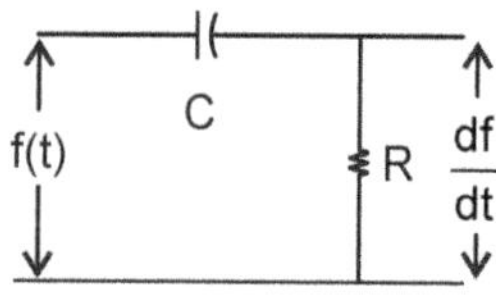

   $$\therefore \quad PDS = \omega^2 |F(\omega)|^2$$

   $$= \omega^2 S(\omega^2)$$

Fig. 6.14

where $S(\omega)^2$ is PDS of f(t). When f(t) is applied to RC Circuit shown output is $\dfrac{df}{dt}$.

4. Find the power spectral density of a periodic signal?

   Ans: Any periodic signal f(t)

   $$f(t) = \sum_{n=-\alpha}^{\alpha} F_n e^{jn\omega_0 t}$$

   and $$F(\omega) = 2\pi \sum_{n=-\alpha}^{\alpha} F_n \delta(\omega - n\omega_0)$$

   The truncated function of f(t) between intervals $-\dfrac{T}{2}$ and $\dfrac{T}{2}$ is

   $$f_T(t) = G_T(t) - f(t)$$

$$\therefore \quad F_T(\omega) = \frac{1}{2\pi} \mathcal{F}\{G_T(t)\} F(\omega)$$

$$= \frac{1}{2\pi} TSa\left(\frac{\omega T}{2}\right) 2\pi \sum F_n \delta(\omega - n\omega_0)$$

$$= T \sum F_n \, Sa\left\{\frac{(\omega - n\omega_0)T}{2}\right\}$$

$$PSD = \lim_{T\to\alpha} \frac{1}{T} S(\omega) = \lim_{T\to\alpha} \frac{|F(\omega)|^2}{T}$$

$$= \lim_{T\to\alpha} T \sum F_n^2 Sa^2\left(\frac{(\omega - n\omega_0)T}{2}\right) = \sum F_n^2 \delta(\omega - n\omega_0)$$

5. Find the power density spectrum of
$$f(t) = 8\cos\left(6\pi t + \frac{\pi}{3}\right)$$

Ans: Auto correlation of f(t)

$$\phi_{11}(t) = \lim_{T\to\alpha} \int_{-T/2}^{T/2} 8\cos(6\pi t + \frac{\pi}{3}) 8\cos\left(6\pi(t-T) + \frac{\pi}{3}\right) dt$$

$$= \lim_{T\to\alpha} \frac{1}{T} \int_{-T/2}^{T/2} 64\cos(6\pi t + \frac{\pi}{3})\cos\left(6\pi(t-T) + \frac{\pi}{3}\right) dt$$

$$= \lim_{T\to\alpha} \frac{32}{T} \int_{-T/2}^{T/2} \cos(12\pi t - 6\pi\lambda + \frac{2\pi}{3}) + \cos 6\pi\lambda \, dt$$

$$=$$

$$\lim_{T\to\alpha} \left\{ \frac{32}{T} \left\{ \left.\frac{\sin(12\pi t - 6\pi\lambda + \frac{2\pi}{3})}{12\pi}\right|_{-T/2}^{T/2} + 32\cos 6\pi\lambda \right\} \right\}$$

$$= 32\cos 6\pi\lambda$$

$$\therefore PSD = \mathcal{F}\{\phi_{11}(t)\} = 32\pi \{\delta(\omega - 6\pi) + \delta(\omega + 6\pi)\}$$

6. Find the PSD function of unit pulse width of Δt?

$$PSD = \lim_{T\to\alpha} \frac{|F(\omega)|^2}{T}$$

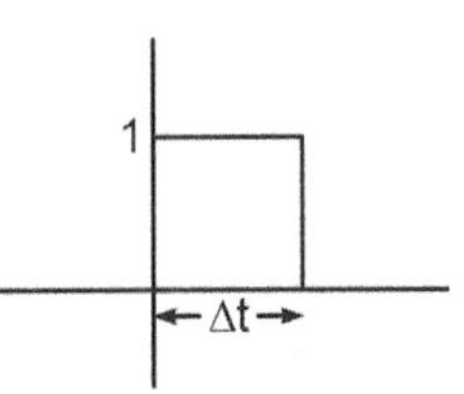

Fig. 6.15

$$F(\omega) = \int_0^{\Delta t} i.\, e^{-j\omega t}\, dt = \left.\frac{e^{-j\omega t}}{-j\omega}\right|_0^{\Delta t}$$

$$= \frac{e^{-j\omega \Delta t} - 1}{-j\omega}$$

$$F(-\omega) = \frac{e^{j\omega \Delta t} - 1}{j\omega}$$

$$\therefore \quad \left|F(\omega)\right|^2 = \frac{2 - e^{-j\omega \Delta t} - e^{j\omega \Delta t}}{\omega^2}$$

$$\text{PSD} = \lim_{T \to \alpha} \frac{1}{T} \frac{2 - e^{-j\omega \Delta t} - e^{j\omega \Delta t}}{\omega^2}$$

$$= \lim_{T \to \alpha} \frac{1}{\Delta t} \frac{2 - e^{-j\omega \Delta t} - e^{j\omega \Delta t}}{\omega^2}$$

7. Find the PSD of $f(t) = A\cos(\omega_0 t + \theta)$

$$\overline{\phi}_{11}(t) = \lim_{T \to \alpha} \frac{1}{T} \int_{-T/2}^{T/2} A\cos(\omega_0 t + \theta) A\cos\big(\omega(t - T) + \theta\big)\, dt$$

$$= \frac{A^2}{2} \cos \omega_0 T$$

$$\therefore \quad \text{PSD} = S_f(\omega) = \frac{\pi A^2}{2}\big\{\delta(\omega + \omega_0) + \delta(\omega - \omega_0)\big\}$$

SOLVED PROBLEMS FROM PREVIOUS PAPERS

1. Find the cross correlation of $\sin \omega t$ and $\cos \omega t$. Nov 2008

Ans:
$$\phi_{11}(t) = \frac{1}{T} \int_{-T/2}^{T/2} \sin \omega t \, \sin(t - T)\, dt$$

$$= \frac{\cos \omega T}{2}$$

$$\phi_{22}(t) = \frac{1}{T} \int_{-T/2}^{T/2} \cos \omega t \, \cos \omega(t - T)\, dt$$

$$= \frac{1}{2T} \int_{-T/2}^{T/2} \cos \omega T + \cos\big(2\omega(t - T)\big)\, dt$$

$$= \frac{1}{2T}\left\{ t\cos\omega T + \frac{\sin\left(2\omega(t-T)\right)}{2\omega}\right\}_{-T/2}^{T/2}$$

$$= \frac{\cos\omega T}{2}$$

2. For the signal shown find auto correlation $g(t) = c\cos(\omega_0 t + \theta_0)$

$$\phi_{11}(T) = \frac{1}{T}\int_{-T/2}^{T/2} c\cos(\omega_0 t + \theta_1)c\cos(\omega_0 t - \omega_c T + \theta)dt$$

$$= \frac{C_2}{2T}\int_{-T/2}^{T/2} \cos\omega T + \cos(2\omega_c t + \omega T + 2\theta)dt$$

$$= \frac{C_2}{2}\cos\omega T$$

3. Determine the cross correlation function of two signals

$$g_1(t) = A\cos(2\pi f_1 t + \theta_1)$$

$$g_2(t) = A\cos(2\pi f_2 t + \theta_2)$$

How does the frequency difference $\left|f_1 - f_2\right|$ affect this cross correlation function?

Ans: $\phi_{12}(t) = \dfrac{1}{T}\displaystyle\int_{-T/2}^{T/2} A\cos(2\pi f_1 t + \theta_1)A\cos\left(2\pi f_2(t-T) + \theta_2\right)dt$

$$= \frac{A^2}{2T}\int_{-T/2}^{T/2}\cos\left(2\pi(f_1 - f_2)t + 2\pi f_2 T + (\theta_1 - \theta_2)\right) + \cos\left(2\pi(f_1 + f_2)t + 2\pi f_2 T + (\theta_1 + \theta_2)\right)dt$$

$$= \frac{A^2}{2T}\left\{\frac{\sin\left(2\pi(f_1 - f_2)\right)t + 2\pi f_2 T}{2\pi(f_1 - f_2)} + +(\theta_1 - \theta_2)\right\} + \left\{\frac{\sin\left(2\pi(f_1 + f_2)\right)t + 2\pi f_2 T}{2\pi(f_1 + f_2)} + +(\theta_1 + \theta_2)\right\}$$

4. Find the output energy of RC low pass filter whose input is $x(t) = \sin c 2\omega t$.   Sep 2007

    Ans: $x(t) = \sin c 2\omega t$

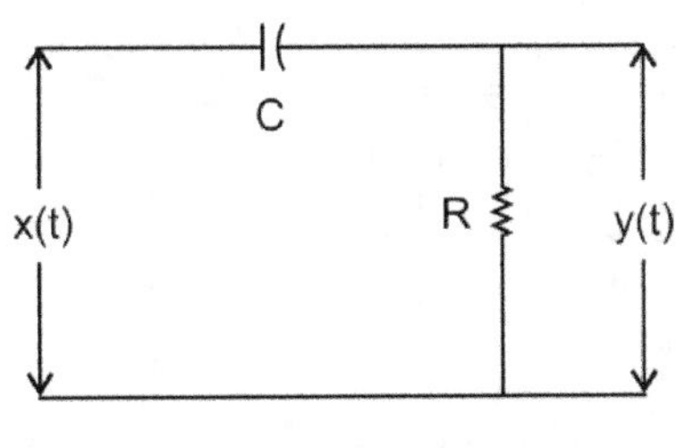

(a)

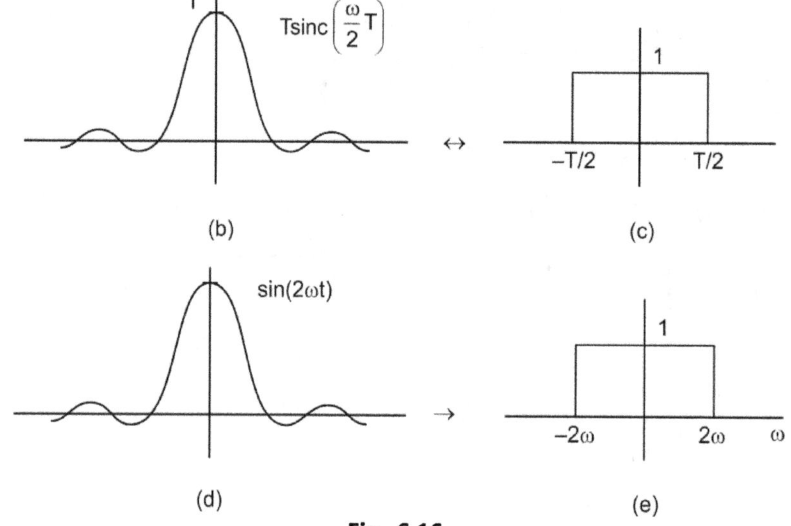

Fig. 6.16

$$\left|H(\omega)\right|^2 = \frac{\omega^2 R^2 C^2}{1+\omega^2 R^2 C^2}$$

$$\left|X(\omega)^2\right| = 1$$

$$S(\omega) = Y(\omega)^2 = \frac{\omega^2 R^2 C^2}{1+\omega^2 R^2 C^2}$$

$$H(j\omega) = \frac{R}{R+\frac{1}{j\omega c}}$$

$$= \frac{j\omega RC}{1+j\omega RC}$$

$$= 1-\frac{1}{1+j\omega RC}$$

$$\left|H(\omega)\right| = \frac{\omega RC\left|90_0^0\right.}{\sqrt{1+(\omega RC)^2}\left|fm^{-1}\left(\dfrac{1}{\omega su}\right)\right.} \qquad \left|\frac{L_e R_e}{\sqrt{1+(\omega rl)^2}}\right| \left|90-tu^{-1}\left(\dfrac{1}{m}\right)\right.$$

OBJECTIVE QUESTIONS

1. Convolution in time domain corresponds to _________ in frequency domain of LTI systems.

2. $x(t) * h(t) \xleftrightarrow{\ f\ }$ _________

3. $x_1(t) * x_2(t) \xleftrightarrow{\ \infty\ }$ _________

4. $f(t) * \delta(t - T) =$ _________

   (a) $f(t) + \delta(t)$ (b) $u(t)$ (c) $f(t) + u(t)$ (d) $f(t - T)$

5. Cross correlation $\phi_{12}(T)$.

   a) $\phi_{21}(t + T)$ b) $\phi_{21}(-T)$ c) $\phi_{21}(t - T)$ d) $\phi_{22}(T)$

6. If the signal f(t) is even function $(f(t) = f(-t))$, then cross correlation and convolution are _________

7. The cross correlation of a signal f(t) with a periodic impulse function $\delta(t) = \delta(t - T_0)$ _________

8. The value of the auto correlation function at the origin is equal to _________ of the signal

9. If $f_T(t) \leftrightarrow f_T(\omega)$, then average auto correlation function $\phi_{11}(T)$ is given by _________

10. The operation of cross correlation is equated in frequency domain to detect a signal buried in noise is same as _________

11. If $P(\omega)$ is average power density function $f(t)$, then $\int_0^{\alpha} P(\omega) d\omega =$ _________

12. If periodic signal S(t) and noise signal n(t) are uncorrelated, then cross correlation $\phi_{sn}(t)$ is _________

13. The energy density function for gate function where $-T < t < T$ is _________

14. The power spectrum of a signal is the fourier transform of its correlation function. This is known as _________

15. _________ method is used to detect a periodic signal in the presence of noise.

16. If $f_1(t)$ and $f_2(t)$ are two functions then cross correlation $\phi_{12}(T)$ is given by _________

17. Average correlation $\phi_{12}(T)$ is _________

18. If $f_T(t) \leftrightarrow F_T(\omega)$, then energy density $S_T(\omega)$ is _________

19. If $y(t) = \dfrac{d}{dt} x(t)$, then auto correlation $\phi_{yy}(T) =$ _________

20. Let S(t) be a periodic signal mixed with a noise signal n(t), then the received signal f(t) is _________

GATE QUESTIONS

1. Let u(t) be the step function, which of the waveform corresponds to convolution of u(t) – u(t–1) with u(t) – u(t – 2)

   GATE 2000

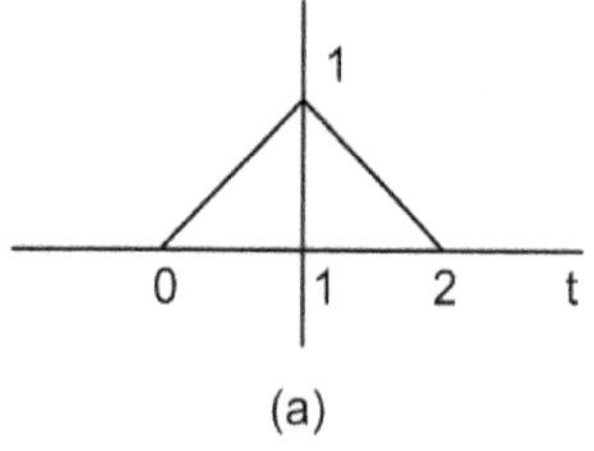

(a)

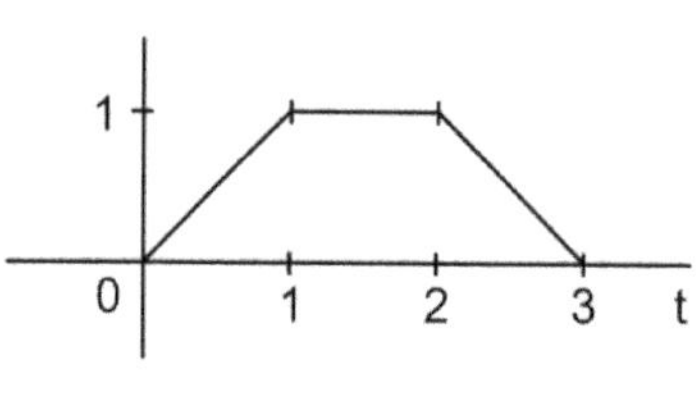

(b)

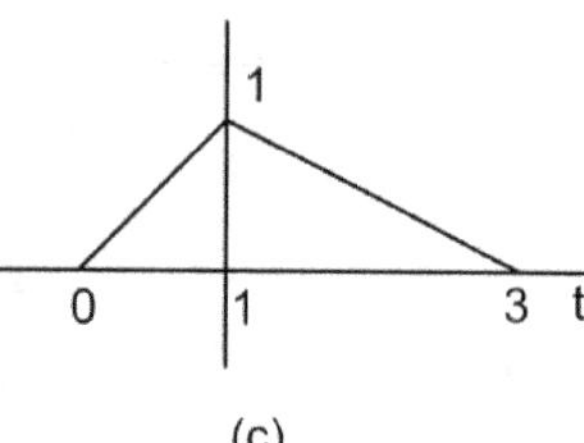

(c)

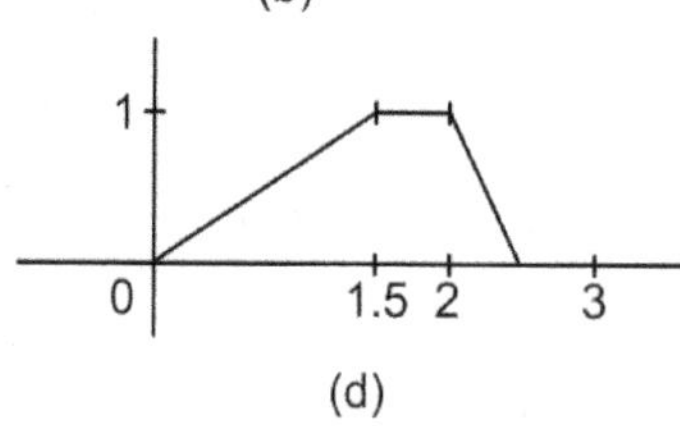

(d)

***Ans:* (b)**

2. If the signal $x(t) = \dfrac{\sin(t)}{\pi t} * \dfrac{\sin(t)}{\pi t}$, * denoting convolution then x(t) is   GATE 2016

   (a) $\dfrac{\sin(t)}{\pi t}$ (b) $\dfrac{\sin 2(t)}{2\pi t}$ (c) $\dfrac{2\sin(t)}{\pi t}$ (d) $\left(\dfrac{\sin(t)}{\pi t}\right)^2$

***Ans:* (a)**

3. Consider a continuous time signal $x(t) = \dfrac{\sin(\pi t / 2)}{(\pi t / 2)} * \sum \delta(t - 10n)$. The nyquist sampling rate is ______ .

   GATE 2015 ***Ans:* 0.4**

4. Consider two real valued signals x(t) band limited to [–500 Hg, 500Hg] and y(t) band limited to [–1kHg, 1 kHg]. For z(t) = x (t) * y(t) the nyquist sampling frequency is ____

   GATE 2014   ***Ans:* 3**

5. The result of convolution x(– t) * δ (– t – t₀) is   GATE 2015

   (a) $x(t + t_0)$ (b) $x(t - t_0)$ (c) $x(-t + t_0)$ (d) $x(-t - t_0)$

***Ans:* (d)**

6. Consider a discrete time signal   GATE 2014

   $x[n] = n$ for $0 \le n \le 10$

   0 otherwise

If y[n] is convolution of x[n] with itself the value of y[4] is ____

(a) 10   (b) 9   (c) 11   (d) 12   ***Ans:* (a)**

7. The sequence $x[n] = 0.5^n\, u[n]$ is convolved with itself to obtain y[n], then $\displaystyle\sum_{n=-\alpha}^{+\alpha} \alpha y[n]$ is ______

(a) 4   (b) 3   (c) 2   (d) 1   ***Ans:* (a)**

8. s(t) is step response and h(t) is impulse response of a system. Then its response y(t) for any input u(t) is

(a) $\dfrac{d}{dt}\displaystyle\int_0^t s(t-\tau)u(\tau)d\tau$ (b) $\displaystyle\int_0^t s(t-\tau)u(\tau)d\tau$

(c) $\displaystyle\int_0^t\int_0^t s(t-\tau)u(\tau_1)dTd\tau$ (d) $\dfrac{d}{dt}\displaystyle\int_0^t h(t-\tau)u(\tau)d\tau$

Hint: $\dfrac{d}{dt}s(t) = h(t)$ ***Ans:* (a)**

9. Impulse response of linear system is $e^{-2t}u(t)$. To produce a response of $t\,e^{-2t}\,u(t)$ the input must be   Gate 1995

(a) $2\,e^{-t}u(t)$ (b) $\dfrac{1}{2}\,e^{-2t}u(t)$ (c) $e^{-2t}u(t)$ (d) $e^{-t}u(t)$

                                                               ***Ans:* (c)**

10. The unit impulse response of a system is given as $c(t) = -4\,e^{-t} + 6e^{-2t}$. The step response of the system for $t \geq 0$ is   Gate 1996

(a) $-3\,e^{-2t} - 4\,e^{-t} + 1$ (b) $-3\,e^{-2t} + 4\,e^{-t} - 1$

(c) $-3\,e^{-2t} - 4\,e^{-t} - 1$ (b) $3\,e^{-2t} + 4\,e^{-t} - 1$ ***Ans:* (b)**

11. If $y(t) = \displaystyle\int_0^t (2+t-\tau)e^{-3(t-\tau)}u(\tau)d\tau$. The transfer function is   Gate 2001

(a) $\dfrac{2e^{-2s}}{s+3}$ (b) $\dfrac{S+2}{(S+3)^2}$ (c) $\dfrac{2S+5}{S+3}$ (d) $\dfrac{2S+7}{(S+3)^2}$

                                                               ***Ans:* (d)**

12. Let s(t) be the step response of a linear system with zero initial conditions. Then the response of the system to an input u(t) is   Gate 2002

  (a) $\displaystyle\int_0^t s(t-\tau)u(\tau)d\tau$ (b) $\displaystyle\frac{d}{dt}\int_0^t\int s(t-\tau)u(\tau)d\tau$

  (c) $\displaystyle\int_0^t s(t-\tau)\left[\int_0^t u(\tau)d\tau\right]dT$ (d) $\displaystyle\int_0^t s(t-\tau)^2 u(\tau)d\tau$ ***Ans: (b)***

13. $x[n] = 0$; $n < -1$, $n > 0$, $x[-1] = -1$, $x[0] = 2$ is the input and

   $y[n] = 0$, $n < -1$, $n > 2$, $y[-1] = -1 = y[1]$

   $y[0] = 3$, $y[2] = -2$ is the output of LTI system

   The system impulse response h[n] is   Gate 2006

  (a) $h[n] = 0$, $n < 0$, $n > 2$, $h[0] = 1$, $h[1] = h[2] = -1$

  (b) $h[n] = 0$, $n < -1$, $n > 1$, $h[-1] = 1$, $h[0] = h[1] = 2$

  (c) $h[n] = 0$, $n < 0$, $n > 3$, $h[0] = -1$, $h[1] = 2$, $h[2] = -1$

  (d) $h[n] = 0$, $n < -2$, $n > 1$, $h[-2] = h[1] = h[-1] = -h[0] = 3$ ***Ans: (a)***

14. A signal A, sin (ω, t + $), is applied to a LTI system. Steady state output is $A_2\, F(\omega_2 t + \theta_2)$. The which of the statements is true   Gate 2007

  (a) F is the periodic function with $\omega_1 = \omega_2$

  (b) F must be sine or cosine with $a_1 = a_2$

  (c) F must be sine with $\omega_1 = \omega_2$ and $\phi = \phi_2$

  (d) F must be sine or cosine with $\omega_1 = \omega_2$ ***Ans: (d)***

15. The impulse response of a causal linear time invariant system is given as h(t). Now consider

   St 1 : Principle of super position holds

   St 2 : $h(t) = 0$ for $t < 0$

   Which one is correct?   Gate 2008

  (a) St 1 is correct and St 2 is false

  (b) St 2 is correct and St 1 false

  (c) Both are wrong

  (d) Both are correct   ***Ans: (d)***

16. A signal $e^{-\alpha t} \sin(\omega t)$ is input to LTI system. Given K and ϕ are constants, the output of the system will be $K\, e^{-\beta t} \sin(vt + \beta)$ where   Gate 2008

   (a) β need not be equal to α and v equal to ω

   (b) v need not be equal to ω but β equal to α

   (c) β equal to α and v equal to ω

   (d) β need not be equal to α and v need not be equal to ω *Ans:* (c)

17. A system having $y(t) = \int\limits_{-\alpha}^{-2t} x(t)d\tau$,the system will be   Gate 2008

   (a) Causal, time invariant and unstable

   (b) Causal, time invariant and stable

   (c) Non causal, time invariant and unstable

   (d) Non causal, time variant and unstable   *Ans:* (d)

18. An LTI system with h (t) produces y (t) when x (t) is applied. Then what is the output when x $(t-\tau)$ is applied   Gate 2009

   (a) $y(\tau)$ (b) $y(2(t-\tau)$

   (c) $y(t-\tau)$ (d) $y(t-2\tau)$ *Ans:* (d)

SUMMARY OF CHAPTER

1. Convolution is a mathematical operation used to express the relation input and output of an LTI system

$$y(t) = x(t) \times h(t)$$

2. Continuous convolution

$$y(t) = \int_{-\alpha}^{\alpha} x(\tau)h(t - \tau)d\tau$$

3. Discrete Convolution

$$y(n) = \sum_{k=-a}^{\alpha} \frac{\alpha}{2}\, x(t)h(n - t)$$

4. Deconvolution is reverse process of convolution.

5. Properties of convolution

   (a) Commutative property

$$x_1(t) * x_2(t)$$

$$x_2(t) * x_1(t)$$

(b)   Distributive property

$$x_1(t) + [x_2(t) + x_3(t)]$$
$$= [x_1(t) + x_2(t)] + [x_1(t) + x_3(t)]$$

(c)   Shifting property

$$x_1(t) * x_2(t) = y(t)$$
$$x_1(t) * x_2(t - t_0) = y(t - t_0)$$

(d)   Associative property

$$x_1(t) * [x_2(t) * x_3(t)] = [x_1(t) * x_2(t) * x_3(t)]$$

(e)   Convolution with impulse

$$x_1(t) * \delta(t) = x(t)$$

6. Correlation is a measure of similarity between two signal

$$= \int_{-\alpha}^{\alpha} x_1(t) x2(t - \tau) d\tau$$

7. Auto correlation is defined as correlation of a signal with itself

$$R_{11}(\tau) = \int_{-\alpha}^{\alpha} x(t) x(t - \tau) dt$$

8. Energy density spectrum can be calculated using

$$E = \int_{-\alpha}^{\alpha} |x(t)|^2 \, df$$

9. Power density spectrum can be calculated using

$$P = \sum_{n=-\alpha}^{\alpha} |C_n|^2$$

10. Parsevals theorem for energy signals states that the total energy in a signal can be obtained by the spectrum of the signal as

$$E = \frac{1}{\varepsilon\pi} \int_{-\alpha}^{\alpha} |x(\omega)|^2 \, d\omega$$

CHAPTER 7

Sampling

Sampling theorem – proof for band limited signals, impulse sampling, Natural and flat sampling, Reconstruction of signals from samples – Aliasing, Band Pass sampling.

7.1 INTRODUCTION

Sampling is the process of measuring the instantaneous values of continuous time signal in a discrete form. When a source generates an analog signal and if that has to be digitized, the signal has to be discredited in time. This discretization of analog signal is called as sampling.

(a) **Sampling:** It is the process of converting a continuous time signal into discrete time signal.

Sampling operation is analogous to operating a switch at fixed time interval T.

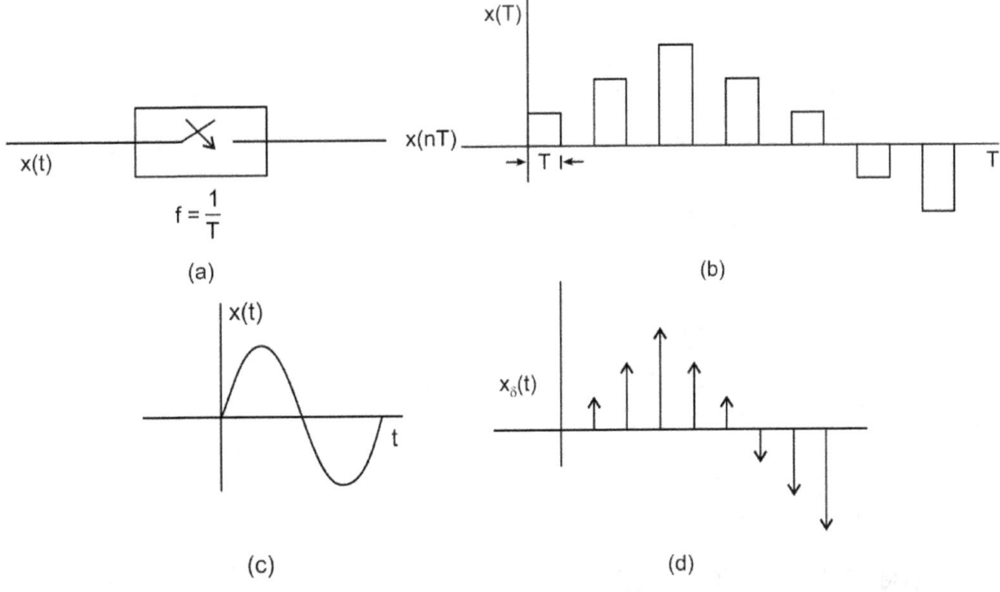

If $P(t) = \delta_T(t)$

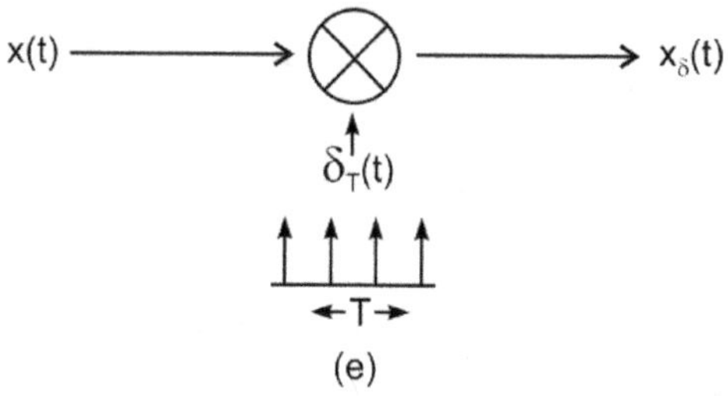

Fig. 7.1

where

$$\delta_T(t) = \sum_{n=-\alpha}^{\alpha} \delta(t-nT)$$

$$x_\delta(t) = \sum_{n=-\alpha}^{\alpha} x(nT)\delta(t-nT)$$

Applying Fourier Transform

$$\mathcal{F}\{x_\delta(T)\} = \mathcal{F}\left\{\sum_{n=-\alpha}^{\alpha} x(nT)\delta(t-nT)\right\}$$

$$= \sum_{n=-\alpha}^{\alpha} x(nT)\,\mathcal{F}\{\delta(t-nT)\}$$

$$\mathcal{F}\{\delta(t-nT)\} = \int_{-\alpha}^{\alpha} \delta(t-nT)e^{-j\omega t}dt = e^{-j\omega nT}$$

$$\mathcal{F}\{x_\delta(T)\} = \sum_{n=-\alpha}^{\alpha} x(nT)e^{-j\omega nT}$$

also we have

$$\sum_{n=-\alpha}^{\alpha} \delta(t-nT) = \delta_T(t) = \frac{1}{T}\sum_{n=-\alpha}^{\alpha} e^{jn\omega_0 t}$$

$$\text{where } \omega_0 = \frac{2\pi}{T}$$

$$\therefore \qquad x_\delta(t) = x(t)\frac{1}{T}\sum_{n=-\alpha}^{\alpha} e^{jn\omega_0 t}$$

$$\mathcal{F}\{x_\delta(t)\} = \frac{1}{T}\sum_{n=-\alpha}^{\alpha} \mathcal{F}\{x(t)e^{jn\omega_0 t}\}$$

$$\mathcal{F}\{x(t)e^{jn\omega_0 t}\} = x\{j(\omega-n\omega_0)\}$$

$$\mathcal{F}\{x_\delta(t)\} = \frac{1}{T}\sum_{n=-\alpha}^{\alpha} x\big(j(\omega-n\omega_0)\big)$$

$$= \frac{1}{T}\sum_{n=-\alpha}^{\alpha} x\left(j(\omega-\frac{2\pi n}{T})\right)$$

∴ fourier transform of the sampled signal is given by an infinite sum of shifted replicas of fourier Transform of the original signal.

Let us consider highest frequency component of $x(t)$ is f_m where [$x(t)$ is band limited]

∴ $x(j\omega) = 0$ for $|\omega| > \omega_m$

$x\left(j(\omega-\frac{2\pi n}{T})\right)$ indicates $x(j\omega)$ is shifted from $\omega = 0$ to $\omega = \frac{2\pi n}{T}$.

Hence $x_0(j\omega)$ is the sum of shifted replicas of $\dfrac{x(j\omega)}{T}$ centering at $\dfrac{2\pi n}{T}$.

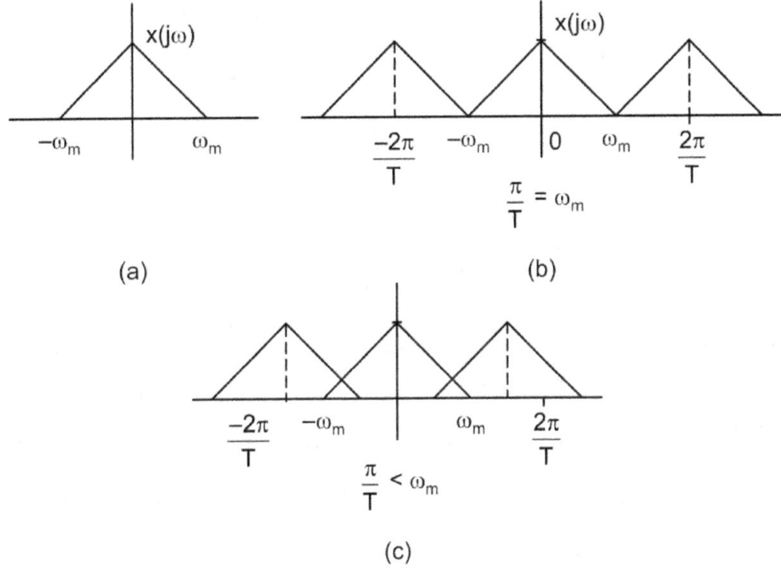

Fig. 7.2

7.2   SAMPLING THEOREM

A band limited signal $x(t)$ with $x(j\omega) = 0$ for $|\omega| > \omega_m$ is uniquely determined from its samples $x(nT)$ if the sampling frequency $f_s \geq 2f_m$.

or

Sampling frequency = Highest frequency in the signal

This is also called as Nyquist Rate, which is equal to $2f_m$. In other words time between samples is no greater than $\dfrac{1}{2f_m}$ secs.

Impulse Sampling: The pulse width of the sampling function must be infinity small. This can be achieved only when sampling function is assumed as infinite train of impulse functions of period T.

7.3 ALIASING

Aliasing happens whenever an analog signal is not sampled at a high enough frequency. It appears itself in the form of spurious low frequencies with a higher frequency but the same sampling rate, a reconstructed signal would appear to be a sine wave of lower frequency (aliased signal).

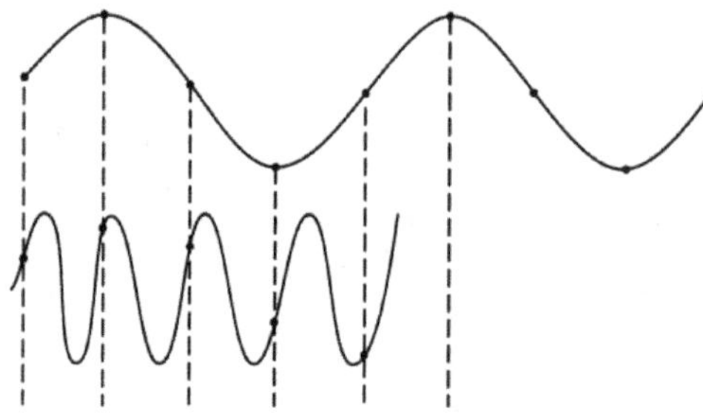

Fig. 7.3

To recover accurately, the signal must be sampled at a rate greater than or equal to two times the highest frequency contained in the signal.

Anti aliasing: When a signal is sampled, with sampling frequency f_s , all the signals with frequency range higher than $\dfrac{\omega_s}{2}$ appear as signal frequencies between 0 and $\dfrac{\omega_s}{2}$ crating aliasing.

Therefore, to avoid aliasing errors caused by the undesired high frequency signals, an analog low pass filter called anti aliasing filter is used prior to sampler.

7.4 SIGNAL RECONSTRUCTION

Let the sampled signal $x_s(t)$ with sampling period T

$$x_s(t) \;=\; \sum_{n=-\alpha}^{\alpha} x(nT)\delta(t-nT)$$

$$\mathscr{F}\{x_\delta(t)\} = \frac{1}{T}\sum x\left[j\left(\omega - \frac{2\pi n}{T}\right)\right]$$

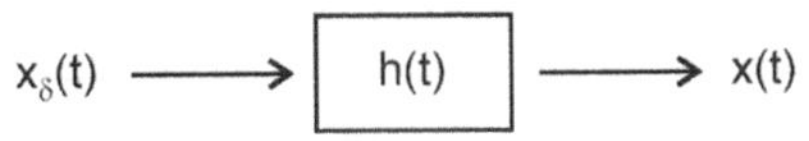

$$x\delta(t) \to \boxed{h(t)} \to x(t)$$

$$x_\delta(t) \longrightarrow \boxed{h(t)} \longrightarrow x(t)$$

Fig. 7.4

$$x(t) = x\delta * h(t) \quad \text{where } h(t) \text{ is impulse response}$$

$$= \left\{\sum_{n=-\alpha}^{\alpha} x(nT)\delta(t-nT)\right\} + h(t)$$

$$= \int_{-\alpha}^{\alpha} \sum_{n=-\alpha}^{\alpha} x(nT)\delta(\lambda - nT)h(t-\lambda)d\lambda$$

$$= \sum_{n=-\alpha}^{\alpha} x(nT)\int_{-\alpha}^{\alpha} \delta(\lambda - nT)h(t-\lambda)d\lambda$$

$$= \sum_{n=-\alpha}^{\alpha} x(nT)h(t-nT)$$

for LPF

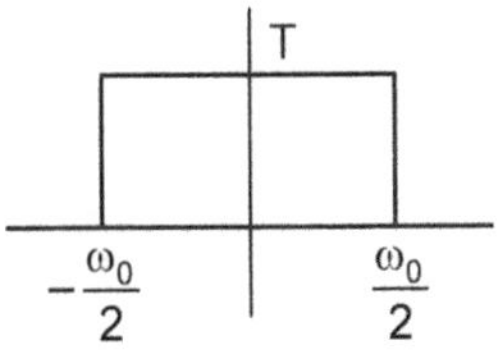

$$H(j\omega) = T \quad \text{for} \quad \omega \le \frac{\omega}{2}m$$

$$h(t) = \frac{1}{2\pi}\int_{-\frac{\omega_m}{2}}^{\frac{\omega_n}{2}} H(j\omega)e^{j\omega t}\,d\omega$$

$$= \frac{T}{2\pi}\int_{-\omega_m/2}^{\omega_n/2} e^{j\omega t}\,d\omega = \frac{T}{\pi T}\sin\frac{\omega_m t}{2}$$

$$= \frac{T}{\pi}\frac{\sin\frac{\pi}{T}}{t}$$

$$h(t-nT) = \frac{T}{\pi}\sin\frac{\frac{\pi}{T}(t-nT)}{t-nT}$$

$$= \sin c\frac{\pi}{T}(t-nT)$$

$$x(t) = \sum x(nT)\sin c\frac{\pi}{2}(t-nt)$$

$$x(t) = \sum x(nT)\delta(t-nT)h(t)$$

$$= \sum x(nT)h(t-nT)$$

Fig. 7.5

LPF response

$$h(t) \;=\; \frac{\omega}{2\pi} Sa\left\{\frac{\omega(t-t_0)}{2}\right\}$$

$$h(t-nT) = \frac{\omega_0}{\pi} \sin c\big(\omega_0(t-t_0)\big)$$

$$= \frac{1}{T}\sin c\,\frac{\pi}{T}(t-nT)$$

$$h(\omega) = G_{\omega_0}(\omega)$$

$$h(t) = \sin c\left(\frac{\omega_0 t}{2}\right) = \sin c\left(\frac{\pi t}{T}\right)$$

$$h(t-nT) = \sin c\,\frac{\pi}{T}(t-nT)$$

$$x(t) = \sum x(nT)\sin c\,\frac{\pi}{T}(t-nT)$$

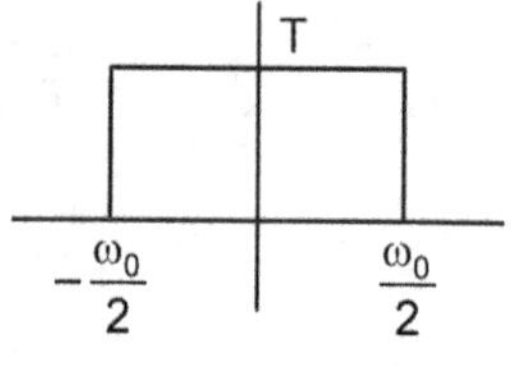

Fig. 7.6

7.5 BAND PASS SAMPLING

Let the frequency range of a band limited signal be

$$\omega_L \le |\omega| \le \omega_H \qquad \text{where} \quad \omega_L > 0$$

This type of signal is called band pass signal .

To sample such a signal, sampling rate $f_s \ge 2f_H$.

As the frequencies are very high the ks sampled signal have spectral gaps, to avoid this

$$\Delta\omega = \omega_H - \omega_L$$

$$\Delta f = f_H - f_L \qquad \text{where} \qquad f_H = k\Delta f$$

$\therefore$ sampling rate is choosen such that

$$f_s = 2\Delta f$$

$$= \frac{2f_H}{k}$$

$$\therefore \; x_\delta(\omega) = \frac{1}{T}\sum x\big(j(\omega - 2k\Delta\omega)\big)$$

The sampled signal is replica of original signal $x(\omega)$. The original signal $x(t)$ can be recovered by passing through a band pass filter with a pass band of $\omega_L \le |\omega| \le \omega_H$.

Example:

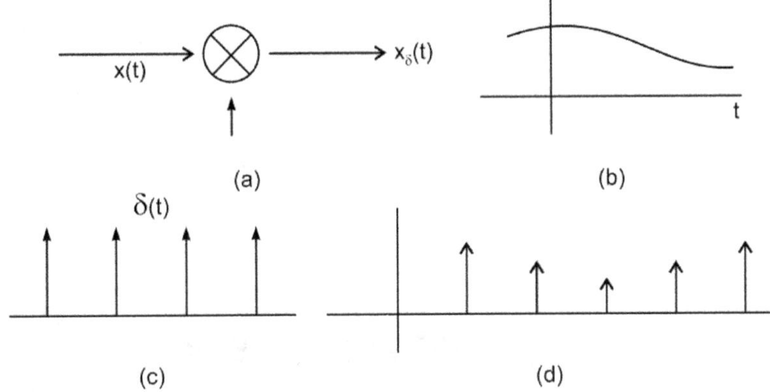

Reconstruction of original signal from samples

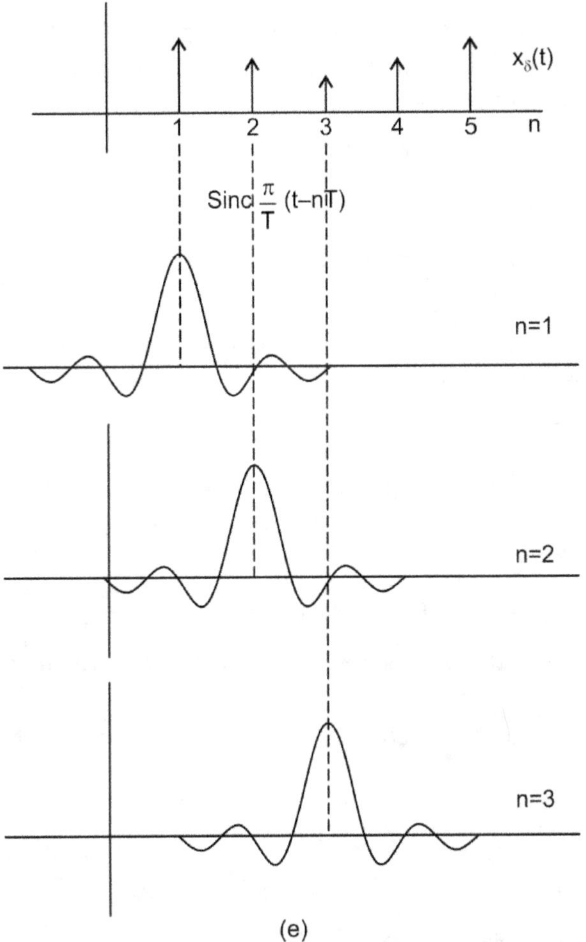

(e)

Fig. 7.7

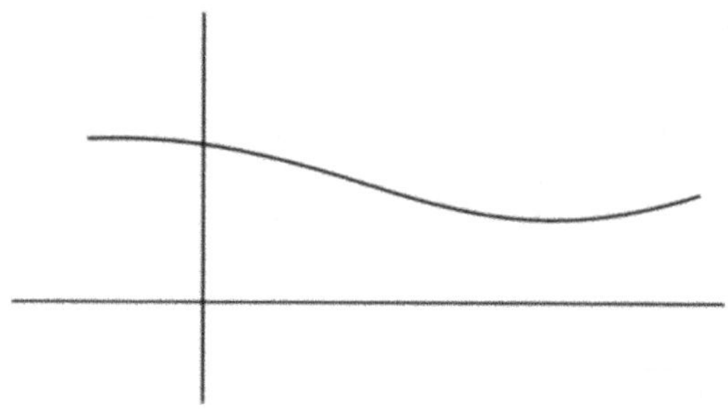

Fig 7.8

SOLVED PROBLEMS

EXERCISE

1. A signal $x(t) = \cos 200\pi t + 2\cos 280\pi t$ is sampled at a sampling rate of 300 Hz and the sampled signal is passed through an LPF whose cutoff frequency is 250 Hz. What frequencies are present in output.

   Ans: $x(t) = \cos 200\pi t + 2\cos 280\pi t$

   $$= \cos 2\pi 100t + 2\cos 2\pi 14t$$

   $$f_{m_1} = 100\,\text{Hz} \qquad f_{m_2} = 140\,\text{Hz}$$

   Sampling frequency $f_3 = 300$ Hz

   Different frequency component are

   $$\left.\begin{array}{l} f_s + f_{m1} = 400 \\ f_s - f_{m1} = 200 \\ f_s + f_{m2} = 440 \\ f_s - f_{m2} = 160 \end{array}\right\} \text{passed through LPF } f_c = 250$$

   $\therefore$ output contains 160 Hz, 200Hz, 100, 140 Hz components.

2. The signal $x(t) = \cos 5\pi t + 0.3\cos 10\pi t$ is sampled. Find maximum interval of sampling.

   $$x(t) = \cos 2\pi \frac{5}{2}t + 0.3\cos 2\pi 5t$$

   $$f_{m1} = \frac{5}{2}$$

   $$f_{m2} = 5\,\text{Hz}$$

   $\therefore$ $f_s = 2f_m = 10\,\text{Hz}$

$\therefore$ Maximum interval $\quad T \;=\; \dfrac{1}{f_s} = 0.1\,\text{sec}$

3. The signal $x(t) = \cos \pi t + 0.5 \cos 10 \pi t$ is sampled with a sampling signal of $\delta t = 5 \Sigma \delta(t - 0.1T)$. Show that $I_T = I_{t+4}$ where $\mathscr{F}_k$ is strength of output samples.

Ans: $\quad f_{m1} = \dfrac{5}{2}$

$\qquad\quad f_{m2} = 5$

$\qquad\quad f_s = 10\,\text{Hg}$

$\qquad\quad x_\delta(t) = \big(\cos 5\pi t + 0.5 \cos 10\pi t\big) 5 \Sigma \delta(t - 0.1T)$

$\qquad\qquad\quad = \; \Sigma \big[5 \cos 5\pi(0.1T) + 2.5 \cos 10\pi(0.1T)\big] \delta(t - 0.1T)$

$\qquad\; I_{T+4} \;=\; \Sigma 5 \cos 5\pi\big(0.1(T+4)\big) + 2.5 \cos 10\pi\big(0.1(T+4)\big)$

$\qquad\qquad\quad = \; \Sigma 5 \cos 0.5\pi T + 2.5 \cos \pi T$

4. The signal $x(t) = 10 \cos 10 \pi t$ is sampled at a rate of 8 samples/sec. Plot the amplitude spectrum

Ans: $\quad x(t) \;=\; 10 \cos 10 \pi t$

$\qquad\quad \mathscr{F}\{x(t)\} \;=\; 10\pi \{\delta(\omega + 10\pi) + \delta(\omega - 10\pi)\}$

$\qquad\quad f_m = 5\,\text{Hz}$

$\qquad\quad f_s = 10\,\text{Hz}$

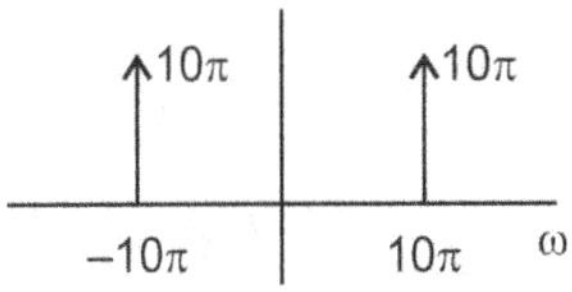

$\uparrow 10\pi \qquad \uparrow 10\pi$

$-10\pi \qquad 10\pi \quad \omega$

Fig. 7.9

but sampling rate is 8 samples/sec

$\qquad \therefore \quad$ signal cannot be recovered and it causes aliasing

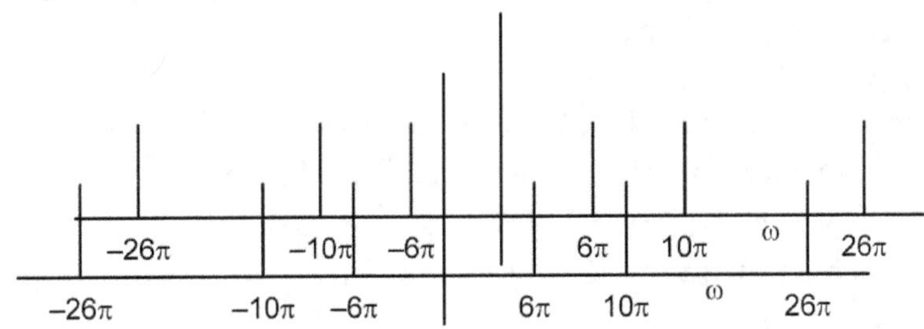

Fig. 7.10

$\qquad f_s = 8$

$\qquad \omega_s = 16\pi$

$\qquad \omega_s - \omega_n = 26\pi$

$\qquad \omega_s - \omega_m = 6\pi$

SOLVED PROBLEMS FROM OLD PAPERS

1. A signal $x(t) = 2\cos 400\pi t + 6\cos 640\pi t$ is sampled at $f_s = 500\,\text{Hz}$. If it is passed through an ideal low pass filter with cutoff frequency 400 Hz, what frequency components appear at output?

Sep 2007, Nov 2009

Ans: $\quad x(t) \;=\; 2\cos 400\pi t + 6\cos 640\pi t$

$\qquad f_{m1} = 200$

$\qquad f_{m2} = 320$

$\qquad$ Nyquist rate $=\; 640\,\text{Hz}$

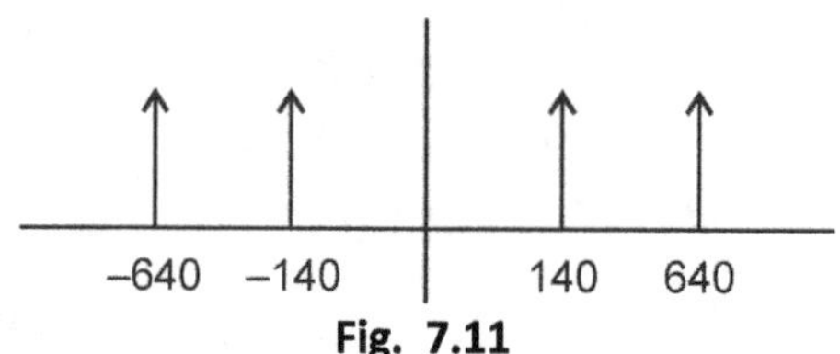

Fig. 7.11

$\qquad$ Sampling $f_s = \; 500\,\text{Hz}$

Low pass filter with cutoff $f_c = \; 400\,\text{Hz}$ is used

$\therefore$ output contains 300, 180, 200, 320, 140 Hz components

2. A rectangular pulse wave shown is sampled at Ts seconds and reconstructed using ideal LPF with cutoff frequency $f_{s/2}$. Sketch the reconstructed wave for $T_s = \frac{1}{6}\,\text{sec}$.

Nov 2009

$\qquad$ Sampling rate $=\; f_s$

$\qquad$ Cutoff frequency $f_c = f_{s/2}$

$\qquad$ Pulse width $\;=\; 1\,\text{sec}$

$\qquad\qquad f_m = \; 1\,\text{Hz}$

$\qquad$ Niquist rate $\;=\; 2\,\text{Hz}$

If $\quad T_s = \frac{1}{6}\,\text{sec}, \quad f_s = 6\,\text{Hz}$

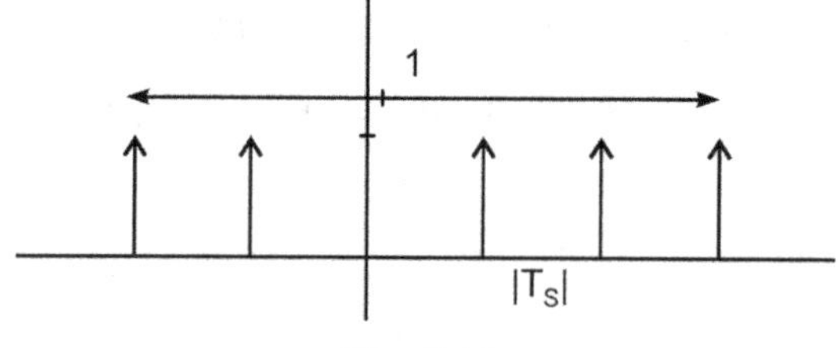

Fig. 7.12

$\therefore \quad f_c = 3\,\text{Hz}$, hence original signal can be reconstructed without any aliasing

3. For the signal $x(t) = 2 + \cos(100\pi t)$ find Nyquist rate and design a reconstruction filter (LPF) to reconstruct original signal if it is sampled at 80 Hz.

Ans: $\quad x(t) = 2 + \cos(100\pi t)\; 2 + \cos(100\pi t)$

$\qquad x(\omega) = 4\pi\,\delta(\omega) + \pi\left\{\delta(\omega + 100\pi) + \delta(\omega - 100\pi)\right\}$

$\qquad f_m = 50\,\text{Hz}$

$\qquad$ Sampling rate $=\; 80\,\text{Hz}$

$\qquad\qquad \omega_s = 160\pi$

$\qquad\qquad \omega_m = 100\pi$

Low pass filter having cutoff frequency 80π is designed to extract the signal.

$$\therefore \quad y(\omega) = 4\pi\delta(\omega) + \pi\big(\delta(\omega+60\pi)+\delta(\omega-60\pi)\big)$$

$$y(t) = 2t\cos(60\pi t)$$

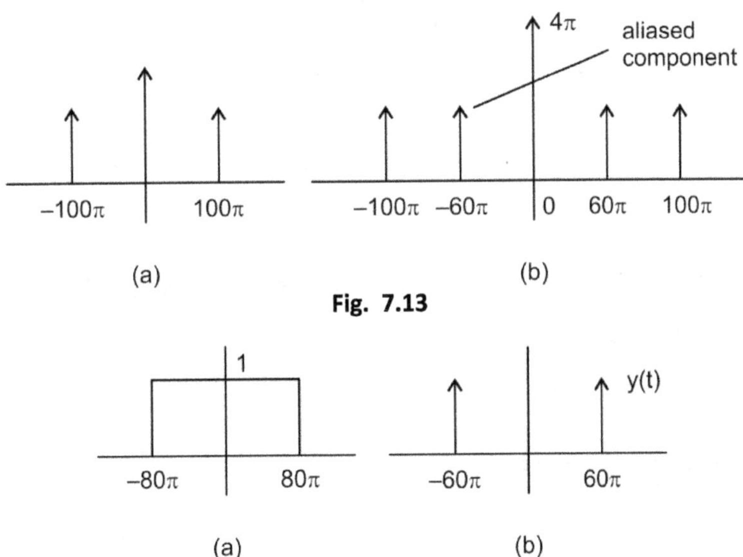

Fig. 7.13

Fig. 7.14

4. Consider a signal $x(t) = 6\cos 2\pi 5t$ sampled at 7 and 14 Hz. Since the highest frequency is 5 Hz explain the effect of sampling at both less than and great than the highest frequency.

Ans: $x(f) = 3\delta(f-5) + 3\delta(f+5)$

The spectrum of sampled $x(t)$ is given by

$$x_s(f) = 3f_s \sum_{n=-\alpha}^{\alpha}\big\{\delta(f-5-nf_s)+\delta(f+5-nf_s)\big\}$$

Let sampling frequency be 7 Hz.

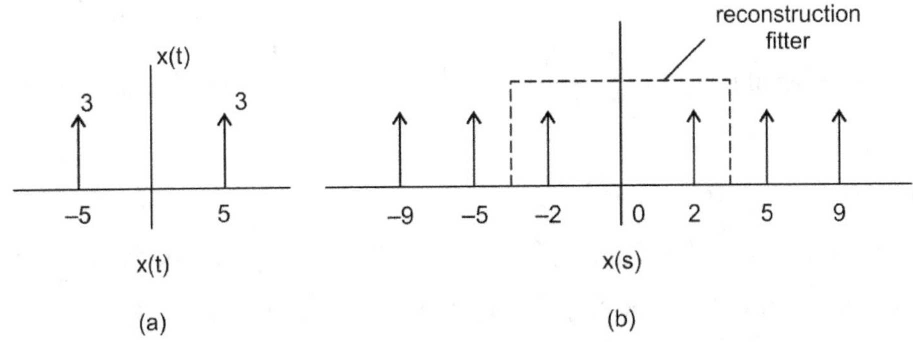

If a low pass filter of cutoff frequency 2 Hz is used the regained signal is $6\cos 2\pi(2)t$ if sampling frequency is 14 Hz.

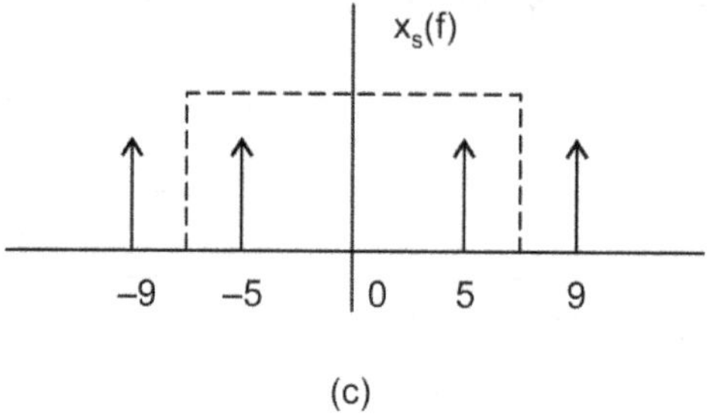

(c)

Fig. 7.15

Low pass filter of cutoff frequency 5 Hz is needed to recover the signal.

$$6\cos 2\pi(5)t$$

(a) 4   (b) (3)   (c) 2   (d) 1

OBJECTIVE QUESTIONS

1. The Nyquist rate of $\sin c(100\pi t)$ is

    (a) 200 Hz   (b) 300 Hz   (c) 100 Hz   (d) 50 Hz

2. The signal $v(t)=\cos 5\pi t+0.5\cos 10\pi t$ is flatteng sampled at Nyquist rate. Let $P(t)$ represent a pulse of unit amplitude extending from $t=0$ to $t=T$. The sampling signal $s(t)=\displaystyle\sum_{k=-\alpha}^{\alpha} p(t-kT)$. The expression for sampled signal.

    (a) $\left(\cos \pi t+0.5\cos 5\pi t\right)\displaystyle\sum_{k=-\alpha}^{\alpha}\delta(t-0.1k)$ (b) $\left(\cos 5\pi t+\cos 10\pi t\right)\displaystyle\sum_{k=-\alpha}^{\alpha}\delta(t-0-k)$

    (c) $\left(\cos 5\pi t+0.5\cos 10\pi t\right)\displaystyle\sum_{k=-\alpha}^{\alpha}\delta(t-0.1k)$ (d) $\left(\cos 5\pi t+0.5\cos \pi t\right)\displaystyle\sum_{k=-\alpha}^{\alpha}\delta(t-k)$

3. A periodic signal is $x(t)=\displaystyle\sum_{k=-\alpha}^{\alpha}\frac{1}{4+j2k}e^{j2400\pi kt}$ is sampled. The minimum sampling rate is

    (a) 1000   (b) 2500   (c) 7200   (d) 5000

4. Given that $x(t)=\cos\left(26000\pi t+\pi/3\right)+\cos\left(16000\pi t+\pi/3\right)+\cos\left(4000\pi t-\pi/3\right)$ and $f_s=10000$ samples/sec. The expression for reconstructed wave.

(a) $\cos\left(16000\pi t + \pi/3\right) + \cos\left(4000\pi t - \pi/3\right)$

(b) $\cos\left(6000\pi t + \pi/3\right) + \cos\left(1000\pi t - \pi/3\right)$

(c) $\cos\left(1000\pi t + \pi/3\right) + \cos\left(2000\pi t - \pi/3\right)$

(d) $\cos\left(6000\pi t + \pi/3\right) + \cos\left(4000\pi t - \pi/3\right)$

5. The Nyquist rate of $\sin c(100\pi t) + \sin c\, 50\pi t$ is

    (a) 200 Hz   (b) 50 Hz   (c) 100 Hz   (d) 300 Hz

6. Impulse sampling is represented as

    (a) $\displaystyle\sum_{n=-\alpha}^{\alpha} \delta(t - nTs)$ (b) $\displaystyle\sum_{n=-\alpha}^{\alpha} \delta(nTs)$

    (c) $\displaystyle\sum_{n=-\alpha}^{\alpha} \delta(-nTs)$ (d) $\displaystyle\sum_{n=-\alpha}^{\alpha} \delta(nt)$

7. If $x(t) = (15 + 30\sin 250\pi t)\cos 1000\pi t$ is sampled, the minimum sampling rate is

    (a) 100 Hz   (b) 1000 Hz   (c) 500 Hz   (d) 1250 Hz

8. A signal $g(t) = \sin c2(5\pi t)$ is sampled at a r ate of 20 Hz. Then it is called as

    (a) Under sampling   (b) Nyquistrate

    (c) Filter   (d) Over sampling

9. A band limited signal which has no spectral components above a frequency from f_m Hz is uniquely determined by its values at uniform intervals less than

    (a) $\dfrac{1}{2f_m}$ (b) $\dfrac{1}{2}$ (c) $\dfrac{1}{f_m}$ (d) f_m

10. If the components of frequencies above $f_s/2$ reappear as components of frequencies below $f_s/2$, then this version is known as

    (a) filtering   (b) convolution   (c) correlation   (d) Aliasing

11. If $x(t) = \cos(16000\pi t) + \cos(4000\pi t) + \cos(1000\pi t)$. If no aliasing occurs for $x(t)$, then sampling rate is

    (a) 8 kHz   (b) 16 kHz   (c) 10 kHz   (d) 5 kHz.

GATE QUESTIONS

1. A continuous time function $x(t)$ is periodic with period T. The function is sampled uniformly with a sampling period T_s, in which of the following cases is sampled signal is periodic?   GATE 2106

    (a) $T = \sqrt{2}T_s$ (b) $T = 1.2\,T_s$ (c) Always   (d) Never

    Ans: (b)

2. Consider the signal $x(t) = \cos(6\pi t) + \sin(8\pi t)$ where t is in seconds. The Niquist sampling rate for the signal $y(t) = x(2t + 5)$ is   GATE 2016

    (a) 8   (b)   12   (c) 16   (d)   32

    Ans: (c)

3. A band limited signal with a maximum frequency of 5 kHz is to be sampled. According to sampling theorem, which frequency is not a valid frequency   GATE 2013

    (a) 5 kHz   (b)   12 kHz   (c) 5 kHz   (d)   20 kHz

    Ans: (a)

4. The signal $\cos\left(10\pi t + \dfrac{\pi}{4}\right)$ is ideally sampled at a sampling frequency of 15 Hz. The sampled signal is passed through a filter with impulse response $\left(\dfrac{\sin \pi t}{\pi t}\right)\cos\left(40\pi t - \dfrac{\pi}{2}\right)$

    The filter output is   GATE 2015

    (a) $\dfrac{15}{2}\cos\left(40\pi t - \dfrac{\pi}{4}\right)$

    (b) $\dfrac{15}{2}\left(\dfrac{\sin \pi t}{\pi t}\right)\cos\left(10\pi t + \dfrac{\pi}{4}\right)$

    (c) $\dfrac{15}{2}\cos\left(10\pi t - \dfrac{\pi}{4}\right)$

    (d) $\dfrac{15}{2}\left(\dfrac{\sin \pi t}{\pi t}\right)\cos\left(40\pi t + \dfrac{\pi}{4}\right)$

    Ans: (a)

5. A band limited signal is sampled at the nyquist rate. The signal can be recovered by passing the samples through   GATE 2001

    (a) RC filter   (b) Energy detector

    (c) PLL   (d) LPF

    Ans: (d)

6. A deterministic signal has the power spectrum given in figure. the minimum sampling rate needed to completely represent signal is   GATE 1997

    (a) 1 kHz   (b)   2 kHz   (c) 3 kHz   (d)   None

    Ans: (b)

7. A signal has frequency components from 300 Hz to 1.8 kHz. The minimum possible
    rate at which the signal has to be sampled is _______   GATE 1991

    (a) 3000 samples/sec   (b) 3600 sample/sec

    (c) 4000 sample/sec   (d) None

    Ans: (b)

8. The nyquist sampling interval for the signal sin c (700t) + sinc (500t)   GATE 2001

    (a) $\dfrac{1}{350}$ sec   (b) $\dfrac{\pi}{350}$ sec   (c) $\dfrac{1}{700}$ sec   (d) $\dfrac{\pi}{175}$ sec

    Ans: (c)

9. The nyquist sampling frequency of signal given by $6 \times 10^4 \sin c^2 (400t) * 10^6$
    $\sin c^3 (100t)$ is   GATE 1999

    (a) 200   (b) 1500   (c) 300   (d) 1000

    Ans: (b)

10. The nyquist sampling rate for the signal $\dfrac{\sin(500\pi t)}{\pi t} \times \dfrac{\sin(700\pi t)}{\pi t}$ is   GATE 2010

    (a) 400   (b) 1200   (c) 600   (d) 1400

    Ans: (b)

11. Increased pulse width in the flat top sampling leads to

    (a) Attenuation of high frequencies   (b) Attenuation of low frequencies

    (c) Greater aliasing   (d) None   *Ans:* (a)

12. A 1.0 kHz signal is flat top sampled at the rate 1800 samples/sec and the samples are
    applied to an ideal rectangular LPF with cut off frequency of 1100 Hz then the output of
    the filter contains   GATE 1994

    (a) only 800 Hz component   (b) 800 Hz and 900 Hz components

    (c) 800 Hz and 1000 Hz components   (d) 800, 900 and 1000 Hz   *Ans:* (c)

13. Flat top sampling of low pass signals   GATE 1995

    (a) gives rise to aperture effects   (b) implies over sampling

    (c) leads to aliasing   (d) Introduces delay distortion   *Ans:* (a)

14. A band limited signal is sampled at the niquist rate. The signal can be recovered by
    passing samples through   GATE 1995

    (a) an RC filter   (b) PLL

    (c) envelop detector   (d) ideal low pass filter   *Ans:* (d)

SUMMARY OF CHAPTER

1. Sampling is the process of converting a continuous signal into discrete time signal.

2. Sampling theorem: A band limited signal $x(t)$ with $x(jw) = 0$ for $|\omega| > |\omega_m|$ is uniquely determined from its samples $x(nT)$ if the sampling frequency $f_0 \geq 2f_m$

3. Aliasing: Aliasing appears whenever an analog signal is not sampled at a high frequency.

4. Antialiasing: When a signal is sampled, with a frequency f_s, all the signals with frequency range higher than $\dfrac{\omega_s}{2}$ appear as signal frequencies between 0 and $\dfrac{\omega_s}{2}$ creating aliasing.

CHAPTER **8**

Laplace Transform

Objectives of the chapter

In this chapter, Laplace transform, evaluation of LT, s-plane and ROC, properties of ROC, convergence of LT, inverse of LT, difference between LT, FT and ZT.

8.1 INTRODUCTION

It is used to analyze a large class of continuous signals that are not absolutely integrable, where FT does not exists for these signals. Laplace transform can be applied to analyse unstable systems. It is also can be called as continuous time generalization of FT.

Laplace transform comes in two varieties: (1) Unilateral (2) bilateral. Laplace transform expresses signals as linear combination of complex exponentials, which are the eigen functions of the differential equations. Impulse responses of LTI systems completely characterize them. Since LT describes the impulse response of LTI system, it best characterizes these systems.

(a) Laplace Transform

Let e^{st} be a complex exponential

where

$S = \sigma + j\,w$ complex frequency

$e^{st} = e^{\sigma t}\cos\omega t + je^{\sigma t}\sin\omega t$

where σ is damping factor

ω is frequency

(b) Eigen function Property of e^{st}

Let us consider an LTI system with an input $x(t) = e^{st}$

$y(t) = h(t) * x(t)$

$$= \int_{-\alpha}^{\alpha} h(T)\,x(t-T)dT$$

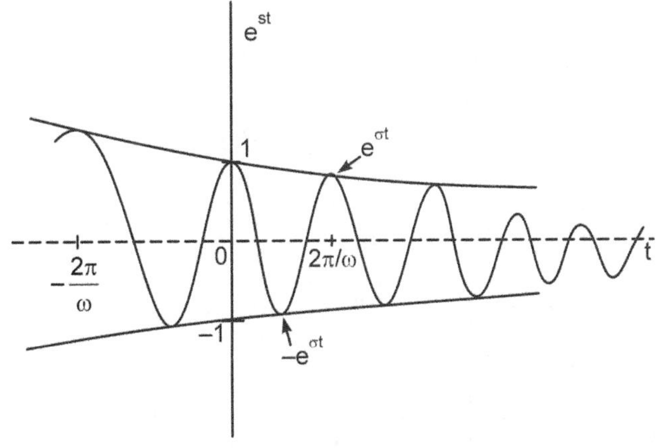

Fig. 8.1

$$= \int_{-\alpha}^{\alpha} h(T) \ e^{s(t-T)} dT$$

$$= e^{st} \int_{-\alpha}^{\alpha} h(T) \ e^{-sT} dT$$

Let us define

$$H(s) = \int_{-\alpha}^{\alpha} h(T) e^{-sT} dT$$

$$\therefore \qquad y(t) = H(s)e^{st}$$

Eigen function is a signal that passes through the system without being modified except for multiplication by a scalar.

Hence e^{st} is eigen function and $H(s)$ is eigen value.

(c) Evaluation of Laplace transform

We have fourier transform

$$\mathcal{F}\{f(t)\} \ = \ F(\omega) = \int_{-\alpha}^{\alpha} f(t)e^{-j\omega t} dt$$

If the signal f(t) is not integrable then fourier transform fails to estimate it. Hence to make it absolutely integrable.

Let $\qquad \phi(t) = f(t)e^{-\sigma t}$

Examples : Find the unilateral LT of following signals

1. $x(t) = \delta(t)$

$$X(t) \ = \ \int_{0}^{\alpha} x(t)e^{-st} dt = \int_{0}^{\alpha} \delta(t)e^{-st} dt = 1$$

2. $x(t) = \ u(t)$

$$X(s) \ = \ \int_{0}^{\alpha} x(t)e^{-st} dt = \int_{0}^{\alpha} e^{-st} dt = \frac{-1}{S} e^{-st} \ \Big|_{0}^{\alpha} = \frac{1}{S}$$

3. $x(t) = \ u(t-2)$

$$X(s) \ = \ \int_{0}^{\alpha} x(t)e^{-st} dt = \int_{0}^{\alpha} u(t-2)e^{-st} dt$$

$$= \ \int_{2}^{\alpha} e^{-st} dt = \frac{1}{-S} e^{-st} \ \Big|_{2}^{\alpha} = \frac{e^{-2s}}{8}$$

4. $x(t) = \cos\omega_0 t$

$$X(s) = \int_0^\alpha \cos\omega_0 t\, e^{-st} dt = \int_0^\alpha \frac{e^{j\omega_0 t} + e^{-j\omega_0 t}}{2} e^{-st} dt$$

$$= \frac{1}{2}\left[\int_0^\alpha e^{-(s-j\omega_0)t} dt + \int_0^\alpha e^{-(s+j\omega_0)t} dt\right]$$

$$= \frac{1}{2}\left[\frac{1}{S-j\omega_0} + \frac{1}{S+j\omega_0}\right]$$

$$= \frac{S}{S^2 + \omega_0^2}$$

5. $x(t) = \sin\omega_0 t$

$$\therefore \mathcal{F}\{\phi(t)\} = \int_{-\alpha}^{\alpha} f(t) e^{-\sigma t} e^{-j\omega t} dt$$

$$= \int_{-\alpha}^{\alpha} f(t) e^{-(\sigma+j\omega)t} dt$$

$$= F(\sigma + j\omega)$$

The inverse fourier transform of $F(\sigma + j\omega)$ is

$$\phi(t) = \frac{1}{2\pi} \int_{-\alpha}^{\alpha} F(\sigma + j\omega) e^{j\omega t} d\omega$$

$$f(t) e^{-\sigma t} = \frac{1}{2\pi} \int_{-\alpha}^{\alpha} F(\sigma + j\omega) e^{(j\omega)t} dt$$

$$f(t) = \frac{1}{2\pi j} \int_{-\alpha}^{\alpha} F(\sigma + j\omega) e^{(\sigma+j\omega)t} d\omega$$

$S = \sigma + j\omega$

where $ds = jd\omega$

limits $\omega = -\alpha$ to α becomes $\sigma - j\infty$ to $\sigma + j\infty$.

Thus

$$\propto\{f(t)\} = \int_{-\alpha}^{\alpha} f(t) e^{-st} dt$$

$$\propto^{-1}\{F(\sigma+j\omega)\} = \frac{1}{2\pi j}\int_{\alpha-j\alpha}^{\alpha+j\alpha} F(s)e^{st}ds$$

This is known as complex FT or bilateral LT where as unilateral LT is given by

$$\propto\{f(t)\} = \int_{0}^{\alpha} f(t)e^{-st}dt$$

8.2 S - PLANE AND ROC

The complex frequency S can be represented graphically known as S-plane. Re(s) i.e., σ is represented on x-axis and Img (s) i.e., jω is on y-axis.

If x(t) is absolutely integrable, then FT is equal to LT when $\sigma=0$. In S-plane, $\sigma=0$ correspond to imaginary axis.

The range of value of S for which the integral converges is referred to as the ROC of LT. ROC consists of those values of $S=\sigma+j\omega$ for which the FT of $x(t)e^{-\sigma t}$.

Properties of ROC

1. ROC of $X_{(s)}$ consists of strips parallel to the $j\omega$ axis in the S plane
2. If $x_{(t)}$ is of finite duration and is absolutely integrable, then ROC is in entire S plane

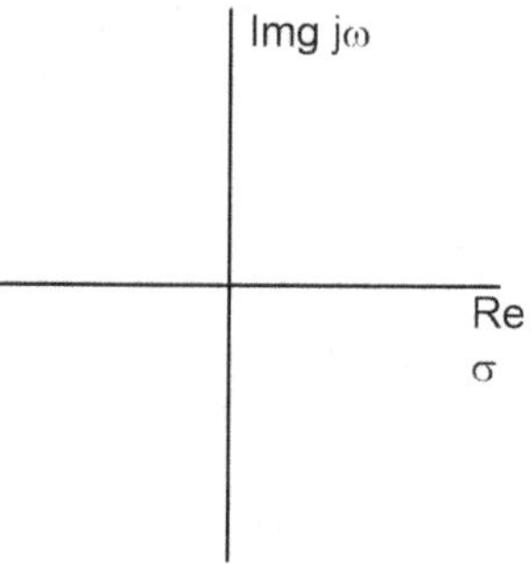

Fig. 8.2

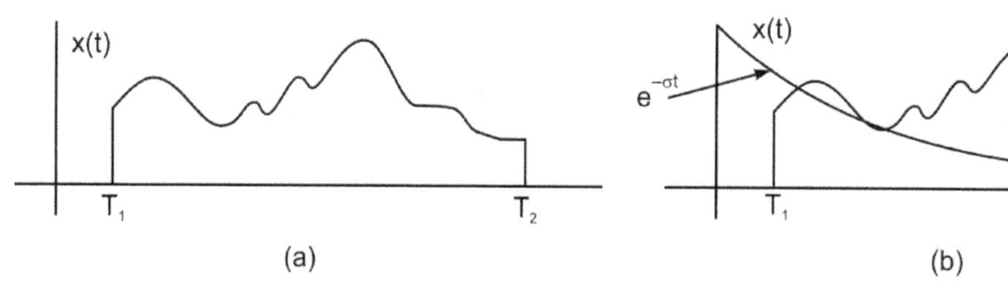

(a)　　　　　　　　　　　(b)

3. If $x_{(t)}$ is right sided and if the line $Re(s)=\sigma_0$ is in ROC, then all values of S for which $Re(s) > \sigma_{0\,x(t)}$ will also be in ROC

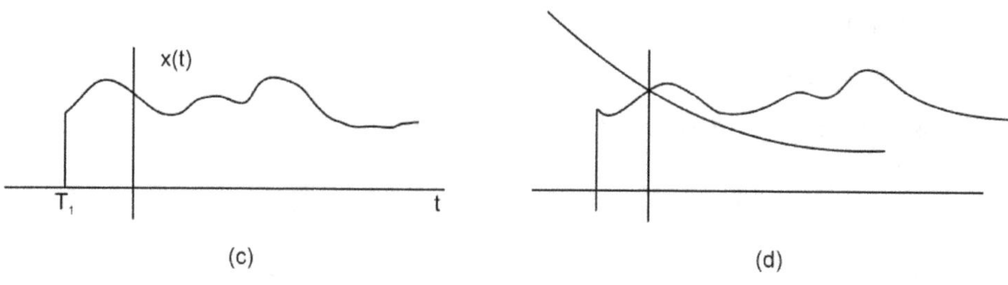

(c)　　　　　　　　　　　(d)

4. If $x(t)$ is left sided and if the line $Re(s) = \sigma_0$ is in ROC, then all values of S for which $Re(s) < \sigma_{0\,x(t)}$ will also be in ROC.

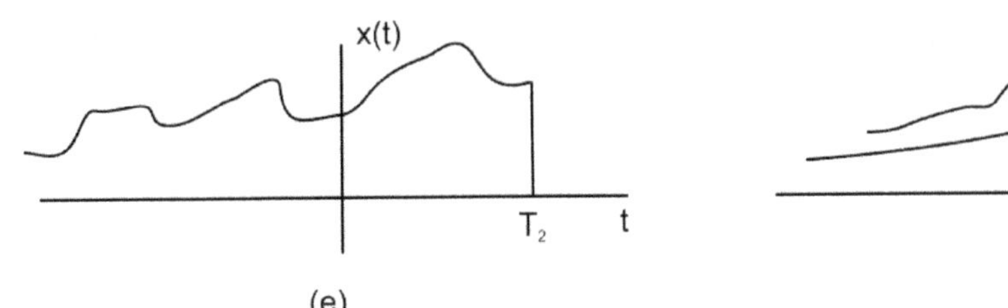

Fig. 8.3

8.3 CONVERGENCE OF LT

LT of signal $x(t)$ exists if

$$X(s) = \int_{-\alpha}^{\alpha} x(t)e^{-st}\,dt$$

substituting $S = \sigma + j\omega$

$$X(s) = \int_{-\alpha}^{\alpha} x(t)e^{-\sigma t}e^{-j\omega t}\,dt$$

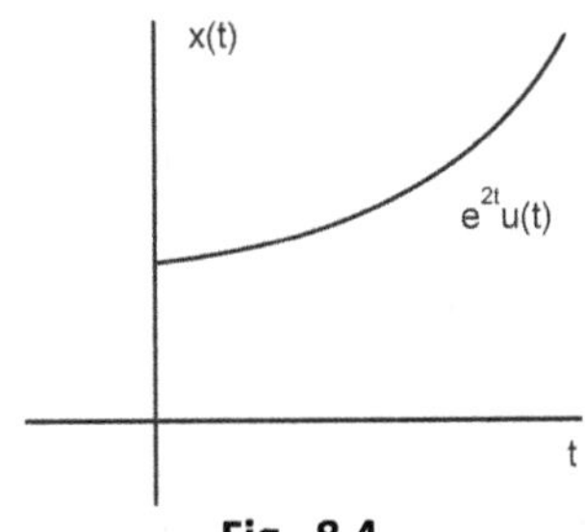

Fig. 8.4

this indicates LT is the $\mathcal{F}T$ of $x(t)e^{-\sigma t}$. LT is absolutely integrability is the condition for convergence.

i.e., $X(s)$ exists if

$$\int_{-\alpha}^{\alpha} \left| x(t)e^{\sigma t} \right| < \alpha$$

The range of σ for which LT converges is known as ROC.

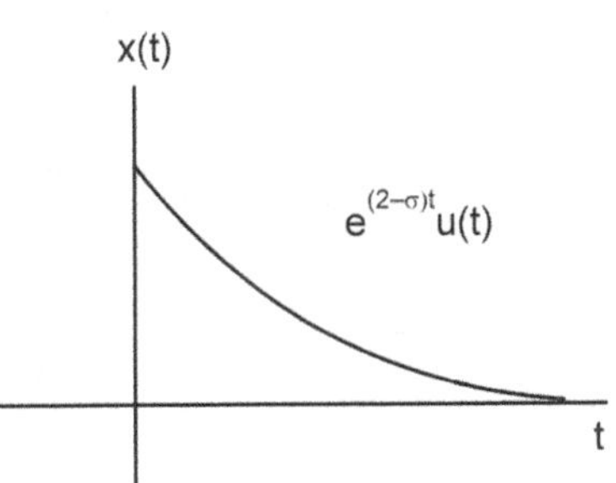

Fig. 8.5

Let us consider a signal

$$x(t) = e^{2t}u(t)$$

FT for $x(t)$ doesn't exists because it is not absolutely integrable.

If we multiply $x(t)$ with $e^{-\sigma t}$

$$x(t) = e^{-\sigma t} = e^{(2-\sigma)t}u(t) \text{ is absolutely integrable} \quad \text{for } \sigma > 2$$

which is a FT of $x(t) = e^{-\sigma t}$ exists for some values of σ.

Ex: Consider $x(t) = e^{2t}u(t)$

which is not absolutely integrable.

If we multiply it with $e^{-\sigma t}$

$$x(t) = e^{(2-\sigma)t}$$

$\therefore$ it is integrable for $\sigma > 2$

Thus LT exists where FT does not exists

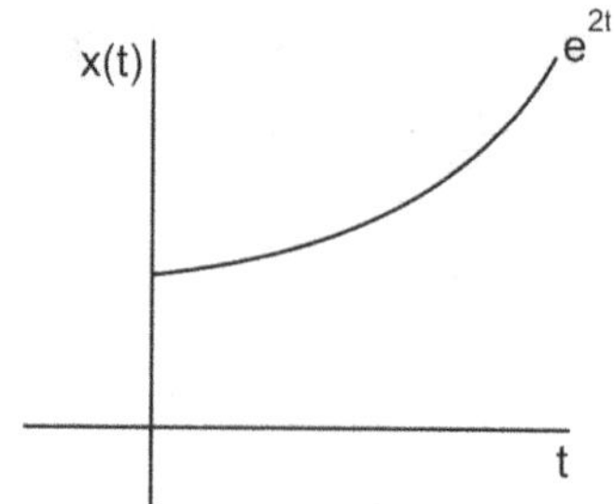

Fig. 8.6

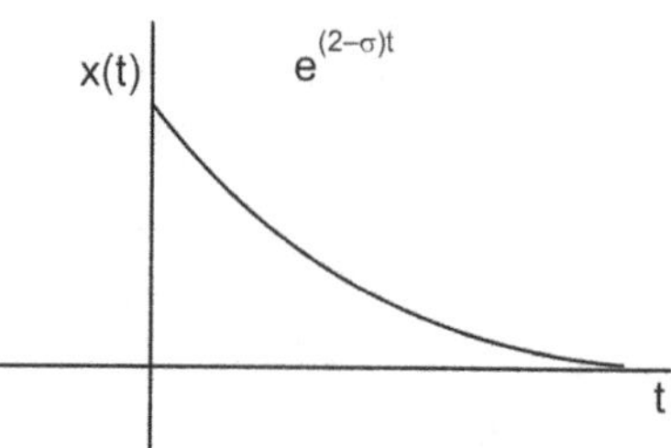

Fig. 8.7

Problems

1. $x(t) = e^{-at}u(t) + e^{-bt}u(-t)$

$$x(s) = \int_0^\alpha e^{-(a+s)t}dt + \int_{-\alpha}^0 e^{-(s+b)t}dt$$

$$= \frac{1}{S+a} + \frac{1}{S+b}$$

$$S > -a \qquad S < -b$$

2. $x(t) = \cos\omega_0 t$

$$x(s) = \int_0^\alpha \cos\omega_0 t\, e^{-st}dt$$

$$= \int_0^\alpha \frac{e^{j\omega_0 t} + e^{-j\omega_0 t}}{2} e^{-st}dt$$

$$= \frac{1}{2}\left\{\frac{1}{S - j\omega_0} + \frac{1}{S + j\omega_0}\right\}$$

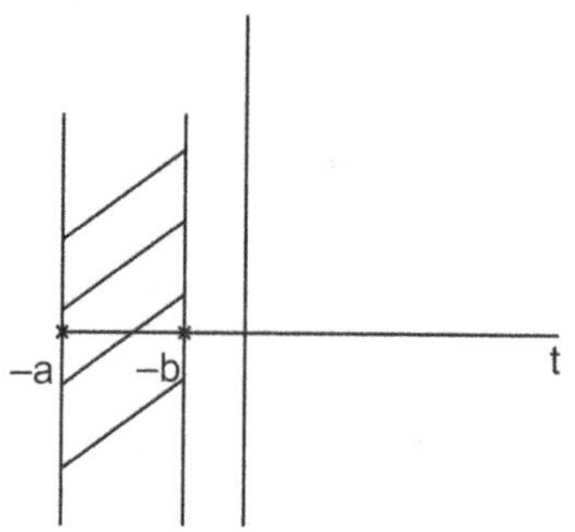

Fig. 8.8

3. $x(t) = t\, e^{-at}$

$$X(s) = \frac{1}{(s+a)^2}$$

4. $x(t) = e^{-at}\cos\omega_0 t$

$$X(s) = \frac{s+a}{(s+a)^2 + \omega_0^2}$$

8.4 PROPERTIES OF LT

1. **Linearity:** $\propto \{ax_1(t) + bx_2(t)\} = ax_1(s) + bx_2(s)$

2. **Scaling:** $\propto \{x(at)\} = \dfrac{1}{|a|} x\left(\dfrac{s}{a}\right)$

3. **Time shifting:** $\propto \{x(t - t_0)\} = e^{-st_0} x(s)$

$$\propto \{x(t - t_0)\} = \int_0^\alpha x(t - t_0)\, e^{-s(t)}\, dt$$

$$= \int_0^\alpha x(p) e^{-s(p + t_0)}\, dp \qquad \begin{aligned} p &= t - t_0 \\ dp &= dt \end{aligned}$$

$$= e^{-st_0} x(s)$$

4. **Frequency Shifting**

$$\propto \{e^{-at} x(t)\} = x(s + a)$$

5. **Differentiation in s-domain**

$$\propto \{-t\, x(t)\} = \frac{d}{ds} x(s)$$

$$\frac{d}{ds} x(s) = \int_0^\alpha x(t) \frac{d}{ds} e^{-st}\, dt$$

$$= \int_0^\alpha x(t) - t e^{-st}\, dt = \propto \{-tx(t)\}$$

Similarly $\quad \propto \{(-t)^n x(t)\} = \dfrac{d^n}{dx^n} x(s)$

6. **Integration in time domain**

If $\qquad x(t) \xleftrightarrow{\ \propto\ } \dfrac{1}{S} \times (s)$

then $\qquad \displaystyle\int_{-\alpha}^{t} x(T)\, dT \xleftrightarrow{\ \propto\ } \dfrac{1}{S} \times (s)$

this is equal to

$$u(t) * x(t) \xleftrightarrow{\ \propto\ } \frac{1}{S} \times (s)$$

7. **Transform of derivatives**

If $\qquad \propto \{x(t)\} = x(s)$

$$\propto \left\{ \frac{d}{dt} x(t) \right\} = S x(s) - x(0^-)$$

Similarly

$$\propto\left\{\frac{d^n}{dt^n}x(t)\right\}=S^n x(s)-S^{n-1}-x(\bar{0})-S^{n-2}\frac{dx}{dt}(\bar{0})-....\frac{d^{n-1}}{dt^{n-1}}x(0^-)$$

8. Transform of Integrables

$$\propto\left\{\int_{-\alpha}^{t}x(t)dt\right\}=\frac{x(s)}{S}+\int_{-\alpha}^{\bar{0}}\frac{x(T)dt}{S}$$

9. Time convolution

$$\propto\left\{x_1(t)*x_2(t)\right\}=x_1(s)x_2(s)$$

10. Initial value theorem

$$\propto\left\{\frac{d}{dt}x(t)\right\}=\int_{0}^{\alpha}\frac{d}{dt}x(t)e^{-st}dt$$

$$=\ Sx(s)-x(0^-)$$

$$\underset{S\to0}{Lt}\ Sx(s)-x(0^-)=0$$

$$\boxed{\underset{S\to0}{Lt}\ Sx(s)=x(0^+)}$$

11.   Final value theorem

$$\underset{S\to0}{Lt}\int_{0}^{\alpha}\frac{dx}{dt}e^{-st}dt=\underset{S\to0}{Lm}Sx(s)-x(0^-)$$

$$\therefore\qquad\underset{t\to0}{Lm}x(t)=\underset{S\to0}{Lm}Sx(s)$$

12.   S – domain shift

$$e^{s_0 t}x(t)\overset{\infty}{\longleftrightarrow}X(s-s_0)$$

$$\propto\left\{e^{s_0 t}x(t)\right\}=\int x(t)e^{s_0 t}e^{-st}dt$$

$$=\ \int x(t)e^{-(s-s_0)t}dt$$

$$=\ X(s-s_0)$$

SOLVED EXERCISE PROBLEMS-1

1. $x(t)=\ e^{-t}(t-2)u(t-2)$

    Using time shifting and S domain shift properties

    Time shift $x(t-t_0)\leftrightarrow e^{-st_0}\times(s)$

Let $x(t) = t\,e^{-at}u(t)$

$$\propto\{x(t)\} = \int\limits_{0}^{\alpha} t \cdot e^{-at} \cdot e^{-st}\,dt$$

$$= \frac{1}{(S+a)^2} = \frac{1}{(S+1)^2}$$

Time shifting property

$$= \frac{e^{-2(s+1)}}{(S+1)^2}$$

2. $x(t) = t^2\,e^{-2t}\,u(t)$

Let $x(t) = t\,e^{-at}$

$$\therefore \propto\{x(t)\} = \frac{1}{(S+a)^2}$$

Using differentiation property

$$\propto\{-t\,x(t)\} = -\frac{d}{ds}\frac{1}{(S+a)^2}$$

$$= +^2\frac{1}{(S+a)^3}$$

$$\therefore \propto\{t^2 e^{-2t}\} = \frac{2}{(S+2)^3}$$

3. $x(t) = tu(t) - (t-1)u(t-1) - (t-2)u(t-2)$

Using time shifting property

$$= \frac{1}{S^2} - \frac{e^{-s}}{S^2} - \frac{e^{-2s}}{S^2}$$

$$= \frac{1 - e^{-s} - e^{-2s}}{S^2}$$

4. $x(t) = e^{-t}u(t) * \cos(t-2)u(t-2)$

Convolution property

$$X(s) = \frac{1}{(S+1)}\frac{e^{-2s}S}{(S^2+1)}$$

5. $f(t) = \dfrac{V}{a}t, \quad 0 \le t \le a$

$\qquad = V, \quad a \le t \le 2a$

$\qquad = 0 \quad t \ge 2a$

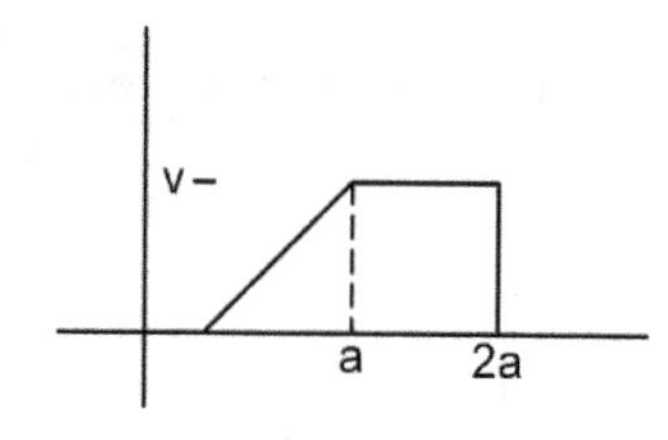

$\mathscr{L}\{f(t)\} = \displaystyle\int_0^a \dfrac{V}{a} te^{-st}dt + \int_a^{2a} Ve^{-st}dt$

$\qquad = \dfrac{V}{as^2}\left\{1 - e^{-as} - ase^{-2as}\right\}$

6. $x(t) = \sin \pi t \quad 0 < t < 1$

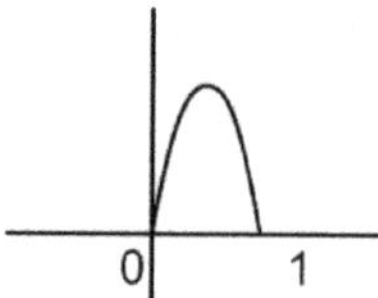 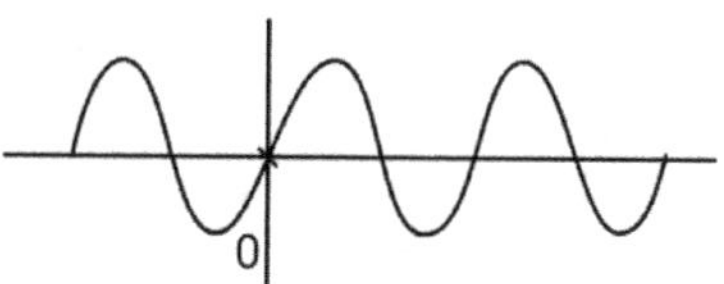

$x(t) = \sin \pi t\, u(t) + \sin \pi(t-1)u(t-1)$

7.

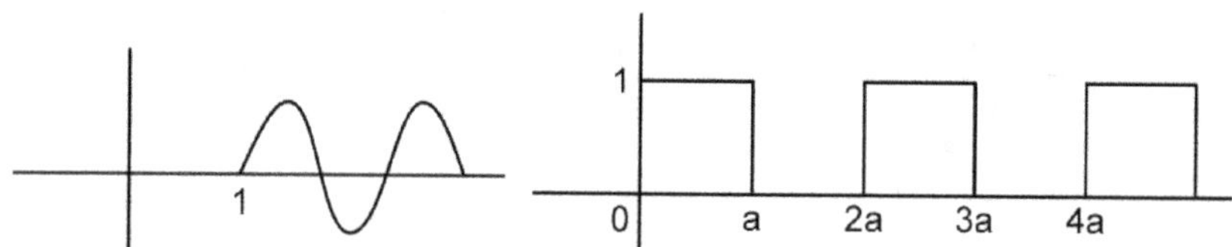

$x(t) = u(t) - u(t-a) + u(t-2a) - u(t-3a) + u(t-4a)\ldots\ldots$

$x(s) = \dfrac{1}{S}\dfrac{e^{-as}}{S} + \dfrac{e^{-2as}}{S} - \dfrac{e^{-3as}}{S} + \ldots\ldots$

$\qquad = \dfrac{1}{S}\left\{\dfrac{1}{1-e^{-2as}} - \dfrac{e^{-as}}{1-e^{-2as}}\right\}$

$\qquad = \dfrac{1}{S}\left\{\dfrac{1}{1+e^{-as}}\right\}$

SOLVED PROBLEMS EXERCISE 2

1. Find LT and ROC of following signals

   (a) $x(t) = u(t-5)$

                                              Ans: $X(s) = \dfrac{e^{-5s}}{s}, \quad \mathrm{Re}(s) > 0$

   (b) $x(t) = e^{5t}(-t+3)$

                                              Ans: $X(s) = \dfrac{e^{-3(s-5)}}{s-5}, \quad \mathrm{Re}(s) < 5$

(c) $x(t) = e^{j\omega_0 t}u(t)$ **Ans:** $X(s) = \dfrac{1}{S - j\omega_0}$, $Re(s) > 0$

(d) $x(t) = \sin 3tu(t)$ **Ans:** $X(s) = \dfrac{3}{S^2 + 9}$, $Re(s) > 0$

(e) $x(t) = e^{-2t}u(t) + 2e^{-3t}u(t)$ **Ans:** $X(s) = \dfrac{2s + 5}{S^2 + 5s + 6}$, $Re(s) > -3$

(f) $x(t) = 3e^{-2t}u(t) - 2e^{-t}u(t)$ **Ans:** $X(s) = \dfrac{3}{S + 2} - \dfrac{2}{S + 1}$, $Re(s) > -1$

(g) $x(t) = e^{-2t}u(t) + e^{-t}\cos 3tu(t)$

$$\textbf{\textit{Ans:}}\ \ X(s) = \frac{1}{S + 2} + \frac{1}{2}\left(\frac{1}{S + (1 - 3j)}\right) + \frac{1}{2}\left(\frac{1}{S + (1 + 3j)}\right),\quad Re(s) > 1$$

(h) $x(t) = \delta(t) - \dfrac{4}{3}e^{-t}u(t) + \dfrac{1}{3}e^{2t}u(t)$ **Ans:** $X(s) = \dfrac{(S - 1)^2}{(S + 1)(S - 2)}$, $Re(s) > 2$

2. Find inverse LT of below functions

(a) $X(s) = \dfrac{3s + 2}{S^2 + 4s + 5}$ **Ans:** $x(t) = 3e^{-2t}\cos tu(t) - 4e^{-2t}\sin tu(t)$

(b) $X(s) = \dfrac{S^2 + S - 2}{S^3 + 3S^2 + 5s + 3}$

$$\textbf{\textit{Ans:}}\ \ x(t) = e^{-t}u(t) + 2e^{-t}\cos\left(\sqrt{2}t\right)u(t) - \frac{1}{\sqrt{2}}e^{-t}\sin\left(\sqrt{2}t\right)u(t)$$

8.5 INVERSE OF LAPLACE TRANSFORM

1. Inversion using Partial fraction Expansion

Inverse LT of each term in the partial fraction expansion is found using the below equations.

$$A_k e^{d_k t}u(t) \xleftrightarrow{\ \infty\ } \frac{A_k}{s - d_k}$$

$$\frac{At^{n-1}}{(n-1)!}e^{d_k t}u(t) \xleftrightarrow{\ \infty\ } \frac{A}{(s - d_k)^n}$$

Example:

1. Find inverse LT of

$$X(s) = \frac{3s+4}{(S+1)(S+2)^2}$$

Ans: Using partial fraction expansion

$$X(s) = \frac{A_1}{S+1} + \frac{A_2}{S+2} + \frac{A_3}{(S+2)^2}$$

Solving by method of residues

$$X(s) = \frac{1}{S+1} - \frac{1}{S+2} + \frac{2}{(S+2)^2}$$

$$\therefore \quad x(t) = e^{-t}u(t) - e^{-2t}u(t) + 2.t\, e^{-2t}u(t)$$

2. $X(s) = \dfrac{-5s-7}{(S+1)(S+2)(S-1)}$

$$= \frac{A}{S+1} + \frac{B}{S+2} + \frac{C}{S-1}$$

$$\therefore \ X(s) = \frac{1}{S+1} + \frac{2}{(S-1)} + \frac{1}{(S+2)}$$

$$x(t) = e^{-t}u(t) - 2\,e^{t}u(t) + e^{-2t}u(t)$$

3. $X(s) = \dfrac{S}{S^2+5s+6}$

$$X(s) = \frac{3}{(S+3)} - \frac{2}{(S+2)}$$

$$X(t) = 3e^{-3t}u(t) - 2e^{-2t}u(t)$$

4. $X(s) = \dfrac{S^2+S-3}{S^2+3S+2}$

$$X(s) = 1 + \frac{1}{(S+2)} - \frac{3}{(S+1)}$$

$$\therefore \ x(t) = \delta(t) + e^{-2t}u(t) - 3e^{-t}u(t)$$

5. $X(s) = \dfrac{2S^3 - 9S^2 + 4s + 10}{S^2 - 3s - 4}$

$$\begin{array}{r} S^2\,3s-4 \end{array} \overline{\left| \begin{array}{l} 2S^3 - 9S^2 + 4s + 10 \\ 2S^3 - 6S^2 - 8s \end{array} \right.}$$

$$\underline{-3S^2 + 12s + 10}$$

$$\underline{-3S^2 + 9s + 12}$$

$$3s - 2$$

$$X(s) \;=\; 2s - 3 + \frac{3s-2}{s^2 - 3s - 4}$$

$$\;=\; 2s - 3 + \frac{1}{S+1} + \frac{2}{S-4}$$

$$\therefore \;\; x(t) \;=\; 2\delta^1(t) - 3\delta(t) + 2e^{4t}u(t) + e^{-t}u(t)$$

8.6 DIFFERENCES BETWEEN LAPLACE, FOURIER AND Z-TRANSFORMS

1. Laplace transform is an extension of continuous time fourier transform.

2. Fourier transform doesn't converge for all signals, in such cases LT and ZT.

3. Z – transform and Laplace Transform are same but first one is for discrete signals and the second is for continuous signals.

4. Correlation between LT and ZT can be achieved by using sampling theorem.

5. Z transforms are mostly used for analysis and synthesis of digital filters.

6. LT are useful for analysis of unstable systems by converting them to a stable system.

7. LT is helpful for analysis of signals that are not absolutely integratble, such as the impulse response of an unstable system.

8. Unilateral LT is useful for solving differential equations with initial conditions.

9. Bilateral LT is useful to study system characteristics such as stability, and frequency response.

10. LT are mainly used to study transient and stability analysis of causal LTI systems described by differential equations.

11. ZT are mainly used to study system characteristics and the derivation of computational structures for implementing discrete time systems.

8.7 SEQUENCE AND THEIR ROC IN Z - TRANSFORMS

Signal   ROC

Finite duration Signals

1. Causal sequence

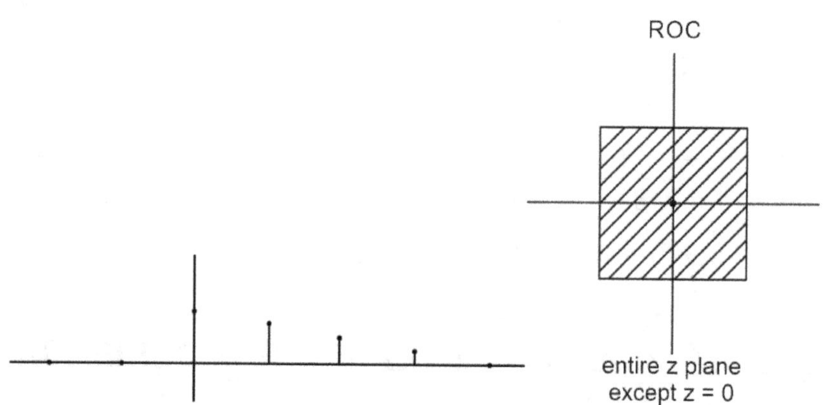

2. Non causal Sequence

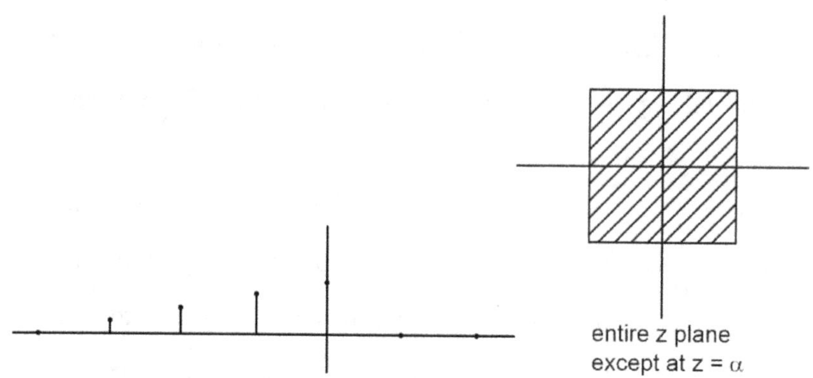

3. Two sided sequence

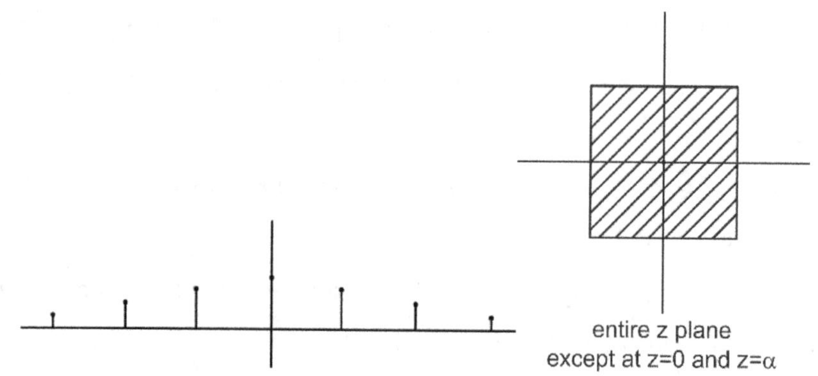

Infinite duration signals

1. Causal sequence

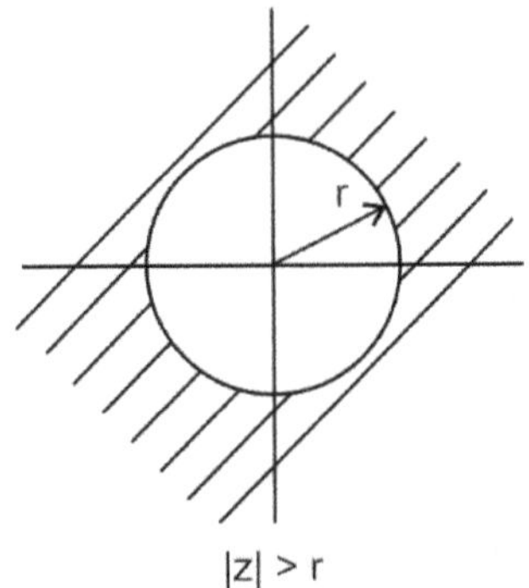

2. Non causal sequence   3. Two sided

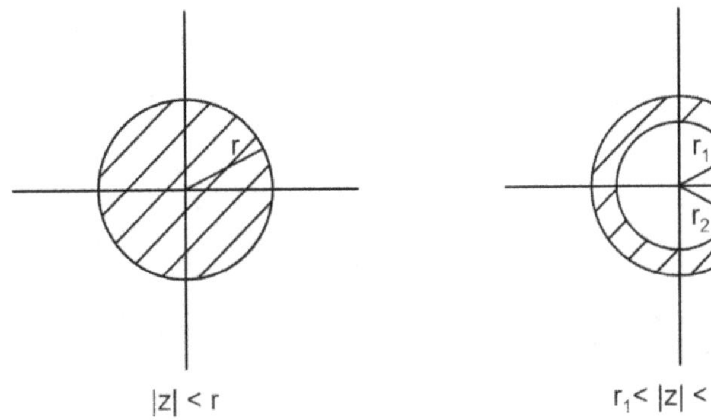

SOLVED PROBLEMS FROM PREVIOUS PAPERS   Nov 2008

1. (a) Find LT of $\sin \omega t$

    (b) Find ILT of   Nov 2009

      (i) $\dfrac{S^2 + 2s + 5}{(S+3)(S+5)^2}$ $R_e(s) > -3$

      (ii) $\dfrac{2s+1}{s+2}$ $R_e(s) > -2$ Nov, Feb 2008, Sep 2007

2. Find ILT of

    (a) $\dfrac{(s+1)^2}{s^2 - s + 1}$ $R_e(s) > \dfrac{1}{2}$

(b) $\dfrac{s^2-s+1}{(s+1)^2}$ $R_e(s)>-1$

3. Find LT of $\dfrac{d^2}{dt^2}[\delta(t)]$ Nov 2008, Sep 2007, Nov, Feb 2008

4. Find LT of

   (a) $e^{-2t}u(t)+e^{-3t}u(t)$

   (b) $e^{-4t}u(t)+e^{-5t}\sin 5tu(t)$

5. The system function of a causal system is $H(s)=\dfrac{s+1}{s^2+2s+2}$.

   Find $y(t)$ for an input $x(t)=e^{-|t|}$

        Feb 2008,

6. Find LT of $0\le t\le 1$ Sep 2007

   (a) $x(t)=1$ $0\le t\le 1$

   (b) $x(t)=t$ $0\le t\le 1$

7. Find the LT of

$$x(t) = e^{-at}\left\{A\cos bt+\left(\dfrac{B-Aa}{b}\right)\sin bt\right\}\cdot u(t) \text{ is where } C=b^2+a^2 \text{ and } a>0$$

$$X(s) = \dfrac{As+B}{s^2+2as+c}$$ Sep 2007

8. If $x(t)$ has following properties

   (a) $x(t)$ is real and even

   (b) $x(s)$ has four poles and no zeros in a finite s plane

   (c) $X(s)$ has pole at $S=\dfrac{1}{2}e^{j\pi/4}$

   (d) $\displaystyle\int_{-\alpha}^{\alpha}x(t)dt=4$

   determine $X(s)$ and ROC   Sep 2007

9. The differential equation of LT is

$$\frac{d^2y(t)}{dt^2} + \frac{6dy(t)}{dt} + 8y(t) = \frac{dx(t)}{dt} + x(t)$$

$$\frac{dy(0)}{dt} = 3, \quad y(\theta) = 1$$

$$x(t) = u(t)$$

Find transfer function

10. Find $x(t)$ if $X(s) = \dfrac{3s^2 + 22s + 27}{(s+1)(s+2)(s2+2s+5)}$ Sep - 2007

11. Use geometric evaluation from pole zero plot to determine the magnitude of FT of the signal whose LT is

$$X(s) \;=\; \frac{s^2 - s + 1}{s^2 + s + 1} \qquad R_e(s) > \frac{1}{2} \qquad\qquad \text{Sep 2007}$$

ANSWERS TO PREVIOUS PAPER PROBLEMS

1. (a)   Find LT of $\sin \omega t$

$$\propto \{x(t)\} = \int_0^\alpha \sin \omega t\, e^{-st} dt$$

$$= \int_0^\alpha \frac{e^{j\omega t} - e^{-j\omega t}}{2j} e^{-st} dt$$

$$= \frac{1}{2j}\int_0^\alpha e^{-(s-j\omega)t} - e^{-(s+j\omega)t} dt$$

$$= \frac{1}{2j}\left\{ \frac{e^{-(s-j\omega)t}}{-(s-j\omega)} + \frac{e^{-(s+j\omega)t}}{(s+j\omega)} \right\}_0^\alpha$$

$$= \frac{1}{2j}\left\{ \frac{1}{s-j\omega} - \frac{1}{(s+j\omega)} \right\}$$

$$= \frac{\omega}{s^2 + \omega^2}$$

(b)   Find ILT of

(i) $\dfrac{s^2+2s+5}{(s+3)(s+5)^2}$ $R_e(s)>-3$

$$x(s)=\frac{A}{s+3}+\frac{B}{s+5}+\frac{C}{(s+5)^2}$$

$$=\frac{-1}{2(s+3)}+\frac{1}{2(s+5)}+\frac{2}{(s+5)^2}$$

$$x(t)=\frac{1}{2}e^{-3t}u(t)+\frac{1}{2}e^{-5t}u(t)+2te^{-5t}u(t)$$

$$A(s+5)(s^2+25+10s)+B(s+3)(s^2+25+10s)+C(s^2+15+8s)=s^2+2s+5$$

$$As^3+Bs^3=0$$

$A+B=0$

$15A+13B+C=1$

$2A+C=1$

$125A+75B+15C=5$

$50A+15C=5$

$10A+3C=1$

$2A+C=1$

$10A+5C=5$

$-2C=-4$

$C=2$

$A=-\dfrac{1}{2}$

$B=\dfrac{1}{2}$

(ii) $\dfrac{2s+1}{s+2}$ $R_e(s)>-2$

$$\begin{array}{r} 2 \\ s+2\,\overline{)\,2s+1} \\ \underline{2s+4} \\ -3 \end{array}$$

$$x(s)=2-\frac{3}{s+2}$$

$$x(t)=2\delta(t)-3e^{-2t}u(t)$$

2. Find ILT of

(a) $\dfrac{(s+1)^2}{s^2-s+1}$ $R_e(s) > \dfrac{1}{2}$

$$\dfrac{\begin{array}{r} 1 \\ \hline s^2-s+1\,\sqrt{\dfrac{s^2+2s+1}{s^2-s+1}} \end{array}}{3s}$$

$$x(s) = \dfrac{s^2+2s+1}{s^2-s+1}$$

$$= 1 + \dfrac{3s}{s^2-s+1}$$

$$= \dfrac{3s}{\left(s-\tfrac{1}{2}\right)^2 + \left(\tfrac{\sqrt{3}}{4}\right)^2}$$

$$= 3\left\{ \dfrac{\left(s-\tfrac{1}{2}\right)}{\left(s-\tfrac{1}{2}\right)^2 + \left(\tfrac{\sqrt{3}}{4}\right)^2} + \dfrac{\tfrac{1}{2}}{\left(s-\tfrac{1}{2}\right)2 + \left(\tfrac{\sqrt{3}}{4}\right)^2} \right\}$$

$$x(t) = \delta(t) + 3\left[e^{\frac{1}{2}t}\cos\left(\tfrac{\sqrt{3}}{4}\right)t + \dfrac{1}{2}e^{\frac{1}{2}t}\sqrt{\tfrac{4}{3}}\sin\left(\tfrac{\sqrt{3}}{4}\right)t \right]$$

$$= \delta(t) + 3e^{\frac{1}{2}t}\cos\left(\tfrac{\sqrt{3}}{4}\right)t + \sqrt{B}\,e^{\frac{1}{2}t}\sin\left(\tfrac{\sqrt{3}}{4}\right)t$$

$$\dfrac{\begin{array}{r} 1 \\ \hline s^2+2s+1\,\sqrt{\dfrac{s^2-s+1}{s^2+2s+1}} \end{array}}{-3s}$$

(b) $\dfrac{s^2-s+1}{(s+1)^2} = 1 - \dfrac{3s}{(s+1)^2}$

$$= \delta(t) - 3(1-t)e^{-t}$$

3. Find LT of $\dfrac{d^2}{dt^2}\{\delta(t)\}$

$$\propto\left\{ \dfrac{d^2}{dt^2}y(t) \right\} = s^2 y(s) - \dfrac{s\,dy(0)}{dt} - y(0)$$

$$\propto\left\{\frac{d^2}{dt^2}\delta(t)\right\} = s^2 - s.\delta^1(0) - \delta(0)$$

$$= s^2$$

As initial condition are not given

4. Find LT of

(a) $e^{-2t}u(t) + e^{-3t}(u(t))$

$$X(s) = \frac{1}{s+2} + \frac{1}{s+3} = \frac{2s+5}{(s+2)(s+3)}$$

(b) $e^{-4t}u(t) + e^{-5t}\sin 5t$

$$X(s) = \frac{1}{s+4} + \frac{5}{(s+5)^2 + 5^2}$$

5. The system function of a causal system is

$$H(s) = \frac{s+1}{s^2 + 2s + 2} \quad \text{and input } x(t) = e^{-|t|}$$

$$x(t) = e^{-|t|} \qquad \therefore \ Y(s) = \frac{1}{S^2 + 2S + 2} = \frac{1}{(S+1)^2 + 1^2}$$

$$Y(t) = e^{-t}\sin t$$

6. Find LT of

(i) $x(t) = 1 \quad 0 \le t < 1$

$$\propto\{x(t)\} = \propto\{1\} = \int_0^1 e^{-st}dt$$

$$= \left.\frac{e^{-st}}{-s}\right/_0^1 = \frac{e^{-s}}{-s} + \frac{1}{s} = \frac{1}{s}\{1 - e^{-s}\}$$

(ii) $x(t) = t \quad 0 \le t \le$

$$= 2 - 1 \quad 1 \le t \le 2$$

$$\propto \{x(t)\} = \int\limits_{0}^{1} t \cdot e^{-st}\,dt + \int\limits_{1}^{2}(2-t)e^{-st}\,dt$$

$$= \left.\frac{e^{-st}}{s^2}\right/_{0}^{1} + \left.\frac{2 \cdot e^{-st}}{-s}\right/_{0}^{2} - \left.\frac{e^{-st}}{s^2}\right/_{1}^{2}$$

$$= \frac{e^{-s}}{s^2} - \frac{1}{s^2} + 2\left(\frac{e^{-2s}-e^{-s}}{-s}\right) - \left(\frac{e^{-2s}-e^{-s}}{s^2}\right)$$

$$= \frac{2e^{-s}}{s^2} - \frac{e^{-2s}}{s^2} - \frac{1}{s^2} + 2\left(\frac{e^{-s}-e^{-2s}}{s}\right)$$

$$= \frac{2}{s^2}\left\{e^{-s}-e^{-2s}-1\right\} + \frac{2}{s}\left\{e^{-s}-e^{-2s}\right\}$$

7. Find the LT of

$$x(t) = \left\{e^{-at}A\cos bt + \left(\frac{B-Aa}{b}\right)e^{-at}\sin bt\right\}u(t)$$

$$X(s) = \frac{A(s+a)}{(s+a)^2+b^2} + \frac{B-Aa}{\not{b}}\frac{\not{b}}{(s+a)^2+b^2}$$

$$= \frac{\not{A}a+As-\not{A}a+B}{(s+a)^2+b^2} = \frac{As+B}{(s+a)^2+b^2}$$

OBJECTIVE QUESTIONS

1. --------------- is useful for analysis of complex signals

2. --------------- is useful to solve differential equations with initial conditions.

3. Laplace transform becomes fourier transform when S = ---------------

4. LT of $u(t)$ = is ------------------

5. LT of $\sin \omega t$ is ---------------------

6. LT of $\cos \omega t$ is -----------------

7. LT of t^n is -----------------------

GATE QUESTIONS AND ANSWERS

1. The Laplace transform of the causal periodic square wave of period T shown in the figure

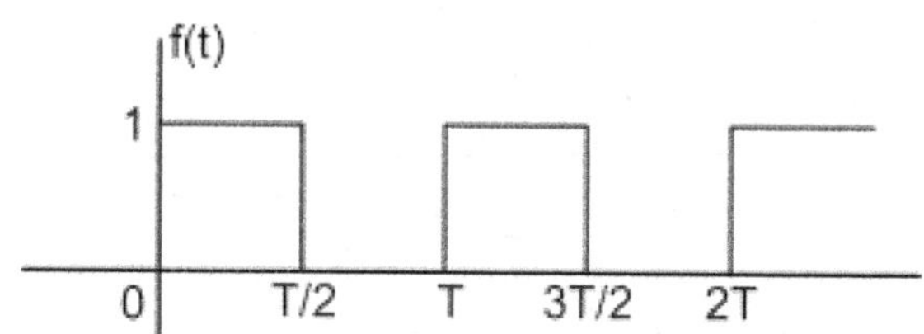

(a) $F(s) = \dfrac{1}{1 + e^{-sT/2}}$

(b) $F(s) = \dfrac{1}{s(1 + e^{-sT/2})}$

(c) $F(s) = \dfrac{1}{s(1 - e^{-sT})}$

(d) $F(s) = \dfrac{1}{(1 - e^{-sT})}$

GATE 2016

2. The bilateral Laplace transform of a function $F(t) = \begin{cases} 1 & \text{if } a \le t \le b \\ 0 & \text{otherwise} \end{cases}$ is

(a) $\dfrac{a - b}{s}$

(b) $\dfrac{e^s (a - b)}{s}$

(c) $\dfrac{e^{-as} - e^{-bs}}{s}$

(d) $\dfrac{e^{s(a-b)}}{s}$ GATE 2015

3. Let $x(t) = \alpha s(t) + s(-t)$ with $s(t) = fe^{-4t} u(t)$. If the bilateral laplace transform of $x(t)$ is
$$X(s) = \frac{16}{s^2 - 16} \quad -4 < \text{Re}\{s\} < 4 \text{ then find } \alpha \qquad \text{GATE 2015}$$

(a) 0   (b) −1   (c) −2   (d) 1

4. The unilateral Laplace transform of f(t) is $\dfrac{1}{s^2 + s + 1}$, find unilateral transform of $g(t) = t$

f(t) is   GATE 2014

(a) $\dfrac{-s}{(s^2 + s + 1)^2}$

(b) $\dfrac{-2s + 1}{(s^2 + s + 1)^2}$

(c) $\dfrac{s}{(s^2 + s + 1)^2}$

(d) $\dfrac{2s + 1}{(s^2 + s + 1)^2}$

5. Given F(s) is the one sided laplace transform of f(t), the laplace transform $\int_0^t f(\tau)d\tau$ is

GATE 2009

(a) $sF(s) - f(0)$ (b) $\dfrac{1}{s}F(s)$ (c) $\int_0^s f(T)dT$ (d) $\dfrac{1}{s}[(s) - f(0)]$

SUMMARY OF CHAPTER

1. Laplace transform is an integral transform that converts a function of a real variable to a function of a complex variable s.

$$F(S) = \int_0^\alpha f(t)e^{-st}dt$$

2. Eigen function is a signal that passes through the system without being modified except for multiplication by a scalar.

3. Representation of complex frequency graphically is called S-plane.

4. If LT is absolutely integrable then we can say it is convergent.

5. LT is used to convert complex differential equations to a simpler form having polynomials.

6. LT of O is O.

CHAPTER 9

Z-Transform

Objectives of the chapter

Difference between continuous and discrete time signals, representation of discrete signals using complex exponential and sinusoidal components, periodicity of discrete time using complex exponential signal. Concept of Z transform of a discrete sequence. Distinction between Fourier and Z transform. ROC in Z transform, Inverse Z transform, properties of Z transforms.

9.1 INTRODUCTION

Z-transform converts a discrete time signal which is a sequence of real or complex numbers into a complex frequency domain representation. It is equivalent to discrete time equivalent of the Laplace transform. Z-domain gives third representation in terms of ratios of polynomials.

It is a basic tool for analysis and syntheses of discrete systems.

We have the sampled signal

$$x_\delta(t) \;=\; \sum x(t)\,\delta(t-nT)$$

$$=\; \sum_{n=0}^{\alpha} x(nT)\delta(t-T)$$

assume $x(t)=0$ for all $t<0$

using laplace transform

$$x_s(s) \;=\; \int_0^{\alpha}\sum_{n=0}^{\alpha} x(nT)\delta(t-nT)e^{-st}\,dt$$

$$=\; \sum_{n=0}^{\alpha} x(nT)\int_0^{\alpha} \delta(t-nT)e^{-st}\,dt\delta$$

$$=\; \sum_{n=0}^{\alpha} x(nT)e^{-snT}$$

Let us define $Z = e^{+ST}$

$$\therefore \ X(z) = \sum_{n=0}^{\alpha} x(nT)z^{-n}$$

we have $S = \sigma + j\omega$

$$\therefore \ z = e^{ST} = e^{\sigma T} \cdot e^{j\omega T}$$

so the magnitude of z is

$$|z| = e^{\sigma T}$$

Right half S plane $\sigma > 0$ representing $|z| >$

Left half S plane $\sigma < 0$ representing $|z| <$

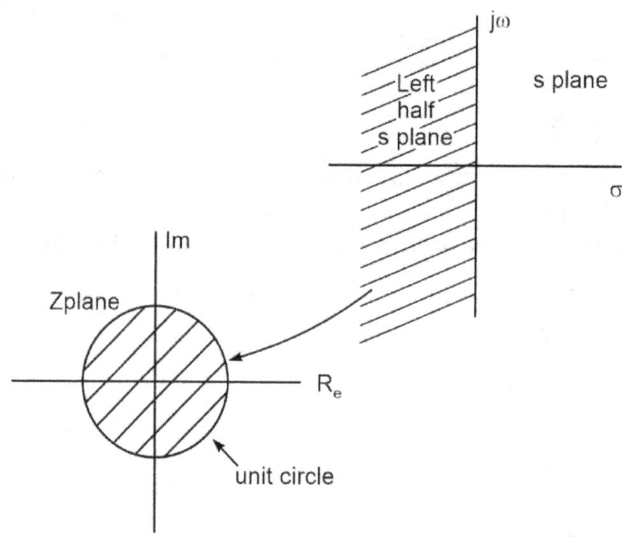

Fig. 9.1

Left half S plane is inside unit circle and right half outside the circle.

Z – transform

For any discrete time sequence $x(n)$

$$x(z) = \sum_{n=-\alpha}^{\alpha} x(n)z^{-n}$$

where $Z = re^{j\omega}$ where r is radius of circle

Discrete and continuous signal

A discrete signal can be generated from continuous signals using A/D converter.

$$x(t) = \sum_{n=-\alpha}^{\alpha} x(n)\delta(t-nT)$$

T is the time between samples.

Discrete time exponential signal

A discrete exponential is defined as $x[n]=Ca^n$ where C and a are complex nos.

If
$$a = e^{j\omega_0}$$
$$x[n] = Ce^{j\omega_0 n}$$

If C and a are real and if

(a) $a > 1$, signal magnitude increases

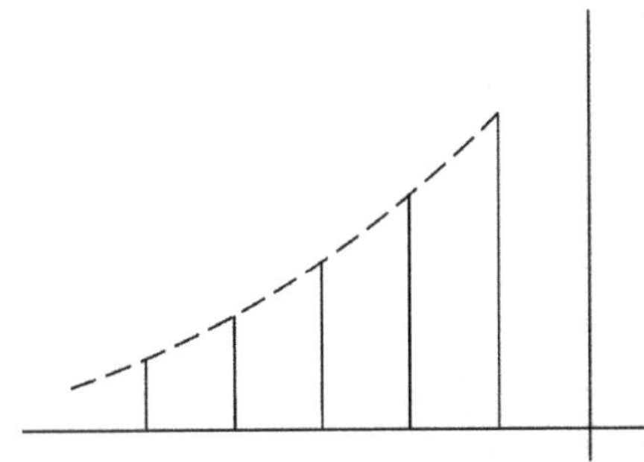

(b)   if $0 < a < 1$ then, magnitude decays exponentially

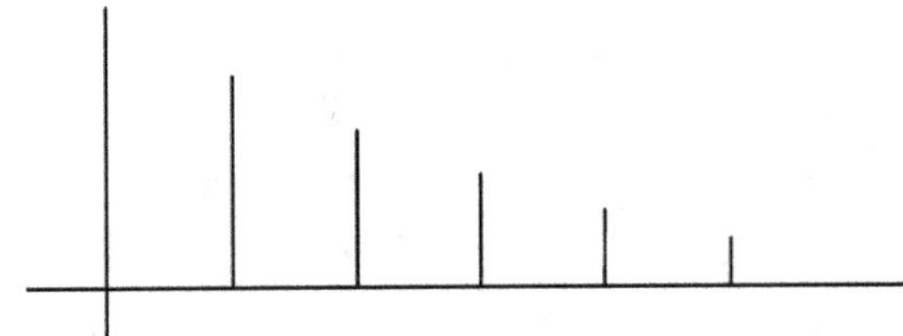

(c)   if $-1 < a < 0$

signal alternates and decays exponentially

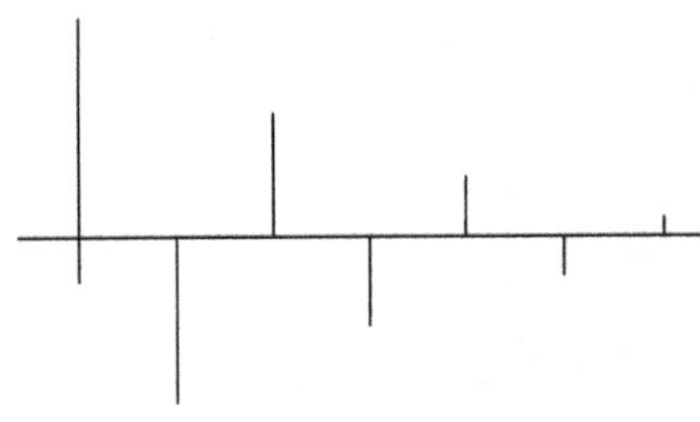

If $C = |C|e^{j\theta}$

$a = |a|e^{j\omega_0}$

$$x[n] = Ca^n = |c||a|^n\, e^{j\theta}.e^{j\omega_0 n}$$

$$= |c||a|^n\, e^{j(\theta+\omega_0 n)}$$

$$= |c||a|^n\, |\{\cos(\omega_0 n+\theta)+j\sin(\omega_0 n+\theta)\}$$

(a) if $|a|=1$, real and imaginary parts are sinusoidal

(b) if $|a|<1$, sinusoidal but decay exponentially

(c) if $|a|>1$, sinusoidal grows exponentially

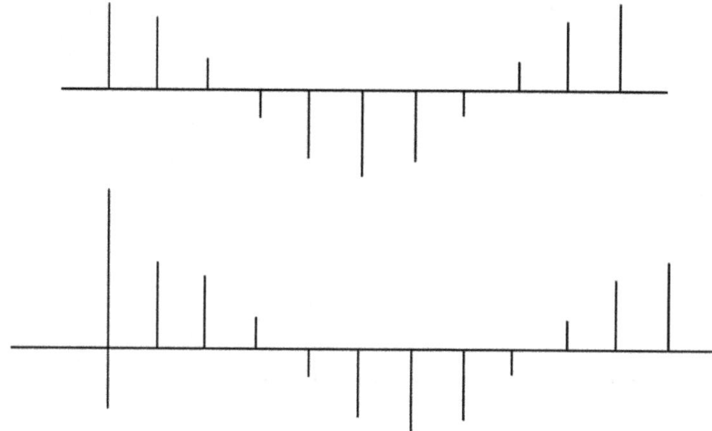

Example:

1. A unit pulse is defined by

$$x(nT) = \begin{cases} 1 & n=0 \\ 0 & n\neq 0 \end{cases}$$

It can be defined by

$$X(n) = 1+0z^{-1}+0z^{-2}+\ldots..$$

we have

$$\delta(n-k)= \begin{cases} 1 & n=k \\ 0 & n\neq k \end{cases}$$

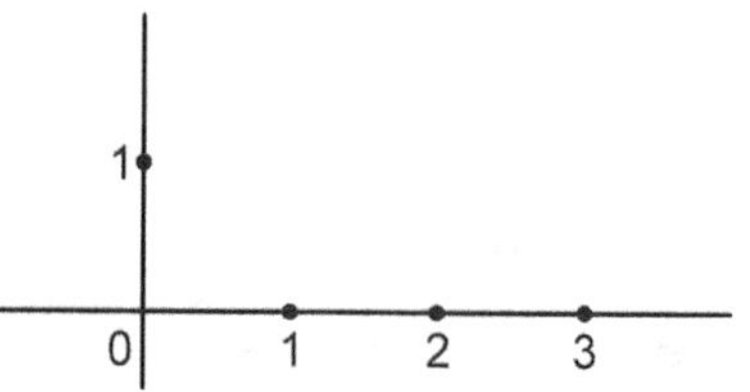

Fig. 9.2

$$x(nT) = \sum_{k=-\alpha}^{\alpha} x(kT)\delta(n-k)$$

2. Unit step is defined by

$$x(nT) = 1 \quad n \geq 0$$

$$x(z) = 1 + z^{-1} + z^{-2} + \ldots$$

$$= \sum_{n=0}^{\alpha} z-n \quad |z| >|$$

$$= \frac{1}{1-z^{-1}}$$

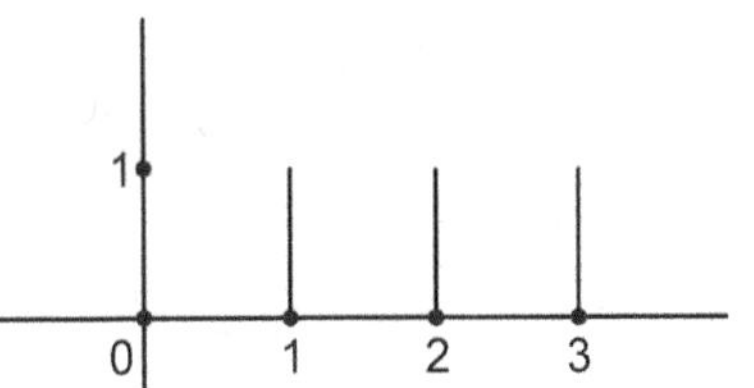

Fig. 9.3

3. Unit exponential sequence

$$x(nT) = e^{-anT} \quad a > 0, n \geq 0$$

$$\therefore \quad X(z) = \sum_{n=0}^{\alpha} e^{-anT} z^{-n}$$

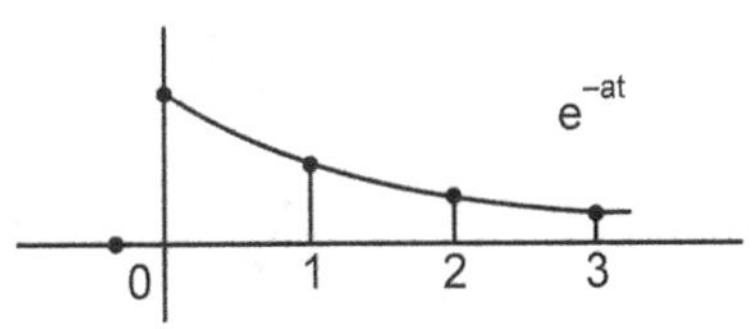

Fig. 9.4

$$= \sum \left(e^{-aT} z^{-1}\right)^n$$

$$= \frac{1}{1-e^{-aT}z^{-1}} \quad |z| > e^{-aT}$$

4. Find the Z transform of

$$x(n) = a^{n-1} u(n-1)$$

If $\quad x(n) = a^n u(n)$

$$X(z) = \frac{1}{1-az^{-1}}$$

Using time shifting property

$$z\{x(n-k)\} = z^{-k} X(z)$$

$$z\{a^{n-1}u(n-1)\} = \frac{z^{-1}}{1-az^{-1}}$$

5. Find Z – transform of

$$x(n) = \frac{-a^n u(n-1)}{n}$$

$$nx(n) = -a\{a^{n-1}u(n-1)\}$$

$$z\{nx(n)\} = -z\frac{d}{dz}x(z)$$

$$= -Z\cdot -a\cdot\frac{d}{dz}\frac{z^{-1}}{1-az^{-1}}$$

$$= -Z\frac{d}{dz}\frac{-az^{-1}}{1-az^{-1}}$$

$$= +Z\frac{d}{dz}\frac{1}{\frac{z}{a}-1}$$

$$= \frac{z}{a}\frac{1}{\left(\frac{z}{a}-1\right)^2}$$

$$= \frac{z}{a}\frac{\left(az^{-1}\right)2}{\left(1-az^{-1}\right)^2} = \frac{az^{-1}}{\left(1-az^{-1}\right)^2}$$

Solved Problems Exercise-1

1. Find the Z – transform of the signal $x(n) = \sin\omega_0 n\ u(n)$

$$x(n) = \sum_{n=-\alpha}^{\alpha} x(n)z^{-n}$$

$$= \sum_{n=-\alpha}^{\alpha} \sin(\omega_0 n)u(n)z^{-n} = \sum_{n=0}^{\alpha}\sin(\omega_0 n)z^{-n}$$

$$= \sum_{n=0}^{\alpha}\left\{\frac{e^{j\omega_0 n}-e^{-j\omega_0 n}}{2j}\right\}z^{-n}$$

$$= \frac{1}{2j}\left\{\sum_{n=0}^{\alpha}e^{j\omega_0 n}z^{-n}-\sum_{n=0}^{\alpha}e^{-j\omega_0 n}z^{-n}\right\}$$

$$= \frac{1}{2j}\left\{\frac{1}{1-e^{j\omega_0}z^{-1}}-\frac{1}{1-e^{-j\omega_0}z^{-1}}\right\}$$

$$= \frac{1}{2j}\left\{\frac{e^{j\omega_0}z^{-1}-e^{-j\omega_0}z^{-1}}{1-e^{j\omega_0}z^{-1}-e^{-j\omega_0}z^{-1}+z^{-2}}\right\}$$

$$= \frac{\sin\omega_0 z^{-1}}{1-2\cos\omega_0 z^{-1}+z^{-2}} \qquad \text{ROC: } |Z|>|$$

2. Find Z – transform, ROC and pole zero locations
   $X(z)$ for

$$x(n) = \left(\frac{2}{3}\right)^n u(n) + \left(-\frac{1}{2}\right)^n u(n)$$

Ans:

$$X(z) = \sum_{n=-\alpha}^{\alpha} \left\{ \left(\frac{2}{3}\right)^n u(n) + \left(-\frac{1}{2}\right)^n u(n) \right\} z^{-n}$$

$$= \sum_{n=0}^{\alpha} \left(\frac{2}{3}\right)^n z^{-n} + \sum_{n=0}^{\alpha} \left(-\frac{1}{2}\right)^n z^{-n}$$

$$X(z) = \frac{1}{1-\frac{2}{3}z^{-1}} + \frac{1}{1+\frac{1}{2}z^{-1}} = \frac{z\left(2z-\frac{1}{6}\right)}{\left(z-\frac{2}{3}\right)\left(z+\frac{1}{2}\right)}$$

poles are at $z = \dfrac{2}{3}$ and $z = \dfrac{-1}{2}$

zeros are at $z = 0$ and $z = \dfrac{1}{12}$

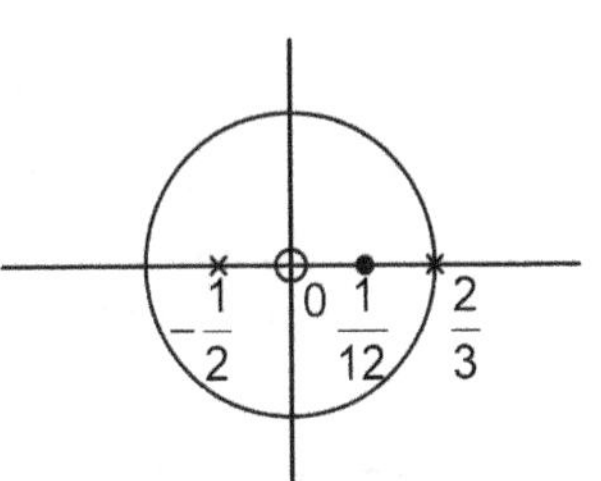

Fig. 9.5

3. Find the Z - transform of $nu(n)$

$$z(u(n)) = \frac{z}{z-1}$$

$$z\{nx(n)\} = -z\frac{d}{dz}x(z)$$

$$= -z\frac{d}{dz}\left(\frac{z}{z-1}\right)$$

$$= = \frac{z}{(z-1)^2} \quad \text{ROC: } |Z|>1$$

4. Find the Z - transform and ROC for the signal
   $$x(n) = a^n u(n)$$

Ans:

$$X(z) = \sum_{n=-\alpha}^{\alpha} x(n)z^{-n}$$

$$= \sum_{n=-\alpha}^{\alpha} a^n u(n)z^{-n} = \sum_{n=0}^{\alpha} a^n z^{-n}$$

$$= \sum \left(a z^{-1}\right)^n = \frac{1}{1-a z^{-1}} \quad \begin{array}{l} |z| > |a| \\ |a z^{-1}| < \end{array}$$

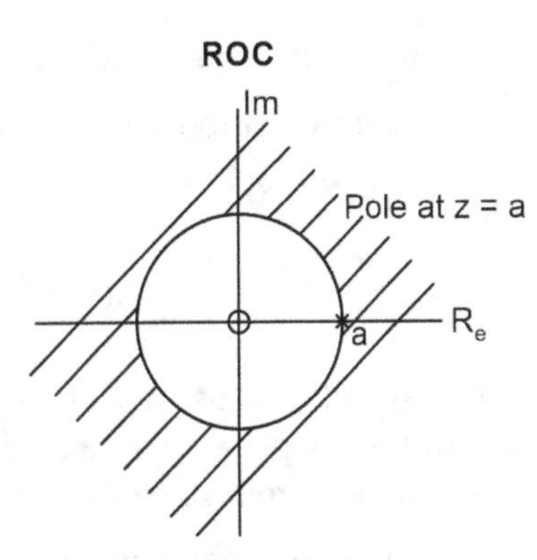

Fig. 9.6

5. Find the Z transform and ROC for

$$x(n) = b^n u(-n^{-1})$$

The act of signals that cause the system output to converge is called ROC. With the z-transform the s plane represents a set of signals. Some of these signals may cause the output of the system to converge while others cause the output to diverge.

9.2 PROPERTIES OF REGION OF CONVERGENCE

1. ROC is a concentric ring

2. ROC cannot contain poles

3. If $x(n)$ is a causal sequence then the ROC is entire Z plane except at $Z = 0$

4. If $x(n)$ is a anti causal sequence then the ROC is entire Z plane except at $z = \alpha$

5. If $x(n)$ is a finite duration two sided sequence then the ROC is entire Z plane except at $z = 0$ and $z = \alpha$

6. ROC of an LTI system contains unit circle

7. ROC is connect region

8. If $x(n)$ is infinite duration two sided sequence then ROC contains circular ring in Z plane, bounded on the interior and exterior by a pole not containing any poles.

ROC

The Z transform $x(z)$ is a ratio of two polynomials

$$x(z) = \frac{b_0 + b_1 z^{-1} + b_2 z^{-2} + \ldots\ldots b_M z^{-M}}{a_0 + a_1 z^{-1} + a_2 z^{-2} + \ldots\ldots\ldots a_N z^{-N}}$$

numerator roots are zeros --- 0

denominator roots are poles --- X

Z transform exists when the infinite sum converges

$$x(z) = \sum_{n=-\alpha}^{\alpha} x(n) z^{-n}$$

The set of signals that cause the system output to converge is called ROC. With the z-transform the s plane represents a set of signals. Some of these signals may cause the output of the system to converge while others cause the output to diverge.

The values of Z for which it converges is called as region of converge, ROC

SOLVED EXERCISE PROBLEMS 2

1. Find the Z-transform and ROC of $x(n) = a^n u(n)$

$$x(z) = \sum x(n)z^{-n}$$

$$= \sum_{n=0}^{\alpha} a^n z^{-n} = \sum_{n=0}^{\alpha} \left(az^{-1}\right)^n$$

$$= \sum \left(\frac{a}{z}\right)^n$$

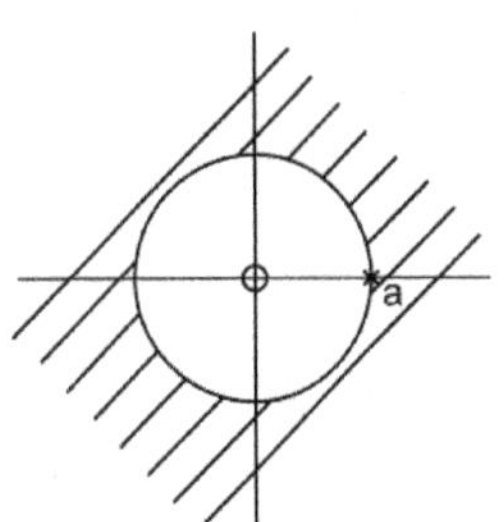

Fig. 9.7

The series converges if $x(n)z^{-n}$ is summala if $\sum \left| x(n)z^{-n} \right| < \alpha$.

The condition for $x(z)$ to be finite is $|z| >$. ROC for $x(z)$ lie outside the unit circle.

$$x(z) = \frac{1}{1-az^{-1}} \qquad \begin{array}{c} \left|az^{-1}\right| < \left| \right. \\ a < z \end{array}$$

2. $\qquad x(n) = -b^n u(-n-1)$

$$x(z) = \sum x(n)z^{-n}$$

$$= \sum -b^n u(-n-1)z^{-n}$$

$$= -\sum_{-\alpha}^{-1} b^n z^{-n}$$

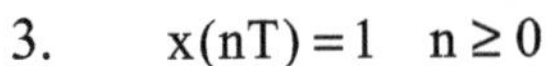

Fig. 9.8

$$= -\sum_{-\alpha}^{-1} \left(bz^{-1}\right)^n = -\sum \left(\frac{b}{z}\right)^n \quad |z| > |b|$$

3. $\qquad x(nT) = 1 \quad n \geq 0$

$$x(z) = \sum_{n=0}^{\alpha} z^{-n}$$

for $|x| < | \quad \sum x^n = \frac{1}{1-x}$

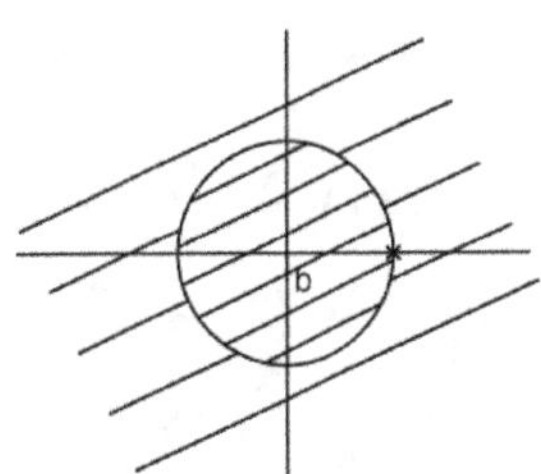

Fig. 9.9

$$x(z) = \sum z^{-n} = \frac{1}{1-z^{-1}} \quad |z| > 1$$

4. $x(nT) = e^{-\alpha nT} \quad \alpha > 0$

$$x(z) = \sum e^{-\alpha nT} z^{-n} = \sum \left(e^{-\alpha T} z^{-1} \right)^n$$

$$x(z) = \frac{1}{1 - e^{-\alpha T} z^{-1}} \quad |z| > e^{-\alpha T}$$

$$K = e^{-\alpha T}$$

$$x(z) = \frac{1}{1 - k z^{-1}} \quad |z| > k$$

$$\frac{1}{1 - k z^{-1}} = \frac{z}{z - k}$$

pole at $z = k$

zero at $z = 0$

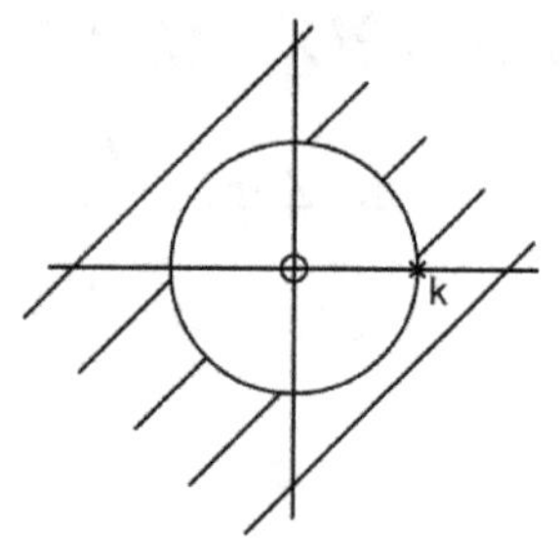

Fig. 9.10

5. $x(nT) = a^n \cos\left(\dfrac{n\pi}{2} \right)$

$$x(n) = \sum a^n \cos\left(\frac{n\pi}{2} \right) z^{-n}$$

$$= \sum a^{2k} (-1)^k z^{-2u}$$

$$x(z) = \sum -\left(a^2 z^{-2} \right)^k$$

$$= \frac{1}{1 + a^2 z^{-2}}$$

6. $x(n) = a^n u(n) - b^n u(-n^{-1})$

$$x(z) = \sum x(n) z^{-n}$$

$$= \sum_{n=0}^{\alpha} a^n z^{-n} - \sum_{n=-1}^{-\alpha} b^n z^{-n}$$

$$x(z) = \frac{1}{1 - a z^{-1}} + \frac{1}{1 + b z^{-1}}$$

$$|a| < |z| \qquad |z| < |b|$$

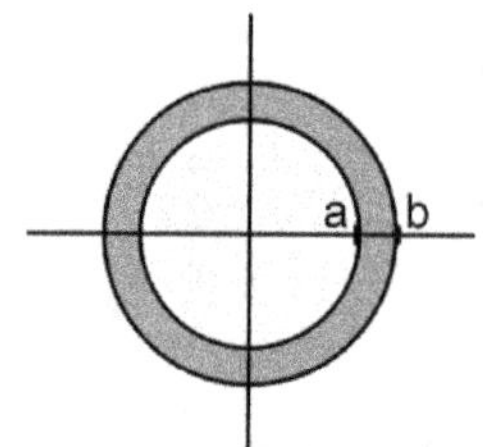

Fig. 9.11

7. Find the ROC and pole zero locations

$$x(n) = \left(\frac{2}{3}\right)^n u(n) + \left(-\frac{1}{2}\right)^n u(n)$$

$$x(z) = \sum \left[\left(\frac{2}{3}\right)^n u(n) + \left(-\frac{1}{2}\right)^n u(n)\right] z^{-n}$$

$$= \frac{1}{1-\frac{2}{3}z^{-1}} + \frac{1}{1+\frac{1}{2}z^{-1}} = \frac{z}{z-\frac{2}{3}} + \frac{z}{z+\frac{1}{2}}$$

$x(z)$ converges for $|z| > \frac{2}{3}$

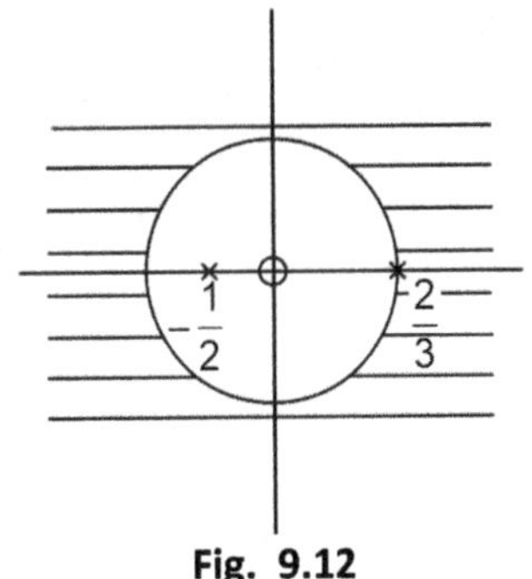

Fig. 9.12

9.3 ROCs OF SOME FUNCTIONS

X(n)	Z – transform	ROC
1. $\delta(n)$	1	All Z
2. $u(n)$	$\dfrac{1}{1-z^{-1}}$	$|z| > 1$
3. $-u(-n-1)$	$\dfrac{1}{1-z^{-1}}$	$|z| < 1$
4. $\delta(n-m)$	x^{-n}	All Z except at $z = 0$
5. a^n	$\dfrac{1}{1-az^{-1}}$	$|z| > |a|$
6. na^n	$\dfrac{az^{-1}}{\left(1-az^{-1}\right)^2}$	$|z| > |a|$
7. $\cos \omega_0 n$	$\dfrac{1-\left(\cos \omega 0\right)z^{-1}}{1-\left(2\cos \omega_0\right)z^{-1}+z^{-2}}$	$|z| > 1$
8. $\sin \omega_0 n$	$\dfrac{\sin \omega_0 n z^{-1}}{1-2\cos \omega_0 z^{-1}+z^{-2}}$	$|z| > 1$

9.4 PROPERTIES OF Z - TRANSFORM

1. Linearity

$$x_1(z) = z\{x_1(n)\} \qquad x_2(z) = z\{x_2(n)\}$$

$$z\{ax_1(n) + bx_2(n)\} = ax_1(2) + bx_2(z)$$

2. Time shifting

$$z\{x(n-m)\} = z^{-m}x(z)$$

$$z\{x(n-m)\} = \sum x(n-m)z^{-n}$$

$$= z^{-m}\sum x(n-m)z^{-(n-m)}$$

$$= z^{-m}x(z)$$

3. Time reversal

If $\quad x(z) = z\{x(n)\}$

$$z\{x(-n)\} = x(z^{-1})$$

4. Multiplication by n

$$z\{xn(n)\} = -\frac{d}{dz}x(z)$$

$$z\{xn(n)\} = \sum x(n)nz^{-n}$$

$$= z\sum x(n)nz^{-n+1}$$

$$= -z\sum x(n)\frac{d}{dz}\left(z^{-n}\right)$$

$$= -z\frac{d}{dz}\sum x(n)z^{-n}$$

$$= -z\frac{d}{dz}x(z)$$

In general $\quad z\{n^k x(n)\} = \left(-z\frac{d}{dz}\right)^k x(z)$

5. Convolution

$$z\{x(n)*h(n)\} = X(z)H(z)$$

$$y(n) = \sum_{k=-\alpha}^{\alpha} x(k)h(n-k)$$

$$y(z) = \sum\sum x(k)h(n-k)z^{-n}$$

$$= \sum x(k)z^{-k}\sum h(n-k)z^{-(n-\alpha)}$$

$$= x(z)\,H(z)$$

6. Time Expansion

$$x_k(n) = \begin{cases} x(n/k) & n \text{ is multiple of } k \\ 0 & n \text{ is not multiple of } k \end{cases}$$

If$\quad z\{x(n)\} = x(z)$

$$z\{x_k(n)\} = x(z^k)$$

7. Conjugation

If$\quad z\{x(n)\} = X(z)$

Then$\quad z\{x^*(n)\} = X^*(z^*)$

Proof:

$$z\{x^*n)\} = \sum_{n=-\alpha}^{\alpha} x^*(n)z^{-n}$$

$$= \left[\sum_{n=-\alpha}^{\alpha} x(n)(z^*)^{-n} \right]^*$$

$$= [X(z^*)]^* = X^*(z^*)$$

8. Correlation

If$\quad x_1(z) \quad = z\{x_1(n)\}$ and $x_2(z) = z\{x_2(n)\}$

Then$\quad z\{\phi_{x1,x2}(l)\} = z\left\{\sum x_1(n)x_2(n-l)\right\}$

$$= X_1(z)X_2(z^{-1})$$

9. Initial value theorem

If$\quad X(z) = z\{x(n)\}$ then

$$x(0) = \lim_{z \to \alpha} X(z)$$

Proof:

$$X(z) = \sum_{n=0}^{\alpha} x(n)z^{-n} = x(0) + x(1)z^{-1} + x(2)z^{-2} + \ldots\ldots$$

As $z \to \alpha$, all the terms vanish except $x(0)$

$$\therefore \qquad \lim_{z \to \alpha} X(z) = \lim_{z \to \alpha} \sum_{n=0}^{\alpha} x(n)z^{-n} = x(0)$$

10.   Final value theorem

If $X(z)\, z\{x(n)\}$ = where the ROC for $X(z)$ includes but is not necessarily confined to $|z| >$ and $(z-1)X(z)$ has no poles or outside the unit circle then

$$x(\alpha) = \lim_{z \to 1}(z-1)X(z)$$

Proof:

$$z\{x(n+1)\} - z\{x(n)\} = \lim_{k \to \alpha}\sum_{n=0}^{k}\{x(n+1) - x(n)\}z^{-n}$$

$$ZX(z) - x(0) - X(z) = \lim_{k \to \alpha}\sum_{n=0}^{k}\{x(n+1) - x(n)\}z^{-n}$$

$$\lim_{k \to 1}(z-1)X(2) - x(0) = \lim_{k \to 0}\{x(1) - x(0)\} + \{x(2) - x(1)\} + ...$$

$$= x(\alpha) - x(0)$$

$$x(\alpha) = \lim_{z \to 1}(z-1)X(z)$$

or

$$x(\alpha) = \lim_{z \to 1}(1 - z^{-1})x(z)$$

9.5 PERIODICITY PROPERTIES

The complex discrete exponential signal

$$x[n] = |C||a|^{n}\, e^{j(\omega_0 n + \theta)}$$

If $\theta = 2\pi n$

$$x[n] = |C||a|^{n}\, e^{j(\omega_0 + 2\pi)n} = e^{j\omega_0 n}$$

Thus the same signal occurs with frequencies $\omega_0 \pm 2\pi$ and so on.

As ω_0 is varied from 0 to π, the signal will oscillate more and more

$$e^{j\omega_0 n} = e^{(j\pi)n} = (-1)^{n}$$

Signal will oscillate rapidly changes at each point.

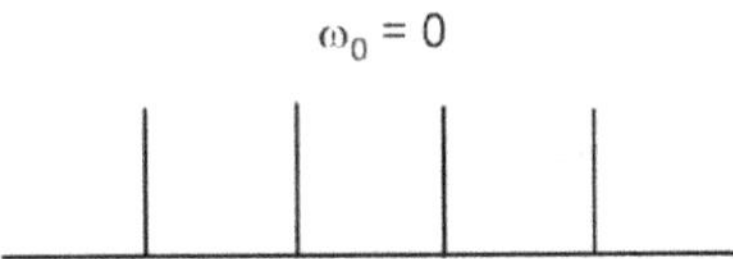

As ω_0 varies from π to 2π, signal oscillations decreases and becomes constant.

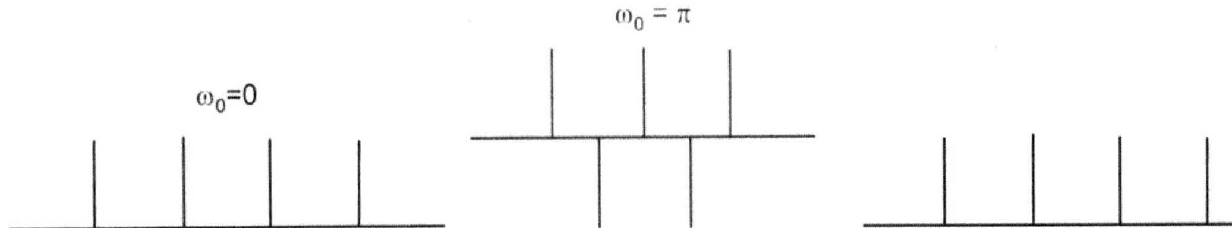

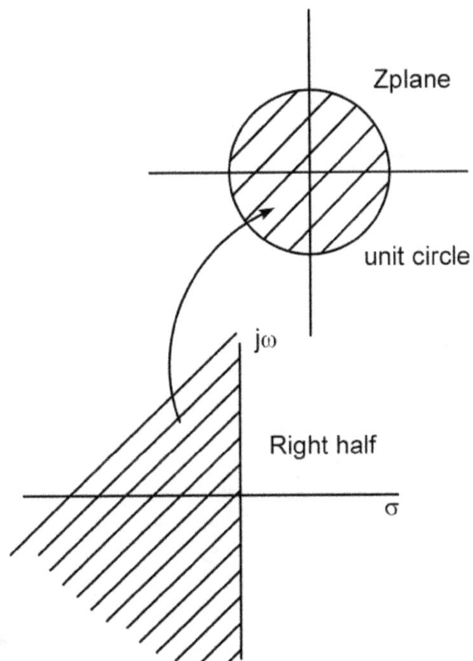

Example: $\quad x(z) = \dfrac{1+3z^{-1}}{1+3z^{-1}+2z^{-2}} \quad |z|>2$

$$x(z) = \frac{z(z+3)}{(z+1)(z+2)}$$

Closed contour C, ROC $|z|>2$

$\therefore$ poles are at $\quad z=-1, z=-2$

$x(n) = $ residue at $Z=-1$ + residue at $z=-2$

$$= \frac{(x+1)(z+3)z^n}{(z+1)(z+2)_{z=-1}} + \frac{(z+2)(z+3)z^n}{(z+1)(z+2)_{z=-2}}$$

$$= 2(-1)^n - (-2)^n$$

for $n > 0$

$$x(z) = \frac{zn(z+3)}{(z+1)(z+2)}$$

$$\underset{z=-1}{\text{Res}} \quad 2(-1)^n$$

$$\underset{z=-1}{\text{Res}} \quad 2(-2)^n$$

$$\therefore \quad x(n) = 2(-1)^n - (-2)^n$$

for $n = 0$ $\quad \dfrac{z+3}{(z+1)(z+2)}$

$$\underset{z=-1}{\text{Res}} = 2 \quad \underset{z=-2}{\text{Res}} = 1$$

$$x(n) = 1$$

$n < 0$ $\quad \dfrac{z^{-1}(z+3)}{(z+1)(z+2)} = \dfrac{z+3}{z(z+1)(z+2)}$

$$\underset{z=-1}{\text{Res}} \quad -2 \qquad \underset{z=-2}{\text{Res}} \quad \frac{1}{-2(-1)}$$

$$= \frac{1}{2}$$

$$\underset{z=0}{\text{Res}} \quad \frac{3}{2}$$

$$x(n) = -2 + \frac{1}{2} + \frac{3}{2} = 0$$

9.6 DIFFERENT METHODS OF EVALUATING Z-TRANSFORMS

(a) **Partial fraction Method**

If $\quad x(z) = \dfrac{a_0 x^m + a_1 z^{m-1} + a_2 z^{m-2} + \ldots a_m}{(z-p_1)(z-p_2) \ldots (z-p_n)}$

$$= A_0 + \frac{A_1}{(z-p_1)} + \frac{A_2}{(z-p_2)} + \ .....$$

$$\text{where } A_0 = \underset{z \to \alpha}{\text{Lt}} \ x(z) \quad \begin{matrix} = a_0 & \text{if } m=n \\ = 0 & m<n \end{matrix}$$

$$A_i = (z - Pi)X(z)\big|_{z=P_1}$$

Ex: $X(z) = \dfrac{z}{z^2 + z + \frac{1}{2}} \qquad z^2 + z + \dfrac{1}{2}$

$$\frac{x(z)}{2} = \frac{z}{z^2 + z + \frac{1}{2}} = \frac{-1 \pm \sqrt{1-2}}{2} = \frac{-1 \pm 2\,1}{2}$$

$$= \frac{A_1}{(z-p_1)} + \frac{A_2}{(z-p_2)}$$

$$P_1 = \frac{-1+j}{2} \qquad\qquad P_2 = \frac{-1-j}{2}$$

$$A_1(z - P_2) + A_2(z - P_1) = 1$$

$$\text{If} \quad z = P_1 \qquad\qquad A_1(P_1 - P_2) = 1$$

$$z = P_2 \qquad\qquad A_2(P_2 - P_1) = 1$$

$$A_1 = -j, \quad A_2 = j$$

$$\frac{x(z)}{z} = \frac{-j}{(z-P_1)} + \frac{j}{(z-P_2)}$$

$$= -j \left\{ \frac{1}{z - \left(\frac{-1+1}{2}\right)} - \frac{1}{z - \left(\frac{-1-j}{2}\right)} \right\}$$

$$X(z) = -j \left\{ \frac{z}{z - P_1} - \frac{z}{z - P_2} \right\}$$

$$X(n) = z^{-1} \left\{ -j \left(\frac{z}{z - P_1} - \frac{z}{z - P_2} \right) \right\}$$

(b)   Convolution Method

If $x(z)$ is $z\{x(n)\}$ and is able to split into

$$x(z) = x_1(z).x_2(z)$$

$$z^{-1}\{x(z)\} = z^{-1}\{x(z)\} \cdot z^{-1}\{x_2(z)\}$$

$$= x_1(n) * x_2(n)$$

Ex: 1 $x(z) = \dfrac{z}{(z-1)^2}$

$$= \dfrac{z}{z-1} \cdot \dfrac{1}{z-1}$$

$$z^{-1}\{x(z)\} = z^{-1}\left\{\dfrac{z}{z-1}\right\} \cdot z^{-1}\left\{\dfrac{1}{z-1}\right\}$$

$$= u(n) * u(n-1)$$

$$= n\, u(n) \qquad\qquad \text{Rays Function}$$

$$= -j\left\{(P_1)^n - (P_2)^n\right\} u(n)$$

Ex: 2 $X(z) = \dfrac{x^2}{z^2+1}$

$$\dfrac{X(z)}{z} = \dfrac{z}{z^2+1}$$

$$= \dfrac{A_1}{(z-P_1)} + \dfrac{A_2}{(z-P_2)}$$

$$P_1 = -j \qquad\qquad P_2 = j$$

$$\dfrac{A_1}{(z+j)} + \dfrac{A_2}{(z-j)}$$

If $\quad z = P_1$

$$A_1(P_1 - P_2) = P_1$$

$$A_1(2j) = j$$

$$A_1 = \dfrac{1}{2}$$

$$z = P_2 \qquad\qquad A_2 = \dfrac{1}{2}$$

$$\dfrac{X(z)}{z} = \dfrac{1}{2}\left\{\dfrac{1}{z+j} + \dfrac{1}{z-j}\right\}$$

$$X(z) = \frac{1}{2}\left\{\frac{1}{z+j} + \frac{1}{z-j}\right\}$$

$$X(n) = z^{-1}\{X(z)\} = \frac{1}{2}\left\{(-j)^n + (j)^n\right\}u(n)$$

Ex 3: $\quad X(z) = \dfrac{3 - \dfrac{5}{6}z^{-1}}{\left(1-\dfrac{1}{4}z^{-1}\right)\left(1-\dfrac{1}{3}z^{-1}\right)}$

$$X(z) = \frac{A}{1-\dfrac{1}{4}z^{-1}} + \frac{B}{1-\dfrac{1}{3}z^{-1}}$$

$$= \frac{1}{1-\dfrac{1}{4}z^{-1}} + \frac{2}{1-\dfrac{1}{3}z^{-1}}$$

$$= \frac{z}{z-\dfrac{1}{4}} + \frac{2z}{z-\dfrac{1}{3}}$$

$$x(n) = \left\{\left(\frac{1}{4}\right)^n + 2\left(\frac{1}{3}\right)^n\right\}u(n)$$

Ex 4: $\quad X(z) = \dfrac{z^{-1} - \dfrac{1}{2}}{\left(1-\dfrac{1}{2}z^{-1}\right)^2}$

$$= \frac{z^{-1}\left(1-\dfrac{z}{2}\right)}{z^{-2}\left(z-\dfrac{1}{2}\right)^2} = \frac{z\left(1-\dfrac{z}{2}\right)}{\left(z-\dfrac{1}{2}\right)^2}$$

$$\frac{x(z)}{z} = \frac{A}{z-\dfrac{1}{2}} + \frac{B}{\left(z-\dfrac{1}{2}\right)^2}$$

$$= \frac{-\dfrac{1}{2}}{z-\dfrac{1}{2}} + \frac{3}{4}\frac{1}{\left(z-\dfrac{1}{2}\right)^2}$$

$$x(z) = -\frac{1}{2}\frac{z}{z-\frac{1}{2}} + \frac{3}{4}\frac{z}{\left(z-\frac{1}{2}\right)^2}$$

$$= \left\{ -\frac{1}{2}\left(\frac{1}{2}\right)^n + \frac{3}{4}n\left(\frac{1}{2}\right)^n \right\}u(n)$$

EX: 5

$$X(z) = \frac{1}{2(z-1)(z+0.5)}$$

$$X(z)z^{n-1} = \frac{z^{n-1}}{2(z-1)(z+0.5)}$$

$$n \geq 1 \quad x(n) = \operatorname*{Res}_{Z=1} x(z) + \operatorname*{Res}_{Z=0..5} X(z)$$

$$\operatorname*{Res}_{Z=1} x(z) = \frac{Z^{n-1}}{2(Z+0.5)}\bigg|_{z=1} = \frac{1}{3}$$

$$\operatorname*{Res}_{Z=-0.5} x(z) = \frac{Z^{n-1}}{2(Z-1)}\bigg|_{z=-0.5} = -\frac{1}{3}\left(-\frac{1}{2}\right)^{n-1}$$

$$x(n) = \frac{1}{3} - \frac{1}{3}\left(-\frac{1}{2}\right)^{n-1}$$

$$n = 0 \quad x(z)z^{n-1} = \frac{z^{-1}}{2(z-1)(z+0.5)}$$

$$= \frac{1}{2z(z-1)(z+0.5)}$$

$$\operatorname*{Res}_{z=0} x(z) = \frac{1}{2-1(0.5)} = -1$$

$$\operatorname*{Res}_{z=1} x(z) = \frac{1}{2z(z+0.5)} = \frac{1}{3}$$

$$\operatorname*{Res}_{z=-0.5} x(z) = \frac{1}{2z(z-1)} = \frac{1}{1.5}$$

$$x(n) = -1 + \frac{1}{3} + \frac{1}{1.5} = 0$$

$$\frac{n<0}{n=-1} \qquad X(z)\,z^{n-1} = \frac{z^{-2}}{2(z-1)(z+0.5)}$$

$$= \frac{1}{2z^2(z-1)(z+0.5)}$$

$$\underset{z=0}{\text{Res}} \qquad X(z) = \frac{1}{(2-1)!}\frac{d^2}{dz^{2-1}}\left\{\frac{1}{2(z-1)(z+0.5)}\right\}$$
$$z=0$$

$$= \frac{d}{dz}\left\{\frac{1}{2(z-1)(z+0.5)}\right\}$$
$$z=0$$

$$= \frac{1}{2}\frac{d}{dz}\left\{\frac{1}{(z-1)} - \frac{1}{(z+0.5)}\right\}$$
$$z=0$$

$$= \frac{1}{2\times1.5}\left\{\frac{1}{(z-1)^2} + \frac{1}{(z+0.5)^2}\right\}$$

$$= \frac{1}{3}\left\{-1+\frac{1}{0.25}\right\} = 1$$

$$\underset{z=1}{\text{Res}} \qquad X(z) = \left\{\frac{1}{2z^2(z-0.5)}\right\} = \frac{1}{3}$$

$$\underset{z=-0.5}{\text{Res}} \qquad X(z) = \left\{\frac{1}{2z^2(z-1)}\right\} = -\frac{1}{0.75}$$

$$X(n) = 1 + \frac{1}{3} - \frac{1}{0.75} = 0$$

$$\therefore \quad X(n) = \frac{1}{3} - \frac{1}{3}\left(-\frac{1}{2}\right)^{n-1} + 0 + 0$$

$$= \frac{1}{3} - \frac{1}{3}\left(-\frac{1}{2}\right)^{n-1}$$

9.7 Z - TRANSFORM VS FOURIER TRANSFORM

Z – transform $X(z)$ of sequence $x(n)$ is

$$X(z) = \sum_{n=-\alpha}^{\alpha} x(n)\, z^{-n}$$

if $x(n) = 0$ for $n < 0$ i.e., it is a causal function

then $$X(z) = \sum_{n=0}^{\alpha} x(n)\, z^{-n}$$

where z is a complex variable $z = re^{j\omega}$

$$X(re^{j\omega}) = \sum_{n=-\alpha}^{\alpha} x(n)\left(re^{j\omega}\right)^{-n}$$

$$= \sum_{n=-\alpha}^{\alpha} x(n) r^{-n} e^{-j\omega n}$$

If $r = 1$ i.e., for $|z| = 1$

$$X(re^{j\omega}) = \sum_{n=-\alpha}^{\alpha} x(n) e^{-j\omega n}$$

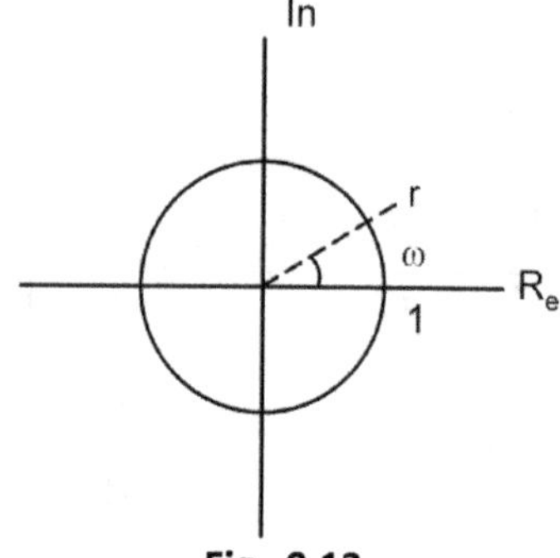

Fig. 9.13

This is F T of the sequence

For any given values of z for which the z – transform converges is called region of convergence. Multiplying by r^{-n} it is possible for z – transform to converge even if F T does not. The sequence is not absolutely integrable for some signals such as $u(n)$. But $u(n)r^{-n}$ is integrable if $|r| > 1$.

If $r = 1$

$$F(x) = F\left(e^{j\omega}\right) = F\{f(n)\}$$

Thus in z plane, within a circle of radius 1, z – transform becomes fourier transform.

Ex: 1. $f(n) \; = \; a^n u(n)$

$$X(n) \; = \; \sum_{n=-\alpha}^{\alpha} a^n u(n)\, z^{-n}$$

$$= \sum_{n=0}^{\alpha} \left(a z^{-1}\right)^{-n} = 1 + az^{-1} + a^2 z^{-2} + \ldots\ldots$$

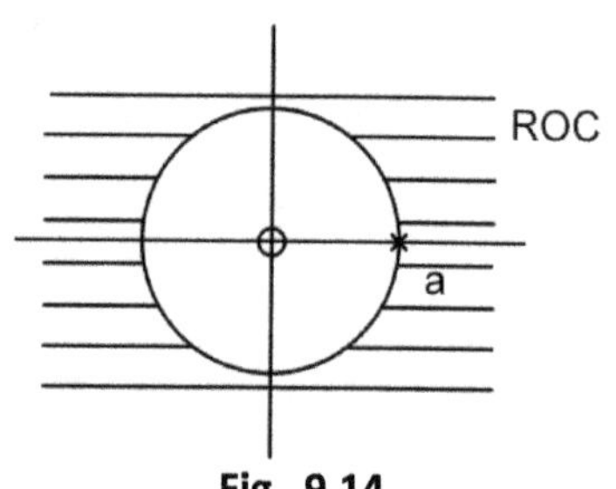

Fig. 9.14

$$= \frac{1}{1-a\,z^{-1}} \qquad \text{for} \quad |z| > |a|$$

$$= \frac{z}{z-a}$$

A zero at $z = 0$

pole at $z = \alpha$

ROC is shaded region

for $|z| > a$

2. Find z – transform of $\delta(nT)$

$$z\{\delta(nT)\} = \sum_{n=0}^{\alpha} \delta(nT)z^{-n} \qquad \text{Assume } T = 1$$

$$= \delta(0) + \delta(T)z^{-1} + \delta(2T)z^{-2} + \ldots\ldots$$

$$= 1 + 0 + 0 = 1$$

$$\therefore \qquad \delta(n) = 1 \quad n = 0$$

$$0 \quad n \neq 0$$

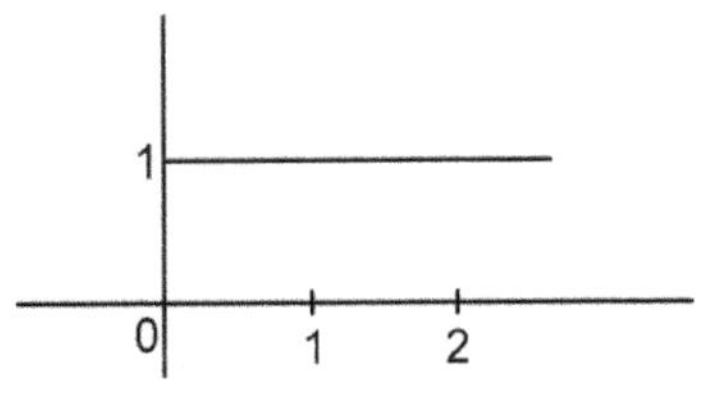

Fig. 9.15

3. Find $z\{\delta(n-1)T\} = z^{-1}$

$$z\{u(nT)\} = \sum_{n=0}^{\alpha} u(nT)z^{-n}$$

$$= u(0) + u(1)z^{-1} + u(2) + z^{-2} + \ldots\ldots\ldots$$

$$= 1 + z^{-1} + z^{-2} + \ldots\ldots$$

$$= \left(1 - z^{-1}\right)^{-1}$$

$$= \frac{1}{1-z^{-1}} = \frac{z}{z-1}$$

4. $x\{u(n-1)T\} \qquad T = 1$

$$= \sum_{n=0}^{\alpha} u(n-1)z^{-n}$$

$$= u(-1) + u(0)z^{-1} + u(1)z^{-2} + \ldots..$$

$$= \quad 0 + \frac{1}{z} + \frac{1}{z^2} + \dots\dots$$

$$= \quad \frac{1}{z}$$

5. $x\{u(nT)e^{-\alpha nT}\} \quad = \quad \sum_{n=0}^{\alpha} e^{-\alpha n}\, z^{-n}$

$$= \quad 1 + e^{-\alpha}z^{-1} + e^{-2\alpha}z^{-2} + \dots\dots$$

$$= \quad \frac{1}{1 - e^{-\alpha}z^{-1}} = \frac{z}{z - e^{-\alpha}}$$

6. $x\{u(nT)\sin\omega nT\} \quad = \quad z\{u(n)\sin\omega n\} \qquad T = 1$

$$= \quad \frac{1}{2j} z\{u(n)e^{j\omega n} - u(n)e^{-j\omega n}\}$$

$$= \quad \frac{1}{2j}\left\{ \frac{z}{z - e^{j\omega}} - \frac{z}{z - e^{j\omega}} \right\}$$

$$= \quad \frac{1}{2j}\left\{ \frac{z^2 - ze^{-j\omega} - z^2 + ze^{j\omega}}{z^2 - z(e^{j\omega} + e^{-j\omega}) + 1} \right\}$$

$$= \quad \frac{\dfrac{z\{e^{j\omega} - e^{-j\omega}\}}{2j}}{z^2 - 2z\cos\omega + 1} = \frac{z\sin\omega}{z^2 - 2z\cos\omega + 1}$$

7. $x\left\{ \left(\dfrac{1}{3}\right)^n \sin\left(\dfrac{\pi}{4}n\right) u(n) \right\}$

$$= \quad \frac{\dfrac{1}{3\sqrt{2}}z}{\left(z - \dfrac{1}{3}e^{j\pi/4}\right)\left(z - \dfrac{1}{3}e^{-j\pi/4}\right)}$$

8. $f(n) = 7\left(\dfrac{1}{3}\right)^n u(n) - 6\left(\dfrac{1}{2}\right)^n u(n)$

$$f(z) = \sum_{n=-\alpha}^{\alpha} \left\{ 7\left(\dfrac{1}{3}\right)^n u(n)z^{-n} - 6\left(\dfrac{1}{2}\right)^n u(n)z^{-n} \right\}$$

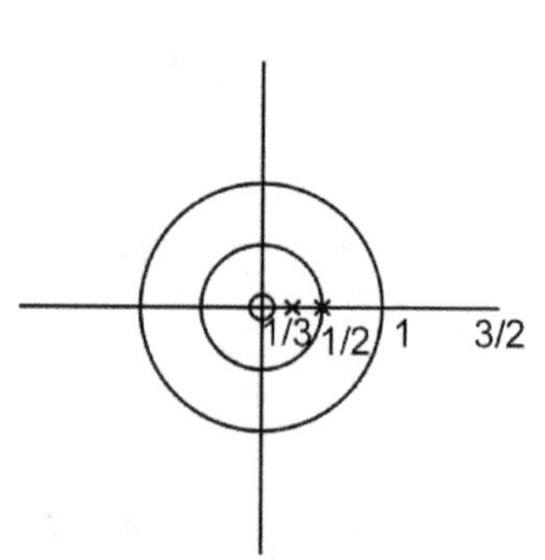

Fig. 9.16

$$= \frac{z}{1-\frac{1}{3}z^{-1}} - \frac{6}{1-\frac{1}{2}z^{-1}} = \frac{z\left(z-\frac{3}{2}\right)}{\left(z-\frac{1}{3}\right)\left(z-\frac{1}{2}\right)}$$

$$|z| > \frac{1}{3}, \quad |z| > \frac{1}{2}$$

9.8 PROPERTIES OF ROC

1. **Finite Length Sequences:** If only a finite number of sequence values are non zero

$$x(z) = \sum_{n=n_1}^{n_2} x(n)z^{-n}$$

Convergence of this requires $|x(n)| < \alpha$ for $n_1 \le n \le n_2$; then z may take on all values except $z = \alpha$ if $n_1 < 0$ and $z = 0$ if $n_2 > 0$.

Finite length of sequences have a region of convergence that is at least $0 < |z| < \alpha$ and it may include either $z = 0$ or $z = \alpha$.

2. **Right Sided Sequences:** A right side sequence is one for which $x(n) = 0$ for $n < n_1$, the z – transform of such a sequence is

$$x(z) = \sum_{n=n_1}^{\alpha} x(n)z^{-n}$$

9.9   INVERSE Z – TRANSFORM

Inverse z – transform

There are four methods often used to find the inverse z – transform. They are

1. Long division method
2. Partial fraction method
3. Residue method
4. Convolution method

1. **Long division method:** z – transform of a sequence $x(n)$ is

$$x(z) = \sum_{n=-\alpha}^{\alpha} x(n)z^{-n}$$

If the sequence $x(n)$ is causal then

$$x(z) = \sum_{n=0}^{\alpha} x(n) z^{-n}$$

If $\qquad X(z) = \dfrac{N(z)}{D(z)} = \dfrac{b_0 + b_1 z^{-1} + \ldots b_M z^{-M}}{1 + a_1 z^{-1} + \ldots a_N z^{-N}}$

If $\;\; X(z)\;$ converges for $|z| > r$ then we can obtain

$$X(z) = x(0) + x(1) z^{-1} + x(2) z^{-2} + \ldots\ldots$$

Which converges for some range of z. Here $x(n)$ is in negative powers of z then the division must be done with highest power of Z.

Ex:

1. $\qquad \dfrac{1 + 3z^{-1}}{1 + 3z^{-1} + 2z^{-2}} \qquad |z| > 2$

$$x(z) = \frac{1 + 3z^{-1}}{1 + 3z^{-1} + 2z^{-2}} = \frac{z(z+3)}{(z+1)(z+2)}$$

$$x(n) = \Sigma \text{ residues of } \frac{z(z+3)}{(z+1)(z+2)} z^{(n-1)} \text{ at poles of same within O}$$

ROC is at $|z| > 2$, it encloses poles at $z = -1$ and $z = -2$

$$\therefore \qquad x(n) = \text{residue of } \frac{(z+3)z^n}{(z+1)(z+2)} \quad \text{at} \quad z = -1$$

$$\text{residue of } \frac{z(+3)z^n}{(z+1)(z+2)} \quad \text{at} \quad z = -2$$

$$\frac{\cancel{(z+1)}(z+3)z^n}{\cancel{(z+1)}(z+2)} + \frac{\cancel{(z+2)}(z+3)z^n}{(z+1)\cancel{(z+2)}} = 2(-1)^n - (-2)^n$$

2. **Partial fraction methods:** Let a rational function $\dfrac{X(z)}{z}$ given by

$$\frac{X(z)}{z} = \frac{N(z)}{D(z)} = \frac{b_0 Z^m + b_1 Z^{m-1} + b_2 Z^{m-2} + \ldots b_m}{Z^n + a_1 x^{n-1} + a_2 Z^{n-2} + \ldots a_n}$$

If $m < n$, then $\dfrac{X(z)}{z}$ is a proper function

If $m \geq n$, then $\dfrac{X(z)}{z}$ is not a proper function, then it can be written as

$$\frac{X(z)}{z} = \delta z^{n-m} + CZ^{n-mf} + \ldots + C_{n-m} \to \frac{N_1(z)}{D(z)}$$

where $\dfrac{N_1(z)}{D(z)}$ is a proper function.

Case 1: When $\dfrac{X(z)}{z}$ has distinct poles

$$\frac{Y(z)}{z} = \frac{C_1}{Z-K_1} + \frac{C_2}{Z-K_2} + \frac{C_3}{Z-K_3} + \ldots$$

where $\qquad C_1 = \left[(z-k_i)\frac{y(z)}{z} \right]_{z=ki} \qquad i = 1,2,3$

Case 2: When $\dfrac{X(z)}{z}$ has l-repeated poles and $(n-l)$ are simple

then $\qquad \dfrac{Y(z)}{z} = \dfrac{C_1}{z-k_1} + \dfrac{C_2}{z-k_2} + \ldots + \dfrac{C_{P_1}}{z-k_p} + \dfrac{C_{P_2}}{(z-k_p)} + \ldots + \dfrac{C_{PL}}{(z-k_p)}$

Where $\qquad C_{pl} = \left[(z-k_p)\dfrac{lX(z)}{z} \right]_{z=k_p}$

Also if z transform has a complex pole

$$\frac{X(z)}{z} = \frac{C_1}{z-k_1} + \frac{C_1^*}{z-k_1^*}$$

3. Residue Method

Inverse z – transform $X(n)$ can be obtained by

$$x(n) = \frac{1}{2\pi j} \oint X(z)z^{n-1}dz$$

where C is a circle in Z plane

This equation can be evaluated by finding sum of all residues of the poles that are inside the circle C.

$\therefore\ x(n) = \Sigma$ residues of $X(z)z^{n-1}$ at the poles inside C

$$= \sum_i (z-z_i)X(z)z^{n-1}\Big|_{z=z_i}$$

If $X(z)$ does not have any poles inside the C, $x(n)=0$

If $X(z)$ has no poles at $z = z_0$ and if the derivative $\dfrac{d}{dz} f(z)$ exists then

$$x(n) = \frac{1}{(k-1)!} \frac{d^{k-1}}{dz^{k-1}} f(z) \bigg|_{z=z_0} \quad \text{must be taken} \quad \text{if } z_0 \text{ is inside C}$$

$$= 0 \quad \text{if } z_0 \text{ is outside C}$$

for $n \geq 0$

$$x(n) = (z - 2) \frac{z \cdot z^{n-1}}{(z - 2)(z - 3)} \bigg|_{z=2}$$

$$= (-2)^n$$

$$\therefore \qquad x(n) = -(2)^n u(n) - (3)^n u(-n-1)$$

4. **Convolution Method:** z-transform $X(z)$ is split into $X_1(z)$ and $X_2(z)$ such that $X_1(z)$ $Y_2(z) = Y(z)$

Signals $x_1(n)$ and $x_2(n)$ are obtained by taking inverse z transform of $x_1(z)$ and $x_2(z)$.

Finally $x(n)$ is obtained by convoluting $x_1(n)$ and $x_2(n)$ in time domain.

$$Z\big[x_1(x) * x_2(n)\big] = x_1(z)x_2(z) = x(z)$$

$$\therefore \qquad x(n) = z^{-1}\big[x(z)\big] = z^{-1}\big[z\{x_1(n) * x_2(n)\}\big]$$

$$\therefore \qquad z^{-1}\big[x(z)\big] = x(n) = x_1(n) * x_2(n) = \sum_{k=0}^{z} x_1(k)x_2(n-k)$$

Solved Problems Exercise - 3

1. Find the z – transform of following

(a) $\qquad x(n) = \{1,\, 2,\, -1,\, 2,\, 3\}$

Given $\qquad x(0) = -1, \quad x(-1) = 2, \quad x(-2) = 1$

$\qquad\qquad x(1) = 2, \qquad x(2) = 3$

It is a non causal signal

$$\therefore \qquad x(z) = 2z^{-1} + 3z^{-2} + z^2 + 2z - 1$$

$$= z^2 + 2z - 1 + 2z^{-1} + 3z^{-2}$$

ROC: Entire Z plane except at

$$Z = 0 \text{ and } z = \alpha$$

(b) $$x(n) = \{1, 2, 3, 4\}$$

assuming it as a causal signal

$$x(0)=1, \quad x(1)=2, \quad x(2)=3, \quad x(3)=4$$

$\therefore$
$$x(z) = 1 + 2z^{-1} + 3z^{-2} + 4z^{-3}$$

ROC: Entire Z- plane except at $z = 0$

2. Find Z transform of

(a) $$\delta(nT)$$

$$x(n) = \delta(nT)$$

$$x(z) = \sum_{n=-\alpha}^{\alpha} \delta(nT)z^{-n}$$

$$= \delta(2T)z^{+2} + \delta(-T)z^{-1} + \delta(0) + \delta(T)z^{-1} + \delta(2T)z^{-2} + \dots$$

If $T = 1$

by definition

$$= \delta(-2)z^{+2} + \delta(-1)z^{+1} + \delta(0) + \delta(1)z^{-1} + \dots\dots$$

$$= \delta(0) = 1$$

$$\delta(0) = 1 \quad n = 0$$
$$0 \quad n \neq 0$$

(b) $$u(nT)$$

$$x(z) = \sum_{n=0}^{\alpha} u(nT)z^{-n}$$

If $T = 1$

$$= u(0) + u(T)z^{-1} + u(2T)z^{-2} + \dots\dots$$

$$u(n) = 1 \quad n \geq 0$$
$$0 \quad n < 0$$

If $T = 1$

$$= 1 + z^{-1} + z^{-2} + \dots\dots$$

$$\left(1 - \frac{1}{2}\right)^{-1} = \frac{z}{z-1}$$

(c) $$x(n) = e^{\alpha nT}u(nT)$$

Let $T = 1$

$$x(z) = \sum_{n=0}^{\alpha} e^{-\alpha n} z^{-n}$$

$$= 1 + e^{-\alpha} z^{-1} + e^{-2\alpha} e^{-2} + \dots\dots$$

$$= \left(1 - \frac{1}{e^{\alpha} z}\right)^{-1}$$

$$= \frac{e^{\alpha} z}{e^{\alpha} z - 1} = \frac{1}{1 - e^{-\alpha} z^{-1}}$$

(d) $x(n) = u(n) - u(n-3)$

$\therefore$ $x(n) = 1$ for $0 \le n \le 2$

$$x(z) = \sum_{n=0}^{2} z^{-n}$$

$$= 1 + z^{-1} + z^{-2}$$

ROC is entire Z plane except at $Z = 0$

3. Find Z- transform of

$$x(n) = 2^{n} u(-n) + \left(\frac{1}{4}\right)^{n} u(n-1)$$

Ans: $x(n) = 2^{n} u(-n) + \left(\frac{1}{4}\right)^{n} u(n-1)$

$$x(z) = \sum_{n=-\alpha}^{\alpha} \left[2^{n} u(-n) + \left(\frac{1}{4}\right)^{n} u(n-1)\right] z^{-n}$$

$$= \sum_{n=-\alpha}^{\alpha} 2^{n} z^{-n} + \sum_{n=1}^{\alpha} \left(\frac{1}{4}\right)^{n} z^{-n}$$

$$= \sum_{n=0}^{\alpha} \left(2^{-1} z\right)^{n} + \sum_{n=0}^{\alpha} \left(\frac{1}{4z}\right)^{n} - 1$$

$$= \frac{1}{1 - 2^{-1} z} + \frac{1}{1 - \dfrac{1}{4z}} - 1 = \frac{\frac{7}{4} z}{\left(z - \dfrac{1}{4}\right)(2 - z)}$$

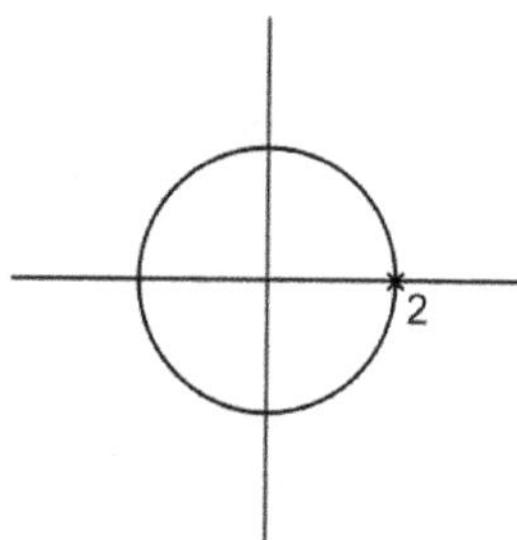

Fig. 9.17

4. Find the Z - transform of

$$x(n) \;=\; 2^n + 3^{-n} \qquad \text{for} \quad n \geq 0$$

$$\qquad\qquad 0 \qquad\qquad \text{for} \quad n \leq 0$$

$$x(z) \;=\; \sum_{n=0}^{\alpha} 2^n z^{-n} + \sum_{n=0}^{\alpha} 3^{-n} z^{-n}$$

$$=\; \frac{1}{1-2z^{-1}} + \frac{1}{1+3^{-1}z^{-1}}$$

$$=\; \frac{z}{z-2} + \frac{z}{z-\frac{1}{3}}$$

$$\text{ROC:} \quad |Z| > 2 \quad \& \quad |Z| > \frac{1}{3}$$

$$\therefore \quad \text{ROC:} \quad |Z| > 2$$

5. Find the Z – transform of

$$x(n) = n(n+3)(0.5)^n u(n)$$

$$x(z) = \sum_{n=0}^{\alpha} n^2 (0.5)^n u(n) z^{-n} \sum_{n=0}^{-n} 3n(0.5)^n u(n) z^{-n}$$

$$\qquad\qquad\quad x_1(z) \qquad\qquad\qquad\qquad x_2(z)$$

$$x_1(z) = \sum_{n=0}^{\alpha} n^2 (0.5)^n z^{-n}$$

$$=\; \sum_{n=0}^{\alpha} n^2 0.5^n z^{-n-1} \, z$$

$$=\; -z \sum_{n=0}^{\alpha} n(-n) 0.5^n z^{-n-1}$$

$$=\; -z \sum_{n=0}^{\alpha} \frac{d}{dz}\left(z^{-n}\right) n \, 0.5^{-n}$$

$$=\; -z \frac{d}{dz} \sum n \, 0.5^{-n} \, z^{-n}$$

$$=\; -z \frac{d}{dz} \sum n \, 0.5^{-n} \, z^{-n-1} \, z$$

$$= +z^2 \frac{d}{dz} \sum \frac{d}{dz} z^{-n} \ 0.5^{-n}$$

$$= z^2 \frac{d^2}{dz^2} \left\{ \sum 0.5^n z^{-n} \right\}$$

$$= z^2 \frac{d^2}{dz^2} \frac{z}{z-0.5} = \frac{z(0.5z+0.25)}{(z-0.5)^3}$$

6. Find z - transform of

$$x(n) = nc_r a^{n-r} u(n)$$

$$x(n) = nc_r a^{n-r} u(n) = \frac{n(n-1)(n-2) \ ... \ (n-r+1)}{r!}$$

$$\therefore \qquad x(z) = \sum_{n=0}^{\alpha} x(n) z^{-n} = \sum_{n=0}^{\alpha} \frac{rp_r}{r!} a^{n-r} z^{-n}$$

$$= \sum_{n=0}^{\alpha} \frac{n(n-1)(n-2) \ ... \ (n-r+1)}{r!} a^{n-r} z^{-n}$$

$$= \frac{1}{a^r r!} \sum_{n=0}^{\alpha} \left(n(n-1)(n-2)... \right) a^n z^{-n}$$

up to $n = r - 1$ the above summation is '0'.

$$\therefore \qquad x(z) = \frac{1}{a^r r!} \sum_{n=r}^{\alpha} n(n-1) \ ... \ (n-r+1) a^n z^{-n}$$

$$= \frac{1}{a^r r!} \left\{ r! \left(az^{-1} \right)^r + \frac{(r+1)!}{1!} \left(az^{-1} \right)^{r+1} + \ ... \ \right\}$$

SOLVED PROBLEMS EXERCISE - 4

1. Find Z – transform of

$$x(n) = r^n \sin \omega n \ u(n)$$

we have

$$z\{ \sin \omega n \ u(n) \} = \frac{\sin \omega z^{-1}}{1 - z \cos \omega z^{-1} + z^{-2}} \qquad\qquad ROC : |z| > 1$$

using multiplication on by exponential sequence property

$$z\{ a^n x(n) \} = X\{ a^{-1} z \}$$

$$z\{r^n \sin \omega n\, u(n)\} = \frac{\sin \omega (r^{-1}z)^{-1}}{1 - 2r \cos \omega\, z^{-1} + r^2 z^{-2}} \qquad \text{ROC: } |z| > |r|$$

2. Find z – transform of $\quad x(n) = n\, u(n)$

$$z\{nx(n)\} = z\frac{d}{dz}\{x(z)\}$$

$$= -z\frac{d}{dz}\left(\frac{z}{z-1}\right)$$

$$= \frac{z}{(z-1)^2}$$

ROC: $|z| > 1$

3. Find the z – transform of $\quad x(n) = a^{n-1} u(n-1)$

We have $\quad x(z) = \dfrac{1}{1 - az^{-1}} \quad$ for $\quad x(n) = a^n u(n)$

using time shifting property

$$z\{x(n-k)\} = z^{-k}x(z) = \frac{z^{-1}}{1 - az^1} = \frac{1}{z-a}$$

ROC: $\quad |z| > |a|$

4. Find $x(\alpha)$, where $x(z)$ is given by

$$x(z) = \frac{z+2}{(z-0.8)^2}$$

$$x(\alpha) = \lim_{z \to 1}(z-1)x(z)$$

$$\lim_{z \to 1}(z-1)\frac{(x+2)}{(z-0.8)^2} = 0$$

5. Find z – transform of

$$x(n) = \frac{1}{2}(n^2 + n)\left(\frac{1}{3}\right)^{n-1} u(n-1)$$

$$= \frac{1}{2}n^2\left(\frac{1}{3}\right)^{n-1} u(n-1) + \frac{1}{2}n\left(\frac{1}{3}\right)^{n-1} u(n-1)$$

we have
$$Z\left\{\left(\frac{1}{3}\right)^{n}u(n)\right\}=\frac{z}{z-\frac{1}{3}}$$

using time shifting property
$$Z\left\{\left(\frac{1}{3}\right)^{n-1}u(n-1)\right\}=\frac{1}{z-\frac{1}{3}}$$

using multiplication property
$$Z\left\{n\left(\frac{1}{3}\right)^{n-1}u(n-1)\right\}=-z\frac{d}{dz}\frac{1}{z-\frac{1}{3}}$$

$$=-z\frac{-1}{\left(z-\frac{1}{3}\right)^{2}}=\frac{z}{\left(z-\frac{1}{3}\right)^{2}}$$

$$Z\left\{n^{2}\left(\frac{1}{3}\right)^{n-1}u(n-1)\right\}=-z\frac{d}{dz}\frac{z}{\left(z-\frac{1}{3}\right)^{2}}$$

$$=-z\frac{\left\{\left(z-\frac{1}{3}\right)^{2}-2z\left(z-\frac{1}{3}\right)\right\}}{\left(z-\frac{1}{3}\right)^{3}}$$

$$=\frac{z\left(z+\frac{1}{3}\right)}{\left(z-\frac{1}{3}\right)^{3}}$$

$$\therefore\quad x(z)=\frac{1}{2}\left\{\frac{z\left(z+\frac{1}{3}\right)}{\left(z-\frac{1}{3}\right)^{3}}+\frac{z}{\left(z-\frac{1}{3}\right)^{2}}\right\}$$

$$=\frac{z^{2}}{\left(z-\frac{1}{3}\right)^{2}}$$

$$\text{ROC:}|z|>\frac{1}{3}$$

6. Find the z – transform of $y(n) = u(n)e^{-\alpha n}\sin\omega$

$$y(n) = \left[e^{-\alpha n}u(n)\right]\left[\sin n\omega u(n)\right]$$

$$f(z) = \frac{z}{z - e^{-\alpha}}$$

$$g(z) = \frac{z\sin\omega}{\left(z - e^{j\omega}\right)\left(z - e^{-j\omega}\right)}$$

using complex convolution

$$\therefore \quad y(n) = \frac{1}{2\pi j}\oint f\left(\frac{z}{v}\right) g(v)v^{1}dv$$

$$= \frac{1}{2\pi j}\oint \frac{z/v}{\dfrac{z}{v} - e^{-\alpha}} \frac{v\sin\omega}{\left(v - e^{j\omega}\right)\left(v - e^{-j\omega}\right)}v^{-1}dv$$

$$= \frac{1}{2\pi j}\oint \frac{-ze^{\alpha}\sin\omega}{(v - ze^{\alpha})(v - e^{j\omega})(v - e^{-j\omega})}dv$$

$$= \operatorname*{Res}_{\upsilon - e^{j\omega}} H(\upsilon) \quad + \quad \operatorname*{Res}_{\upsilon - e^{-j\omega}} H(\upsilon)$$

$$= -z\,e^{\alpha}\sin\omega\left\{\frac{1}{\left(e^{j\omega} - ze^{\alpha}\right)\left(e^{j\omega} - e^{-j\omega}\right)} + \frac{1}{\left(e^{-j\omega} - ze^{\alpha}\right)\left(e^{-j\omega} - e^{j\omega}\right)}\right\}$$

$$= \frac{ze^{\alpha}\sin\omega}{z^{2} - 2ze^{-\alpha}\cos\omega + e^{-2\alpha}}$$

7. Find the inverse transform of $x(z) = \dfrac{z}{(z-1)^{2}}$

$$x(z) = \frac{z}{z-1} + \frac{1}{z-1}$$

using convolution property

for $\qquad \dfrac{z}{z-1}, \quad x(n) = u(n)$

$$x(n) = u(n) * u(n-1)$$

$$= \sum u(x)\,u(n-1-x)$$

$$= n\,u(x)$$

8. Find $x(n)$ for $x(z) = \dfrac{1+3z^{-1}}{1+3z^{-1}+2z^{-2}}$; $|z| > 2$

$$x(z) = \dfrac{z(2+3)}{(z+1)(z+2)}$$

$$= \dfrac{z}{(z+1)} \times \dfrac{z+3}{z+2}$$

$\therefore \quad x_1(n) = -1^n u(n)$

$\quad x_2(n) = -2^n u(n) + 3. - 2^{n-1} u(n-1)$

$\therefore \quad x(n) = -1^n u(n) * \left\{ -2^n u(n) + 3. - 2^{n-1} u(n-1) \right\}$

$$= \sum_{k=0}^{n} -1^k u(k) - 2^{n-k} u(n-k) + 3\sum_{k=0}^{n-1} -1^k u(k) - 2^{n-k-1} u(n-k-1)$$

$$= -2^n \sum_{k=0}^{n} -1^k . - 2^{-k} + 3. - 2^{n-1} \sum_{k=0}^{n-1} -1^k - 2^{-k}$$

$$= -2^n \sum_{k=0}^{n} \left(\frac{1}{2}\right)^k + 3. - 2^{n-1} \sum_{k=0}^{n-1} \left(\frac{1}{2}\right)^k$$

$$= -2^n \left\{ \frac{1-0.5^{n+1}}{1-0.5} \right\} + 3. - 2^{n-1} \left\{ \frac{1-0.5^n}{1-0.5} \right\} \quad \text{as} \quad \sum_{k=0}^{n} a^k = \frac{1-a^{n+1}}{1-a}$$

$$= (-1)^n .2 - (-2)^n$$

$$n(n) = \left\{ 2(-1)^n - (-2)^n \right\} u(n)$$

PREVIOUS PAPER QUESTIONS

1. Find the inverse z – transform of

(a) $X(z) = \log\dfrac{1}{(1-az^{-1})}$, $|z| > |a|$

(b) $X(z) = \log\dfrac{1}{(1-a^{-1}z)}$ $|z| < |a|$

2. Find the z – transform of $\qquad$ Nov 08

$$z[n] = \left(\frac{1}{2}\right)^{n} u[n] + \left(\frac{1}{3}\right)^{n} u[-n-1]$$

3. A finite sequence is defined as $x[n] = \{5, 3, -2, 0, 4, -3\}$. Find $X[z]$

4. Consider the sequence $x[n] = \begin{cases} a^{n} & 0 \le n < N-1 \\ 0 & \text{otherwise} \end{cases}$

   Find $X[z]$

5. Find the z – transform of $x(n) = \cos(n\omega)u(n)$

6. Find inverse z – transform of

   (a) $X(n) = \dfrac{z}{2z^{2} - 3z + 1}$ $\qquad |z| < \dfrac{1}{2}$

   (b) $X(n) = \dfrac{z}{2z^{2} - 3z + 1}$ $\qquad |z| > 1$

   (c) $X(n) = \dfrac{z}{z(z-1)(z-2)^{2}}$ $\qquad |z| > 2$ $\qquad$ Feb 2008   Sep 2007

7. Find z – transform of the following sequence $\qquad$ Feb 2008

   (a) $\qquad x[n] = a^{-n}u[-n-1]$

   (b) $\qquad x[n] = u[-n]$

   (c) $\qquad x[n] = a^{n}u[-n-1]$

8. Find inverse z – transform of

$$x[z] = \frac{1}{1 - 1.8z^{-1} + 0.8z^{-2}}$$

   $\qquad$ Sep 2007

9. Find z transform of $x(n) = \sin(n\omega)u(n)$ $\qquad$ Sep 2007

10. Prove that final value of $x(n)$ for

$$x(z) = \frac{z^{2}}{(z-1)(z-0.2)}$$ is 1.25 and its initial value is unity $\qquad$ Sep 2007

11. Given $X(z) = \dfrac{1}{1-az^{-1}}$ $|z| > |a|$

    find x(a) using long division method

12. Find z transform of $a^n \cos n\pi / 2$ May 2005

13. Find z- transform of   May 2005

$$x(n) = \left\{4(5^n) - 3(4^n)\right\} u(n)$$

14. Find z – transform of

(a) $\dfrac{1}{3}^n u(-n)$

(b) $\dfrac{1}{3}^n \left(u(-n) - u(n-8)\right)$

15. Given $x(z) = \dfrac{z}{(z-1)^3}$ find $x(n)$ using contour integration method   May 2005

1. Find the inverse z – transform of below functions

(a) $x(z) = \log \dfrac{1}{(1-az^{-1})}$, $|z| > |a|$

$$\dfrac{d}{dz} x(z) = \dfrac{d}{dz} \log (1-az^{-1})^{-1}$$

$$= \dfrac{1}{\left(1-az^{-1}\right)^2} \cdot az^{-2} = \dfrac{az^{-2}}{\left(1-az^{-1}\right)^2}$$

$$-z\dfrac{d}{dz} x(z) = \dfrac{-az^{-1}}{\left(1-az^{-1}\right)^2}$$

$$z\{nx(n)\} = -z\dfrac{d}{dz} x(z) = \dfrac{-az^{-1}}{\left(1-az^{-1}\right)^2}$$

$$= z\left\{-na^n u(n)\right\}$$

$$nx(n) = -na^n u(n)$$

$$x(n) = -a^n u(n)   \qquad |z| > |a|$$

(b)
$$x(z) = \log\frac{1}{\left(1-a^{-1}z\right)}$$

$$\frac{d}{dz}x(z) = \frac{1}{1-a^{-1}z} = \frac{-a\,z^{-1}}{\left(1-a\,z^{-1}\right)}$$

$$= \frac{d}{dz}\log\left(\frac{-a\,z^{-1}}{1-a^{-1}z}\right)$$

2. Find the z – transform of

$$x[n] = \left(\frac{1}{2}\right)^n u[n] + \left(\frac{1}{3}\right)^n u[-n-1]$$

we have

$$z\{a^n u(n)\} = \frac{1}{a-az^{-1}} \quad \text{and} \quad z\{-a^n u(-n-1)\} = \frac{1}{1-az^{-1}}$$

$$\therefore \qquad x(z) = \frac{1}{1-\frac{1}{2}z^{-1}} - \frac{1}{1-\frac{1}{3}z^{-1}}$$

$$= \frac{2z}{(2z-1)} - \frac{3z}{(3z-1)} \qquad \text{ROC: } |z|$$

3. A finite sequence is defined as

$$x[n] = \{5, 3, -2, 0, 4, -3\} \quad \text{find} \quad x[z]$$

we have

$$x(z) = \sum_{n=-\alpha}^{\alpha} x(n)z^{-n}$$

$$x(0) = 0 \qquad\qquad x(-1) = -2$$

$$x(1) = 4 \qquad\qquad x(-2) = 3$$

$$x(2) = -3 \qquad\qquad x(-3) = 5$$

4. Consider the sequence $x[n] = \begin{cases} a^n & 0 \le n \le N-1 \\ 0 & \text{otherwise} \end{cases}$

find $\qquad x(z)$

$$x(z) = \sum_{n-0}^{N-1} a^n z^{-n}$$

$$= 1 + az^{-1} + a^2 z^{-2} + \ldots a^{N-1} z^{-(N-1)}$$

$$= \frac{1}{1 - az^{-1}} \qquad \text{if } |az^{-1}| <$$

or

$$|a| < |2|$$

5. Find the z – transform of $\quad x(n) = \cos(n\omega)\, u(n)$

$x(n) = \cos(n\omega)\, u(n)$

$$x(z) = \sum_{z=0}^{\alpha} \cos(n\omega) z^{-n} = \sum_{n=0}^{\alpha} \frac{e^{j\omega n} + e^{-j\omega n}}{2} z^{-n}$$

$$= \frac{1}{2}\left\{ \sum_{n=0}^{\alpha} e^{j\omega n} z^{-n} + \sum_{n=0}^{\alpha} e^{-j\omega n} z^{-n} \right\}$$

$$= \frac{1}{2}\left\{ \sum \left(e^{j\omega} z^{-1} \right)^n + \sum \left(e^{-j\omega} z^{-1} \right)^n \right\}$$

If $\quad e^{j\omega} z^{-1} < |$ and $\quad e^{-j\omega} z^{-1} < |$ then $\quad |z| > 1$

$$= \frac{1}{2}\left\{ \frac{1}{1 - e^{j\omega} z^{-1}} + \frac{1}{1 - e^{-j\omega} z^{-1}} \right\}$$

$$= \frac{1}{2}\left\{ \frac{1 - e^{-j\omega} z^{-1}}{1 - e^{j\omega} z^{-1}} + \frac{1 - e^{j\omega} z^{-1}}{1 - e^{-j\omega} z^{-1}} \right\}$$

$$= \frac{1}{2}\left\{ \frac{2 - \left(e^{j\omega} z^{-1} + e^{j\omega} z^{-1} \right)}{1 + z^{-2} - e^{j\omega} z^{-1} - e^{-j\omega} z^{-1}} \right\}$$

$$= \frac{1 - \cos\omega z^{-1}}{1 - 2\cos\omega z^{-1} + z^2}$$

6. Find the inverse Z transform of $\qquad$ Feb   2008

$$x(z)= \frac{z}{2z^2 - 3z +1} \qquad |z|<\frac{1}{2}$$

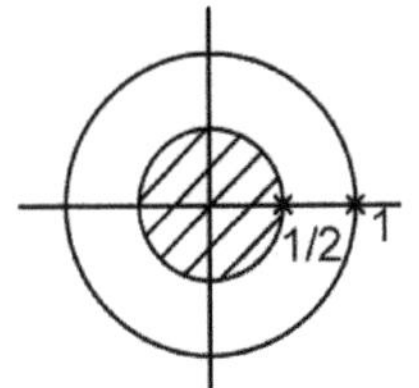

$$= \frac{z}{(z-1)(2z-1)}$$

$$= \frac{z^{-1}}{\left(1-z^{-1}\right)\left(2-z^{-1}\right)} = \frac{A}{1-z^{-1}} + \frac{B}{2-z^{-1}}$$

$$= \frac{1}{1-z^{-1}} - \frac{2}{2-z^{-1}} = \frac{1}{1-z^{-1}} - \frac{1}{1-\frac{1}{2}z^{-1}}$$

$$x(n) \; = u(n)-\left(\tfrac{1}{2}\right)^{n} u(n)$$

Since $\quad |z|<\dfrac{1}{2}$

$$x(n) \; = -\left(\frac{1}{2}\right)^{n} u(n)$$

7. Find Z transform of the following

(a) $\quad x(n) \; = -a^{-n}u[-n-1]$

$$x(z) \; = \frac{1}{1-a\,z^{-1}} \qquad |z|<|a|$$

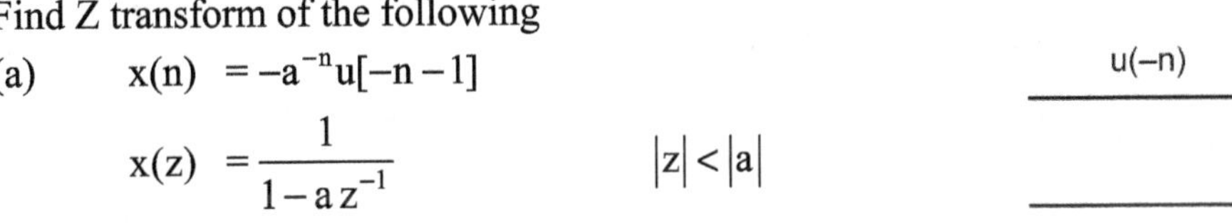

(b) $\quad x(n) \; = u(-n)$

$$x(z) \; = \sum_{n=-\alpha}^{0} z^{-n} = \sum_{n=0}^{\alpha} z^{n}$$

$$x(z) \; = \frac{1}{1-z} \qquad |z|<1$$

(c) $\quad x(n) \; = a^{-n}u(-n-1)$

$$x(z)= \sum_{n=-\alpha}^{-1} a^{-n}z^{-n}$$

$$= \sum_{+1}^{\alpha} a^{+n}z^{n}$$

$$= \frac{1}{1-az^{-1}} -1 = \frac{a\,z^{-1}}{1-a\,z^{-1}}$$

8. Find the inverse z – transform of

$$x(z) = \frac{1}{1 - 1.8z^{-1} + 0.8z^{-2}}$$

$$= \frac{z^2}{z^2 - 1.8z + 0.8} = \frac{z^2}{(z - 0.8)(z - 1)}$$

using long division method

$$
\begin{array}{r}
1 + 1.8z^{-1} + 2.44z^{-2} \\
\hline
z^2 \qquad\qquad\qquad\qquad\quad \\
\end{array}
$$

$$z^2 - 1.8z + 0.8$$

$$1.8z - 0.8$$

$$z^2 - 1.8z + 0.8 \quad\quad 1.8z - 3.24 + 1.44z^{-1}$$

$$2.44 - 1.44z^{-1}$$

$$2.44 - 4.39z^{-1} + 1.95z^{-2}$$

$$2.95z^{-1} - 1.95z^{-2}$$

$$x(z) \;=\; 1 + 1.8z^{-1} + 2.44z^{-2} + \ldots$$

$$x(n) \;=\; \{1,\; 1.8,\; 2.44,\; \ldots\}$$

9. Find the z – transform of $x(n) = \sin(n\omega)u(a)$ \hfill Sep 2007

$$x(z) \;=\; \sum_{n=-\alpha}^{\alpha} x(n)z^{-n}$$

$$= \sum_{n=-\alpha}^{\alpha} \sin(n\omega)\, u(n) z^{-n}$$

$$= \sum_{n=0}^{\alpha} \left\{ \frac{e^{j\omega n} - e^{-j\omega n}}{2j} \right\} z^{-n}$$

$$= \frac{1}{2j} \left\{ \sum_{n=0}^{\alpha} \left(e^{j\omega n} - e^{-j\omega n} \right) z^{-n} \right\}$$

$$= \frac{1}{2j} \left\{ \frac{1}{1 - e^{j\omega}z^{-1}} - \frac{1}{1 - e^{-j\omega}z^{-1}} \right\}$$

$$= \frac{1}{2j} \left\{ \frac{e^{j\omega}z^{-1} - e^{-j\omega}z^{-1}}{1 - e^{j\omega}z^{-1}\, e^{-j\omega}z^{-1} + z^{-2}} \right\}$$

$$= \frac{\sin\omega z^{-1}}{1 - 2\cos\omega z^{-1} + z^{-2}} \qquad \text{ROC:}\quad |z| < 1$$

10. Prove that final value of x(n) for

$$x(n) = \frac{z^2}{(z-1)(z-0.2)} \quad \text{is 1.25 and its initial value is 1}$$

Sep 2007

$$x(z) = \frac{z^2}{(z-1)(z-0.2)}$$

according to initial value theorem

$$x(0) = \lim_{z \to \alpha} x(z)$$

$$= \frac{1}{\left(1-\frac{1}{z}\right)\left(1-\frac{0.2}{z}\right)} = 1$$

according to final value theorem

$$x(\alpha) = \lim_{z \to 1}(z-1)x(z)$$

$$= \frac{z^2}{(z-1)(z-0.2)} \times (z-1)$$

$$= \frac{1}{1-0.2} = \frac{1}{0.8} = 1.25$$

11. Given $x(z) = \dfrac{1}{1-az^{-1}}$ $|z| < |a|$, find $x(n)$ using long division method

$$\therefore \quad x(z) = 1 + az^{-1} + a^2z^{-2} + a^3z^{-3} + \dots$$

$$x(n) = \left\{1,\ a,\ a^2,\ a^3,\ \dots\right\}$$

12. Find Z transform of $a^n \cos \frac{\pi}{2} a(n)$

May 2005

$$x(z) = \sum_{n=-\alpha}^{\alpha} a^n \cos \frac{\pi}{2} z^{-n} u(n)$$

$$= \sum_{n=-\alpha}^{\alpha} \frac{e^{jn\pi/2} + e^{-jn\pi/2}}{2} a^n z^{-n} u(n)$$

$$= \frac{1}{2}\left\{\sum_{n=0}^{\alpha} e^{jn\pi/2} a^n z^{-n} + \sum_{n=0}^{\alpha} e^{-jn\pi/2} a^n z^{-n}\right\}$$

$$= \frac{1}{2}\left\{\left(\frac{1}{1-ae^{j\pi/2}z^{-1}}\right)+\left(\frac{1}{1-ae^{-j\pi/2}z^{-1}}\right)\right\}$$

$$= \frac{1}{2}\left\{\frac{2-a\left(e^{j\pi/2}+e^{-j\pi/2}\right)}{\left(1-a(e^{j\pi/2}z^{-1}-e^{-j\pi/2}z^{-1})+a^2z^{-2}\right)}\right\}$$

$$= \frac{a^n}{2}\left\{\frac{1-a\cos\pi/2}{1-2a\cos\pi/2\,z^{-1}+a^2z^{-2}}\right\}$$

$$= \left\{\frac{1}{1+a^2z^{-2}}\right\} = \frac{z^2}{z^2+a^2}$$

13. Find z – transform of

$$x(n) = \left\{4(5^n)-3(4^n)u(n)\right\}$$

May 2005

as we have for $a^n u(n)$

$$x(z) = \sum_{n=0}^{\alpha} a^n z^{-n} = \frac{1}{1-az^{-1}}$$

$$4.5^n = \frac{4}{1-5z^{-1}}$$

$$3.4^n = \frac{3}{1-4z^{-1}}$$

$$\therefore \quad x(z) = \frac{4}{1-5z^{-1}}-\frac{3}{1-4z^{-1}}$$

14. Find z – transform of

$$\left(\frac{1}{3}\right)^n u(-n)$$

$$x(n) = \left(\frac{1}{3}\right)^n u(-n)$$

$$\therefore \quad x(z)= \sum_{n=-\alpha}^{0}\left(\frac{1}{3}\right)^n z^{-n}$$

$$= \sum_{n=0}^{\alpha} \left(\frac{1}{3}\right)^n z^{+n}$$

$$= \sum_{n=0}^{\alpha} (3z)^n$$

$$x(z) = \frac{1}{1-3z} \qquad \text{ROC:} \quad \begin{aligned} &|3z| < | \\ &|z| < \frac{1}{3} \end{aligned}$$

15. Given $x(z) = \dfrac{z}{(z-1)^3}$ find $x(n)$ using contour integration method by using complex integral (contour integration) $\hfill$ May 2005

$$x(n) = \frac{1}{2\pi j} \oint_c x(z) z^{n-1} dz$$

for $n \geq 0$

$$x(n) = \frac{1}{2\pi j} \oint_c \frac{z}{(z-1)^3} z^{n-1} dz$$

$$= \frac{1}{2\pi j} \oint_c \frac{z^n}{(z-1)^3} dz$$

$$= \text{residue of} \frac{z^n}{(z-1)^3} \text{ at the pole } z=1 \text{ of multiplicity } 3$$

$$\therefore \quad \frac{1}{2\pi j} \oint_c \frac{z^n}{(z-1)^3} dz = \left. \frac{1}{(k-1)!} \frac{d^{k-1}}{dz^{k-1}} x(z) \right|_{z=1}$$

$$= \left. \frac{1}{2!} \frac{dz}{dz^2} z^n = \frac{1}{2} n \frac{d}{dz} z^{n-1} \right|_{z=1}$$

$$= \left. \frac{1}{2} n(n-1) z^{n-2} \right|_{z=1}$$

$$= \frac{1}{2} n(n-1)$$

16. Let $g(t) = p(t) * p(t)$ where * denotes convolution and $p(t) = u(t) - u(t-1)$

    (a) Impulse response of filter matched to the signal $s(t) = g(t) - \delta(t-2) * g(t)$ is

$$p(t) \;=\; u(t) - u(t-1)$$

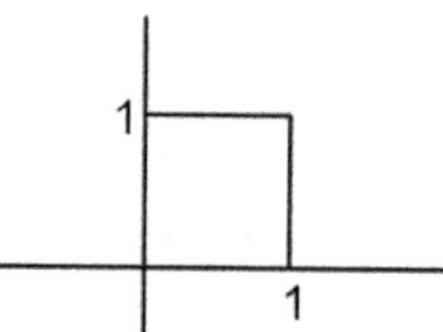

$$g(t) \;=\; p(t) * p(t)$$

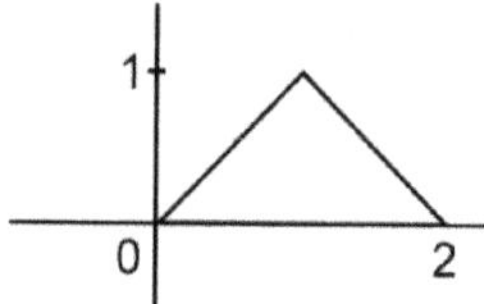

$$\delta(t-2) * g(t) \;=$$

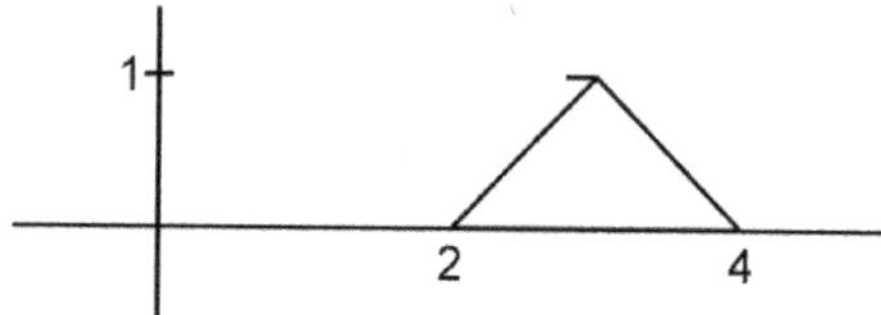

$$g(t) - \delta(t-2) * g(t)$$

$$y(t) \;=\; -s(t)$$

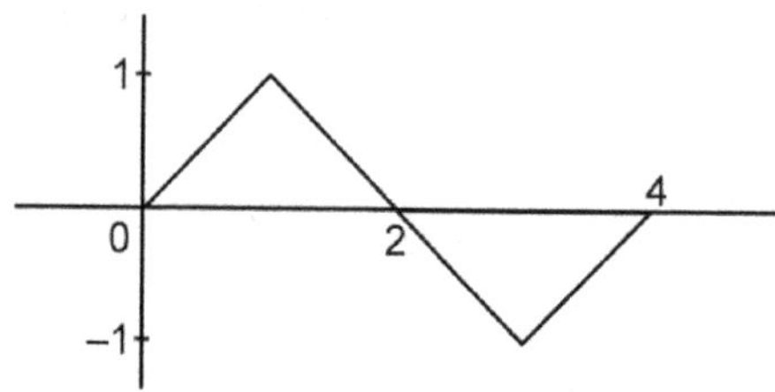

    (b) An AM signal is given by

$$X(t) \;=\; 100\big((p(t) + 0.5g(t))\big)\cos \omega t$$

for $0 \le t \le 1$

$$p(t) = 1$$

$$g(t) = t$$

$\therefore \quad X(t) \;=\; \big(100(1 + 0.5t)\big)\cos \omega t$

modulating signal is 0.5 t

modulation index = 0.5

17. The frequency response of LTI system is

$$H(f) = \frac{5}{1 + j10\pi f}, \text{ what is step response}$$

$$H(s) = \frac{5}{1 + 5s} \times (s) = \frac{1}{s}$$

$$Y(s) = \frac{1}{s}\frac{5}{1 + 5s} = \frac{1}{s} - \frac{5}{1 + 5s}$$

$$g(t) = 1 - 5.e^{-\frac{1}{5}t}$$

18. The impulse response of LTI system is $h(t) = e^{-2t}u(t)$

frequency response of system is

$$H(\omega) = \int_0^\alpha e^{-2t}e^{-j\omega t}dt = \frac{1}{2 + j\omega}$$

19. As LTI system is

$$\frac{d^2}{dt^2}y(t) + \frac{4dy(t)}{dt} + 3y(t) = \frac{2dx(t)}{dt} + 4x(t)$$

assume input $x(t) = e^{-2t}u(t)$, find $y(t)$

$$y(s)s^2 + 4sy(s) + 3y(s) = 2sx(6)4x(s)$$

$$\frac{y(s)}{x(s)} = \frac{2s + 4}{s^2 + 4s + 3}$$

given $x(s) = \dfrac{1}{s + 2}$

$$y(s) = \frac{2(s + 2)}{s^2 4s + 3} \cdot \frac{1}{(s + 2)} = \frac{2}{s^2 + 4s + 3}$$

$$= \frac{1}{s + 1} - \frac{1}{s + 3}$$

$$y(t) = \left(e^{-t} - e^{-3t}\right)u(t)$$

20. $x(t) = e^{-2t}u(t) + \delta(t - 6)$, $h(t) = u(t)$, find $y(t)$

$$x(s) = \frac{1}{s + 2} + e^{-6s}$$

$$h(s) = \frac{1}{s}$$

$$\therefore \quad y(s) = \frac{1}{s(s+2)} + \frac{e^{6s}}{s}$$

$$y(t) = 0.5(1 - e^{-2t})u(t) + u(t-6)$$

OBJECTIVE QUESTIONS

1. ZT of impulse function is -------------------
2. ZT of unit step function is -------------------
3. ZT of $\sin \omega_0 n$ is -------------------
4. ZT of $\cos \omega_0 n$ is ----------------
5. Discrete time system is stable if the poles are ----------- unit circle
6. For a stable system $|z|$ is ------------------
7. For unstable system $|z|$ is ------------------
8. According to initial value theorem $x(0) = $ ------------------
9. The correlation of two sequences $x_1(1) * x_2(-1)$ is given by --------------
10. One side Z – transform is given by ---------------
11. According to final value theorem $\lim_{n \to \alpha} x(n) = $ --------------
12. Z – transform is most widely used for designing of ------------- filters
13. $F(z)$ becomes ---------------- at the poles
14. ---------------------- cannot lie in the ROC of Z – plane
15. -------------------- determines the boundary of ROC of Z plane
16. Type of sequence can be determined by -------------
17. In LT left half S – plane corresponds to ----------- in Z – plane
18. Right half of S – plane corresponds to ---------- in Z – plane
19. Single sided Z – transform is used for analyses of -------------- systems
20. For right hand sided sequence, the ROC is entire Z – plane except ------------------
21. For left hand sided sequence, the ROC is entire Z – plane except ----------------
22. Convolute of $x(n) = (1, 2, 1)$ and $h(n) = (1, 1, 1)$ is ------------------
23. Z – transform of $\delta(n-m)$ is ----------------
24. Z – transform of $r^n x(n)$ is ------------------
25. Z – transform of $\delta^m(n)$ is -------------------

ANSWERS TO OBJECTIVE QUESTIONS

1. 1

2. $\dfrac{1}{1-Z^{-1}}$

3. $\dfrac{\sin\omega_0 2^1}{1-2Z^{-1}\cos\omega_0 - Z^{-2}}$

4. $\dfrac{1-\cos\omega_0 Z^{-1}}{1-2\cos\omega_0 Z^{-1} + Z^{-2}}$

5. $|Z|<1$ (with unit circle)

6. $|Z|<1$

7. $|Z|>1$

8. $\lim\limits_{z\to\infty} x(z)$

9. $x_1(Z)x_2(Z^{-1})$

10. $x(Z) = \sum\limits_{n=\partial}^{\alpha} x(n)Z^{-n}$

11. $\lim\limits_{Z\to 1}\left(1-Z^{-1}\right)x(z)$

12. Digital

13. Infinite

14. Poles

15. Poles

16. ROC

17. Inside the ring

18. Outside the ring

19. Casual

20. $Z=\alpha$

21. (1, 3, 4, 3, 1)

23. $\dfrac{1}{Z^n}$

24. $X(Z^{-1}\alpha)$

25. S^n

(A) GATE QUESTIONS AND ANSWERS ON Z-TRANSFORMS

1. Consider the sequence $x|n| = a^n u[n] + b^n u[n]$, where $u[n]$ denotes unit stip and $0<|a|<|b|<1$. The region of convergence of z transform of $x[n]$ is ______   GATE 2016

    (a) $|z|>|a|$

    (b) $|z|>|b|$

    (c) $|z|<|a|$

    (d) $|a|<|z|<|b|$ *Ans:* (b)

2. The ROC of z transform of discrete time signal $x|n| = (2.0)^{|n|}$, $-\alpha < n < +\alpha$ is ______
                                                                    GATE 2016

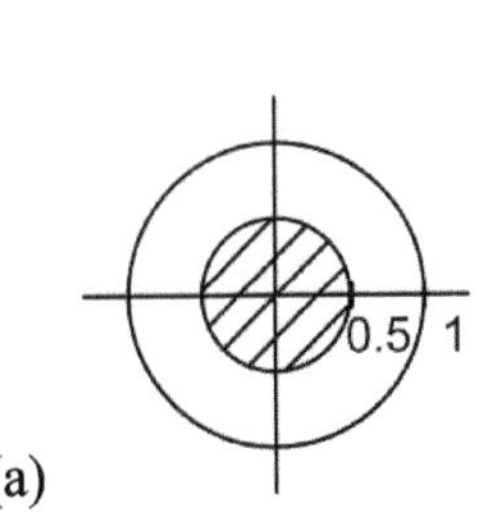

(a)

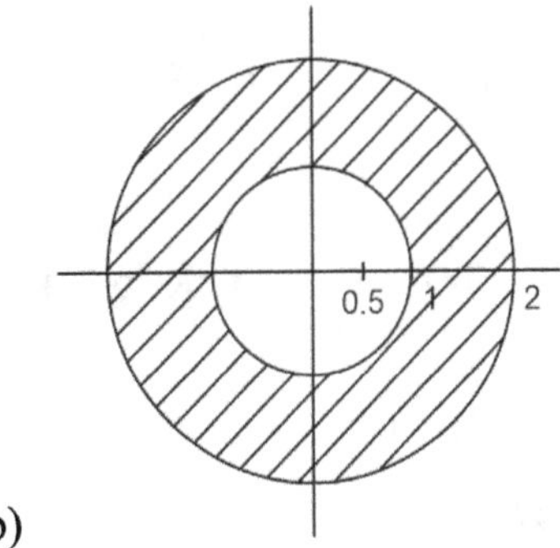

(b)

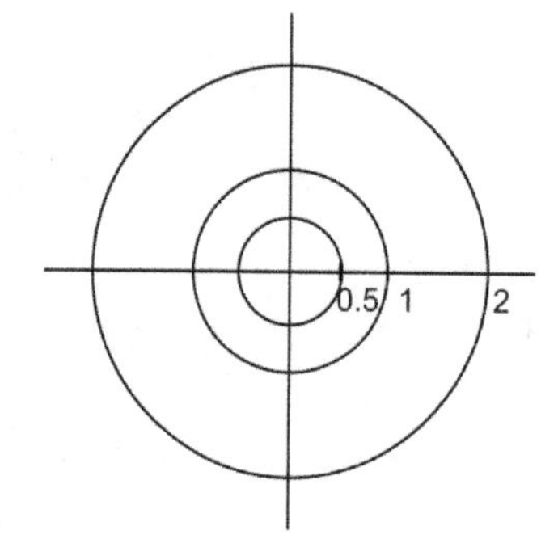

(c)

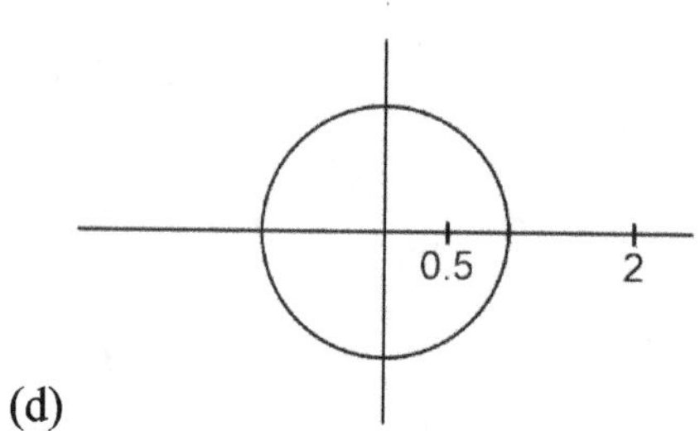

(d)

Ans: (d)

3. Two casual discrete time signals x[n] and y[n] are related by $y[n] = \sum_{m=0}^{n} x[m]$. If z transform of y[n] is $\dfrac{2}{z(z-1)^2}$ the value of x[2] is _____   GATE 2015

 (a) 0   (b) 1   (c) 2   (d) 3

Ans: (a)

4. The Z transform of the sequence x[n] is given by $X(z) = \dfrac{1}{(1-2z^{-1})^2}$ with ROC $|z| > 2$, then x[2] is _____   GATE 2014

 (a) 10   (b) 11   (c) 12   (d) 13

Ans: (c)

5. Consider Z-transform $X(z) = 5z^2 + 4z^{-1} + 3; \ 0 < |z| < \alpha$. The inverse z transform x[n] is

GATE 210

 (a) $5\delta[n+2] + 3\delta[n] + 4\delta[n-1]$ (b) $5\delta[n-2] + 3\delta[n] + 4\delta[n-1]$

 (c) $5u[n+2] + 3u[n] + 4u[n-1]$ (d) $5u[n-2] + 3u[n] + 4u[n+1]$

Ans: (a)

6. Z-transform of time function $\sum_{k=0}^{\alpha} \delta(n-k)$ is   GATE 1998

 (a) (z–1)/z   (b) z/(z–1)

 (c) $z/(z-1)^2$ (d) $(z-1)^2/z$

Ans: (c)

7. z-transform of the function $f(nT) = a^{nT}$ is   GATE 1999

(a) $\dfrac{z}{z-a^T}$

(b) $\dfrac{z}{z-a^{-T}}$

(c) $\dfrac{z}{z+a^T}$

(d) $\dfrac{z}{z+a^{-T}}$

Ans: (a)

8. The region of convergence of z-transform of unit step function is   GATE 2001

(a) $|z| > 1$

(b) $|z| < 1$

(c) Real $|z| > 0$

(d) Real $|z| < 0$

Ans: (a)

9. The Z transform of a system is $\dfrac{z}{z-0.2}$. If ROC is $|z| < 0.2$, the impulse response of the system is   GATE 2007

(a) $(0.2)^n u[n]$

(b) $(-0.2)^2 u[n]$

(c) $(0.2)^2 u[-n] - 1$

(d) $-(0.2)^n u[-n] - 1$

Ans: (d)

10. The region of convergence of z-transform of the sequence $\left(\dfrac{5}{6}\right)^n u[n] - \left(\dfrac{5}{6}\right)^n u[-n-1]$

must be   GATE 2005

(a) $|z| < \dfrac{5}{6}$

(b) $|z| > \dfrac{5}{6}$

(c) $\dfrac{5}{6} < |z| < \dfrac{5}{6}$

(d) $\dfrac{5}{6} < |z| < \varsigma$

Ans: (c)

11. If the region of convergence of $x_1[n] + x_2[n]$ is $\dfrac{1}{3} < |z| < \dfrac{2}{3}$ then the region of convergence of $x_1[n] - x_2[n]$ is   GATE 2006

(a) $\dfrac{1}{3} < |z| < 3$

(b) $\dfrac{2}{3} < |z| < 3$

(c)　$\dfrac{3}{2} < |z| < 3$

(d)　$\dfrac{1}{3}|z| < \dfrac{2}{3}$

Ans: (d)

12. Two discrete time systems with impulse responses $h_1[n] = \delta[n-1]$ and $h_2[n] = \delta[n-2]$ is connected in cascade. The overall impulse response of the cascade system is

GATE 2006

(a)　$\delta[n-1] + \delta[n-2]$

(b)　$\delta[n-4]$

(c)　$\delta[n-3]$

(d)　$\delta[n-1]\delta[n-2]$

Ans: (c)

13. A discrete time signal $x[n] = d = [n-3] + d[n-3]$ has z transform $X[z]$. If $Y[z] = X[-z]$ is z transfer of another signal $y[n]$, then

GATE 2016

(a)　$y[n] = x[n]$

(b)　$y[n] = x[-n]$

(c)　$y[n] = -x[n]$

(d)　None

Ans: (d)

14. Let $x[n] = \left(-\dfrac{1}{9}\right)^n u[n] - \left(-\dfrac{1}{3}\right)^n u(-n-1)$, the ROC of z-transform is　GATE 2014

(a)　$|z| > \dfrac{1}{9}$

(b)　$|z| < \dfrac{1}{3}$

(c)　$\dfrac{1}{3} > |z| < \dfrac{1}{9}$

(d)　does not exist

Ans: (c)

15. If $x[n] = \left(\dfrac{1}{3}\right)^{|n|} - \left(\dfrac{1}{2}\right)^n u[n]$, then ROC of z transform is　GATE 2006

(a)　$\dfrac{1}{3} < |z| < 3$

(b)　$\dfrac{1}{3} < |z| < \dfrac{1}{2}$

(c)　$\dfrac{1}{2} < |z| < 3$

(d)　$\dfrac{1}{3} < |z|$

Ans: (b)

(B) GATE QUESTIONS
CONVOLUTION & CORRELATION (LTI SYSTEMS)

1. Auto correlation of an energy signal has even symmetry.

2. A system having a unit impulse as an impulse response $h(n)$ is excited by a signal $s(n) = a^n u(n)$. Find $y(n)$.

   Given Impulse response of $h(n) = \delta(n) * h(n) = \delta(n) = h(n)$

   $$h(n) = 1 \quad n = 0$$
   $$0 \quad\quad n \neq 0$$

   $$x(n) \;=\; a^n u(n)$$

   $$\therefore \quad y(n) \;=\; x(n) * h(n) = a^n u(n)$$

3. An input signal $A\,e^{-at}u(t)$ with $a > 0$ is applied to a causal filter, the impulse response is $A\,e^{-at}$. Find filter output.

   $$x(t) \;=\; A\,e^{-at}u(t)$$
   $$\text{impulse response } h(t) \;=\; \delta(t) * h(t)$$
   $$=\; A\,e^{-at}$$
   $$y(t) \;=\; x(t) * h(t)$$
   $$X(s) \;=\; \frac{A}{s+a}$$
   $$H(s) \;=\; \frac{A}{s+a}$$
   $$Y(s) \;=\; X(s)\,Y(s) = \frac{A^2}{(s+a)^2}$$
   $$y(t) \;=\; \alpha^{-1}\{Y(s)\} = A^2 t\,e^{at}\,u(t)$$

4. The power spectral density of deterministic signal is $\left(\dfrac{\sin(f)}{f}\right)^2$. AC in time domain is rectangular functions.

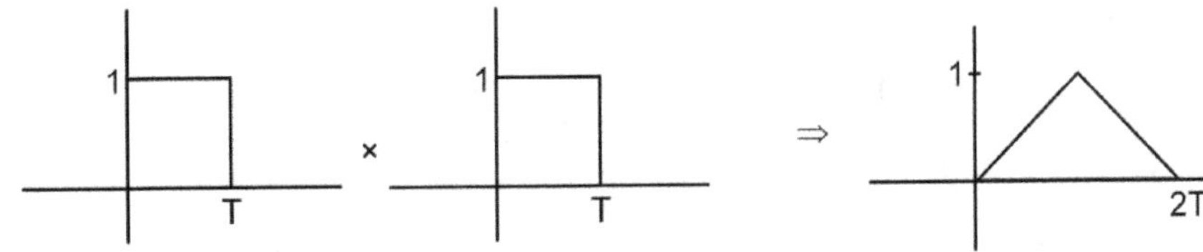

5. The unit impulse response of LTI system is $u(t)$. Find the response of system for an excitation of $e^{-at}u(t)$, $a>0$.

$$\text{Impulse response} \quad h(t) \;=\; u(t)$$

$$\therefore \quad H(s) \;=\; \frac{1}{s}$$

$$\text{Input response} \quad X(t) \;=\; e^{-at}$$

$$X(s) \;=\; \frac{1}{s+a}$$

$$\therefore \quad Y(s) \;=\; \frac{1}{s}\frac{1}{(s+a)} = \frac{1}{a}\left\{\frac{1}{s} - \frac{1}{s+a}\right\}$$

$$y(t) \;=\; \frac{1}{a}\left\{1-e^{-at}\right\}u(t)$$

6. Consider a rectangular pulse $g(t)$, $|t| \le \dfrac{T}{2}$. Fourier transform of $g(t)$ is sine function. Find output when it is convolved with itself.

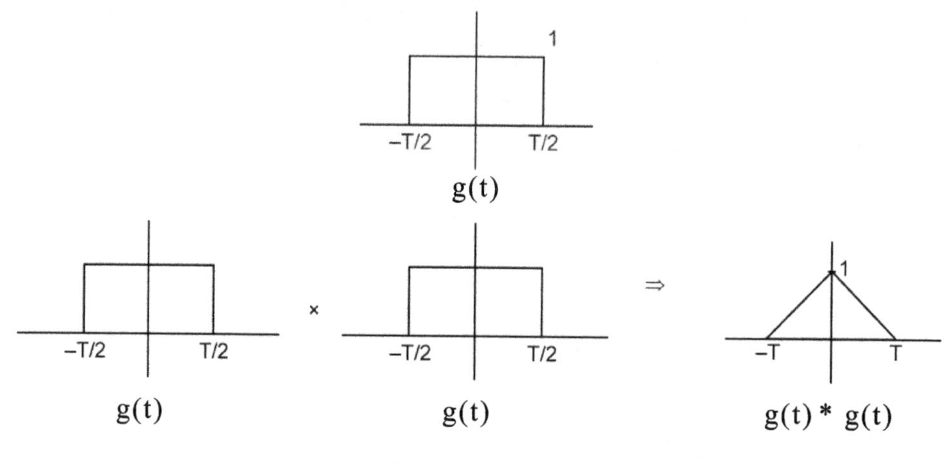

$$g(t) = \left(2\,\frac{\sin \omega t}{\omega t}\right)^2$$

7. If in a system $y(t) = t\,x(t)$, then it is linear but not time invariant

$$y_1(t) \;=\; t\,x_1(t)$$

$$y_2(t) \;=\; t\,x_2(t)$$

$$y(t) \;=\; t\{x_1(t) + x_2(t)\}$$

$$\;=\; t\,x_1 t + x_2(t)$$

$\therefore$ it is linear and it is increasing with t hence it is time variant

8. In an LTI system if impulse response is e^{-2t} what is the response of e^{-3t}.

$$h(t) = e^{-2t}$$

$$x(t) = e^{-3t}$$

$$H(s) = \frac{1}{s+2}$$

$$X(s) = \frac{1}{s+3}$$

$$Y(s) = \frac{1}{s+2} - \frac{1}{s+3} = \frac{1}{s+2} - \frac{1}{s+3}$$

$$y(t) = \left(e^{-2t} - e^{-3t}\right)u(t)$$

9. Find convolution of $u(t) - u(t-1)$ with $u(t) - u(t-2)$.

$$x_1(t) = u(t) - u(t-1)$$

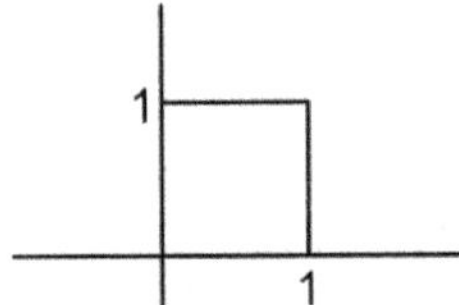

$$x_2(t) = u(t) - u(t-2)$$

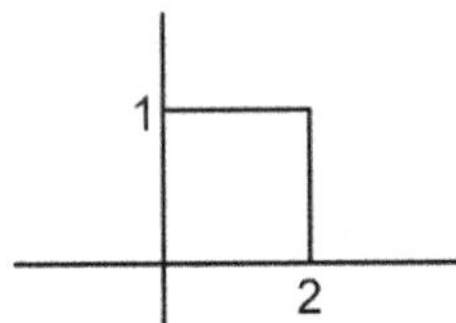

$$g(t) = x_1(t) * x_2(t)$$

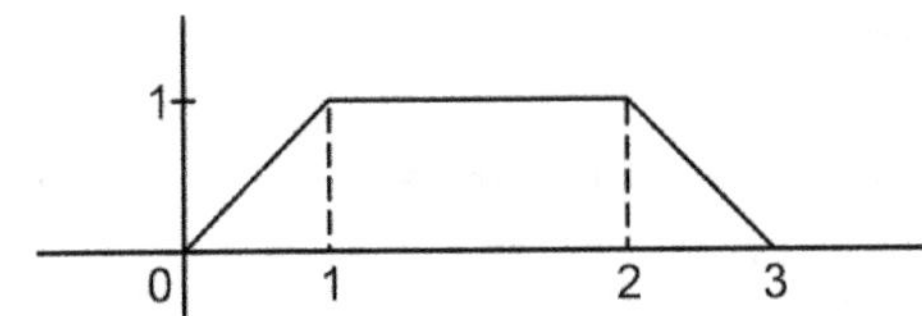

$$= \text{Linear increase} \qquad 0 \le t \le 1$$

$$= \text{const} \qquad 1 \le t \le 2$$

$$= \text{Linear Decrease} \qquad 2 \le t \le 3$$

10. If transfer function of a system is $H(s) = \dfrac{1}{s^2(s-2)}$. The impulse response of system is

$$H(s) \;=\; H_1(s)H_2(s) = \frac{1}{s^2} \cdot \frac{1}{s-2}$$

$$\therefore \; h(t) \;=\; t \cdot e^{2t}$$

11. Convolution of $X(t+5)$ with $\delta(t-T)$ is

$$y(t) \;=\; \int x(T)\delta(t-T)dT$$

Using time shifting property

$$x(t-7+5) \;=\; x(t-2)$$

$$T \;=\; t+5$$

$$T-7-T = t-7-t-5 = -2$$

12. A system is described by the following diff. equation.

$$\frac{d^2y}{dt^2} + \frac{3dy}{dt} + 2y = x(t)$$

if $x(t) = 2u(t)$,   what is output.

$$s^2 y(s) + 3y(s) + 2y(s) = \frac{2}{3}$$

$$y(s) = \frac{2}{s(s^2+3s+2)} = \frac{2}{s(s+1)(s+2)}$$

$$= \frac{1}{s} - \frac{2}{s+1} + \frac{1}{s+2}$$

$$y(t) = \left(1 - 2e^{-t} + e^{-2t}\right)$$

13. A rectangular pulse train s(t) is convolved with $\cos^2\left(4\pi \times 10^3 t\right)$, the output is

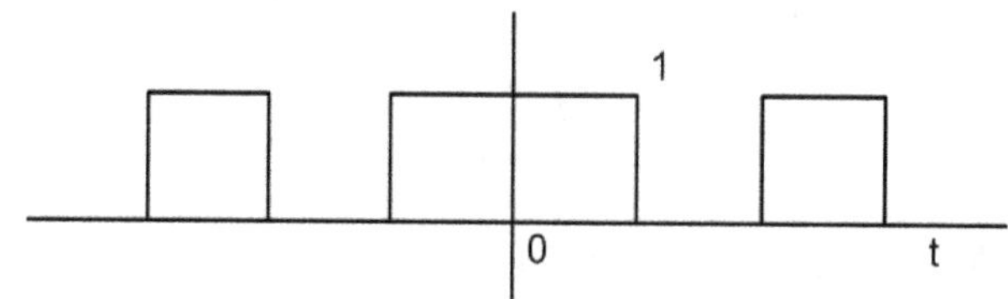

14. A causal system having the transfer function $H(s) = \dfrac{1}{s+2}$ is excited with $10u(t)$. The time at output reaches 99% of steady state is

$$H(s) = \frac{1}{s+2}$$

$$x(t) = \frac{10}{s}$$

$$y(s) = \frac{10}{s(s+2)} = \frac{5}{s} - \frac{5}{s+2}$$

$$y(t) = 5(1-e^{2t}); \quad y(\alpha) = 5$$

$$0.99 \times 5 = 5(1-e^{-2t})$$

$$t = 2.4 \text{ sec.}$$

15. The unit impulse response of a system is $h(t) = e^{-t}$, $t > 0$, for this system the steady state value of the output for unit step is?

$$h(t) = e^{-t}$$

$$x(t) = u(t)$$

$$H(s) = \frac{1}{s+1}$$

$$X(s) = \frac{1}{s}$$

$$Y(s) = \frac{1}{s(s+1)} = \frac{1}{s} - \frac{1}{s+1} = (1-e^{-t})$$

Steady state $\quad \lim_{t \to \alpha}(1-e^{-t}) = 1$

(C) GATE QUESTIONS FOURIER TRANSFORMS

1. Fourier transform of $e^{-at} \cos at$

$$\mathscr{F}T\{\cos at\} = \frac{s}{s^2 + a^2}$$

$$\mathscr{F}T\{e^{-at} \cos at\} = \frac{(s-a)}{(s-a)^2 + a^2}$$

2. $\mathcal{F}T\{g(t-2)\} = e^{-j4\pi f}G(f)$

$\mathcal{F}T\left\{g\left(\frac{t}{2}\right)\right\} = 2G(2f)$

3. $F(s) = \dfrac{\omega}{s^2+\omega^2}$ find $\lim\limits_{t\to\alpha} f(t)$

$f(t) = \sin\omega t$ its value is between -1 and 1

we can't determine $t=\alpha$, it is available for $t = \dfrac{2\pi}{\omega}$ and $\dfrac{-2\pi}{\omega}$

4. $\mathcal{F}T$ of $\dfrac{dx(t)}{dt} \, j\,2\pi\,f \times (f)$

5.   If $f(t) \leftrightarrow F(s)$

$f(t-T) = e^{sT}F(s)$

6.   FT $G(\omega)$ of a signal $g(t)$ in figure (a) is given as $G(\omega) = \dfrac{1}{\omega^2}\left(e^{j\omega} - j\omega e^{j\omega} - 1\right)$. Find FT of signals in (b) & (c).

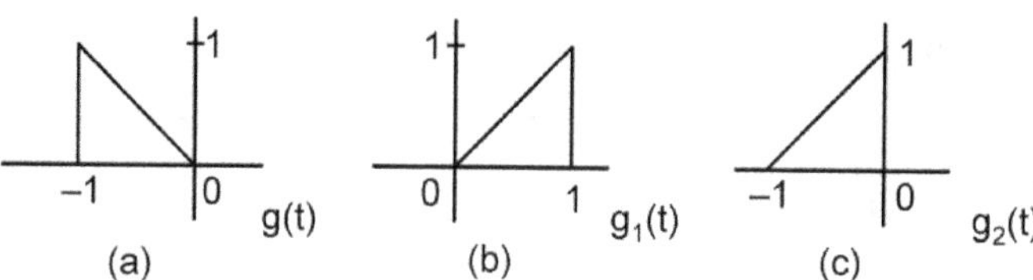

In   fig(b) $g_1(t) = g(-t)$

$\therefore$ $G_1(\omega) = G(-\omega) = \dfrac{1}{\omega^2}\left(e^{-j\omega} + j\omega e^{-j\omega} - 1\right)$

In   fig(c) $g_2(t) = g_1(t-1) = e^{-j\omega}G_1(\omega)$

7.   FT of $e^{-t}u(t)$ is $\dfrac{1}{1+j2\pi f}$, what is FT of $\dfrac{1}{1+j2\pi t}$ duality property

$F(t) \leftrightarrow F(\omega)$

$F\left\{\dfrac{1}{1+j2\pi t}\right\} = e^{f}u(-f)$

8. The inverse fourier transform of $X(3f + 2)$

$$x(f - a) \leftrightarrow e^{-j2\pi at} x(t)$$

$$x(af) \leftrightarrow \frac{1}{a} x\left(\frac{t}{a}\right)$$

$$\frac{1}{3} x\left(\frac{t}{3}\right) e^{+j4\pi\, t/3}$$

9. The sequence $y(n) = x\left(\dfrac{n}{2} - 1\right)$ for n even

$$\qquad\qquad 0 \qquad\qquad \text{for} \quad \text{n odd}$$

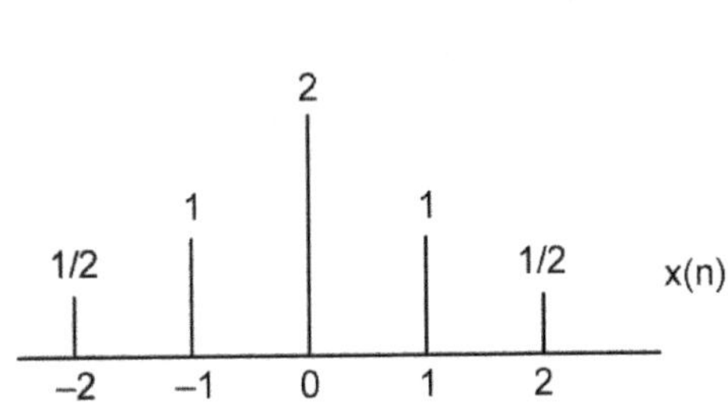

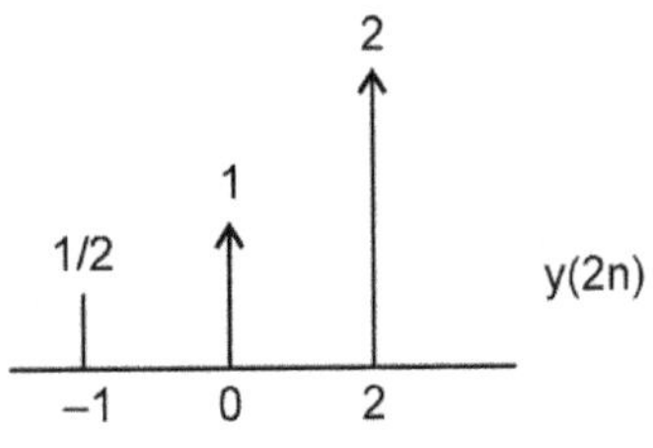

$\mathcal{F}$T of   y(2n) is

$$y(2n) \;=\; x(n-1)$$

$$y(j\omega) \;=\; 2e^{-j\omega} + \frac{1}{2} e^{j\omega} + 1 + e^{-j2\omega} + \frac{1}{2} e^{-j3\omega}$$

$$\qquad\quad =\; e^{-j\omega}\left(2 + \cos 2\omega + 2\cos\omega\right)$$

10. $\mathcal{F}$T $X(5t - 3)$ is

$$X(j\omega) \;=\; \frac{1}{5} \times \left(\frac{j\omega}{5}\right) e^{-j3\omega/5}$$

11. The fourier transform of a signal h(t) is H(jω)

$$H(j\omega) = (2\cos\omega)(\sin 2\omega)\,/\,\omega. \text{ The value of } h(0) \text{ is}$$

Let

$$\frac{\sin 2\omega}{\omega} \quad \longleftrightarrow \quad h^1(t)$$

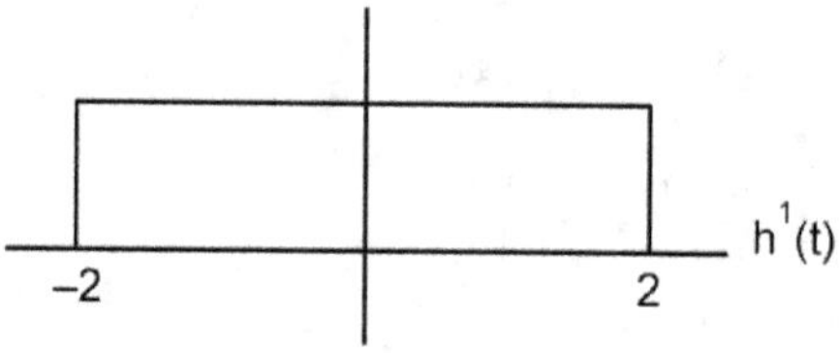

$$2\cos\omega\left(\frac{\sin 2\omega}{\omega}\right) = \left(e^{j\omega}+e^{-j\omega}\right)\frac{\sin 2\omega}{\omega} = e^{j\omega}\frac{\sin 2\omega}{\omega}+e^{-j\omega}\frac{\sin 2\omega}{\omega}$$

$$h(t) = h^1(t-1)+h^1(t+1)$$

Answers to Objective Questions

Chapter 1

1. $E = \int_{-\alpha}^{\alpha} \left| x(t)^2 \right| dt$

2. Energy signal

3. Zero

4. $x(t) = x(t + T)$

5. e^{-jwx}

6. $w_0 = 2\pi f_0$

7. $\int_{-a}^{a} x_0(t)dt = 0$

8. Fundamental time period

9. Continuous

10. Zero

11. Finite

12. Negative

13. $\sum_{-\alpha}^{\alpha} \delta(n)dn - 1$ at $n = 0$

14. $t = 1$

15. $n \geq -1$

16. Rectangular signal

17. Ramp

18. 1

19. Aperiodic

20. $2A^2$

21. Infinite

22. $x(t-1)$

23. yes

24. Even

25. (a) t u (t)

    (b) u (t) – u(t–1)

    (c) u (t) – u (–t)

    (d) u (t + a) – u (t – a)

26. Power signal

27. 8π

28. α

29. a, d are energy

    c, e are power

30. a, b, c

Chapter 2

1. Magnitude, direction

2. $|a||b|\cos\theta$

3. $|a||b|\sin\theta$

4. Zero

5. Orthogonal

6. Orthogonal

7. One

Chapter 3

1. d

2. a

3. a

4. 0.5

5. d

6. 2

7. b, it is non periodic

8. c, $e^{-|t|}$ is decaying exponentially hence it is not periodic

9. b

10. b

11. b

Chapter 4

1. 1

2. $2/j\omega$

3. $\sin c\left(\dfrac{\omega}{\pi}\right)$

4. 1

5. Sic c^2

6. $e^{-j\omega t_o} f(\omega)$

7. $j\omega f'(\omega)$

8. $\delta(t) = 1$

9. $X(-f)$

Chapter 5

1. Stable

2. Time invariant

3. $\int_{-\alpha}^{\alpha} |h(T)| dT < \alpha$

4. Disfortionless

5. Impulse Response

6. Memory less

7. Impulse response is absolutely integrable

8. Every bounded 1/P results every bounded O/P

9. Invertible

10. Memory

11. Super position

12. Present and past

13. Non linear

14. b

15. b

16. c

17. a

18. d

19. d

20. c

21. c

22. b

23. a

24. b

Chapter 6

1. Multiplication

2. $X(w) \times h(w)$

3. $\int_{-\alpha}^{\alpha} x_1(T)) x_2(t-T) dT$

4. $f(t-T)$

5. $\phi_{21}(-T)$

6. equal

7. Original waves

8. Energy

9. $\mathop{\mathrm{lt}}\limits_{T\to\alpha}\dfrac{1}{2T}\left|F_T(\omega)\right|^2$

10. Filtering

11. $\mathop{\mathrm{lt}}\limits_{T\to\alpha}\dfrac{1}{T}\displaystyle\int_{-T/2}^{T/2} f^2(t)\,dt$

12. 0

13. $\dfrac{A^2T^2}{\pi}\,Sa^2\left(\dfrac{\omega T}{2}\right)$

14. Wiener – Kinchine theorem

15. Correlation

16. $\displaystyle\int_{-\alpha}^{\alpha} f_1(t)f_2(t-T)\,dt$

17. $\mathop{\mathrm{lt}}\limits_{T\to\alpha}\displaystyle\int_{-T/2}^{T/2} f_1(t)f_2(t-T)\,dt$

18. $\dfrac{1}{\pi}\left|F_T(\omega)\right|^2$

19. $-\dfrac{d_2}{dt^2}\phi_{xx}(T)$

20. $S(t)+n(t)$

Chapter 7

1. c

2. c

3. b

4. a

5. c

6. a

7. d

8. a

9. a

10. d

11. b

Chapter 8

1. Laplace transform

2. Unilateral LT

3. $S=j\omega$

4. $1/S$

5. $\dfrac{\omega}{S^2+\omega^2}$

6. $\dfrac{S}{S^2+\omega^2}$

7. $\dfrac{n1}{S^n+1}$

Gate Questions and Answers

1. *Ans:* All

2. *Ans:* c

3. *Ans:* b

4. *Ans:* d

5. *Ans:* b

Chapter 9

1. 1

2. $\dfrac{1}{1-z^{-1}}$

3. $\dfrac{\sin\omega_0 z^{-1}}{1-2z^{-1}\cos\omega_0+z^{-2}}$

4. $\dfrac{1-\cos\omega_0 z^{-1}}{1-2\cos\omega_0 z^{-1}+z^{-2}}$

5. $|z|<1$ (within unit circle)

6. $|z|<1$

7. $|z|>1$

8. $\displaystyle\lim_{z \to \alpha} x(z)$

9. $x_1(z)x_2(z^{-1})$

10. $\displaystyle x(z) = \sum_{n=0}^{\alpha} x(n)z^{-n}$

11. $\displaystyle\lim_{z \to 1}\left(1 - z^{-1}\right)x(z)$

12. Digital

13. Infinite

14. Poles

15. Poles

16. ROC

17. Inside the ring

18. Outside the ring

19. Causal

20. $z = 0$

21. $z = \alpha$

22. $(1, 3, 4, 3, 1)$

23. $\dfrac{1}{z^m}$

24. $X(z^{-1}r)$

25. s^m

Appendix

Some useful Mathematical Functions

1. Power series

(a) $\quad e^x = 1 + x + \dfrac{x^2}{2!} + \dfrac{x^3}{3!} + \dots \dfrac{x^n}{n!}$

(b) $\quad \sin x = x - \dfrac{x^3}{3!} + \dfrac{x^5}{5!} - \dfrac{x^7}{7!} + \dots$

(c) $\quad \cos x = 1 - \dfrac{x^2}{2!} + \dfrac{x^4}{4!} - \dfrac{x^6}{6!} + \dots$

(d) $\quad \tan x = x - \dfrac{x^3}{3} + \dfrac{2x^5}{15} - \dfrac{17x^7}{315} + \dots$

(e) $\quad (1+x)^n = 1 + nx + \dfrac{n(n-1)}{2!}x^2 + \dfrac{n(n-1)(n-2)}{3!}x^3 + \dots$

(f) $\quad \dfrac{1}{1-x} = 1 + x + x^2 + x^3 + \dots$

2. Sums

(a) $\quad \displaystyle\sum_{km}^{n} r^k = \dfrac{r^{n+1} - r^m}{r-1}$

(b) $\quad \displaystyle\sum_{k=0}^{n} k = \dfrac{n(n+1)}{2}$

(c) $\quad \displaystyle\sum_{k=0}^{n} k^2 = \dfrac{n(n+1)(2n+1)}{6}$

(d) $\quad \displaystyle\sum_{k=0}^{n} kr^k = \dfrac{r + [n(r-1)-]r^{n+1}}{(r-1)^2}$

3. Trigonometric Identities

   (a) $e^{\pm jx} = \cos x \pm j\sin x$

   (b) $\cos x = \dfrac{1}{2}(e^{jx} + e^{-jx})$

   (c) $\sin x = \dfrac{1}{2j}(e^{jx} - e^{-jx})$

   (d) $\cos\left(x \pm \dfrac{\pi}{2}\right) = \mp\sin x$

   (e) $\sin\left(x \pm \dfrac{\pi}{2}\right) = \pm\cos x$

   (f) $2\sin x \cos x = \sin 2x$

   (g) $\cos 2x - \sin^2 x = \cos 2x$

   (h) $\cos^2 x = \dfrac{1}{2}(1 + \cos 2x)$

   (i) $\sin^2 x = \dfrac{1}{2}(1 - \cos 2x)$

   (j) $\cos^2 x = \dfrac{1}{4}(3\cos x + \cos 3x)$

   (k) $\sin^3 x = \dfrac{1}{4}(3\sin x - \sin 3x)$

   (l) $\tan(x + y) = \dfrac{\tan x + \tan y}{1 \mp \tan x \tan y}$

   (m) $\sin x \sin y = \dfrac{1}{2}(\cos(x - y) - \cos(x + y))$

   (n) $\cos x \cos y = \dfrac{1}{2}(\cos(x - y) + \cos(x + y))$

   (o) $a\cos x + b\sin x = c\cos(x + \theta)$ $\theta = \tan^{-1} b/a$

                                                                  $c = \sqrt{a^2 + b^2}$

4. Indefinite integrals

   (a) $\displaystyle\int u\, dv = uv - \int v\, du$

   (b) $\displaystyle\int f(x)\, g(x)\, dx = f(x)\, g(x) - \int f(x) g(x) dx$

   (c) $\displaystyle\int \sin ax\, dx = \dfrac{1}{a}\cos ax$

(d) $\int \cos^2 ax\,dx = \dfrac{1}{a}\sin ax$

(e) $\int \sin^2 ax\,dx = \dfrac{x}{2} - \dfrac{\sin 2ax}{4a}$

(f) $\int \cos^2 ax\,dx = \dfrac{x}{2} + \dfrac{\sin 2ax}{4a}$

(g) $\int x \sin ax\,dx = \dfrac{1}{a^2}(\sin ax - ax\cos ax)$

(h) $\int x \cos ax\,dx = \dfrac{1}{a^2}(\cos ax + ax\sin ax)$

(i) $\int x^2 \sin ax\,dx = \dfrac{1}{a^3}(2ax\,s\sin ax + 2\cos ax - a^2x^2\cos ax)$

(j) $\int x^2 \cos ax\,dx = \dfrac{1}{a^3}(2ax\cos ax - 2\sin ax + a^2x^2\sin ax)$

(k) $\int \sin ax\,\sin bx\,dx = \dfrac{\sin(a-b)x}{2(a-b)} - \dfrac{\sin(a+b)x}{2(a+b)}$

(l) $\int \sin ax\,\cos bx\,dx = -\left[\dfrac{\cos(a-b)x}{2(a-b)} + \dfrac{\cos(a+b)x}{2(a+b)}\right]$

(m) $\int \cos ax\,\cos bx\,dx = \dfrac{\sin(a-b)x}{2(a-b)} + \dfrac{\sin(a+b)x}{2(a+b)}$

(n) $\int e^{ax}\,dx = \dfrac{1}{a}e^{ax}$

(o) $\int xe^{ax}\,dx = \dfrac{e^{ax}}{a^2}(ax-1)$

(p) $\int x^2 e^{ax}\,dx = \dfrac{e^{ax}}{a^3}(a^2x^2 - 2ax + 2)$

(q) $\int \dfrac{1}{x^2+a^2}\,dx = \dfrac{1}{a}\tan^{-1}\dfrac{x}{a}$

(r) $\int \dfrac{x}{x^2+a^2}\,dx = \dfrac{1}{2}\ln(x^2+a^2)$

5. Common derivatives

(a) $\dfrac{d}{dx}f(u) = \dfrac{d}{du}f(u)\dfrac{du}{dx}$

(b) $\dfrac{d}{dx}(uv) = \dfrac{udv}{dx} + v\dfrac{du}{dx}$

(c) $\dfrac{d}{dx}\left(\dfrac{u}{v}\right) = \dfrac{v\dfrac{du}{dx} - u\dfrac{du}{dx}}{v^2}$

(d) $\dfrac{dx^n}{dx} = nx^{n-1}$

(e) $\dfrac{d}{dx} = \ln(ax) = \dfrac{1}{x}$

(f) $\dfrac{d}{dx} = \log(ax) = \dfrac{\log c}{x}$

(g) $\dfrac{d}{dx}e^{bx} = be^{bx}$

(h) $\dfrac{d}{dx}a^{bx} = b(\ln a)a^{bx}$

(i) $\dfrac{d}{dx}\sin ax = a\cos ax$

(j) $\dfrac{d}{dx}\cos ax = -a\sin ax$

(k) $\dfrac{d}{dx}\tan ax = \dfrac{a}{\cos^2 ax}$

(l) $\dfrac{d}{dx}(\sin^{-1} ax) = \dfrac{a}{\sqrt{1-a^2x^2}}$

(m) $\dfrac{d}{dx}(\cos^{-1} ax) = \dfrac{-a}{\sqrt{1-a^2x^2}}$

(n) $\dfrac{d}{dx}(\tan^{-1} ax) = \dfrac{a}{1+a^2x^2}$